Design and Analysis of Algorithms

The text introduces readers to different paradigms of computing in addition to the traditional approach of discussing fundamental computational problems and design techniques in the random access machine model. Alternate models of computation including parallel, cache-sensitive design and streaming algorithms are dealt in separate chapters to underline the significant role of the underlying computational environment in the algorithm design. The treatment is made rigorous by demonstrating new measures of performances along with matching lower bound arguments.

The importance of greedy algorithms, divide-and-conquer technique and dynamic programming is highlighted by additional applications to approximate algorithms that come with guarantees. In addition to several classical techniques, the book encourages liberal use of probabilistic analysis and randomized techniques that have been pivotal for many recent advances in this area. There is also a chapter introducing techniques for dimension reduction which is at the heart of many interesting applications in data analytics as well as statistical machine learning. While these techniques have been known for a while in other communities, their adoption into mainstream computer science has been relatively recent.

Concepts are discussed with the help of rigorous mathematical proofs, theoretical explanations and their limitations. Problems have been chosen from a diverse landscape including graphs, geometry, strings, algebra and optimization. Some exposition of approximation algorithms has also been included, which has been a very active area of research in algorithms. Real life applications and numerical problems are spread throughout the text. The reader is expected to test her understanding by trying out a large number of exercise problems accompanying every chapter.

The book assumes familiarity with basic data structures, to focus on more algorithmic aspects and topics of contemporary importance.

Sandeep Sen has been a faculty member in the Department of Computer Science and Engineering, Indian Institute of Technology Delhi, India, since 1991 where he is currently a professor. His research spans randomized algorithms, computational geometry, dynamic graph algorithms and models of computation. He is a Fellow of the Indian National Science Academy and the Indian Academy of Sciences.

Amit Kumar is a faculty member in the Department of Computer Science and Engineering, Indian Institute of Technology Delhi, India. His research lies in the area of combinatorial optimization, with emphasis on problems arising in scheduling, graph theory and clustering. He is a Fellow of Indian Academy of Sciences, and is a recent recipient of the Shanti Swarup Bhatnagar Award for Mathematical Sciences.

Design and Analysis of Algorithms
A Contemporary Perspective

Sandeep Sen
Amit Kumar

CAMBRIDGE
UNIVERSITY PRESS

CAMBRIDGE
UNIVERSITY PRESS

University Printing House, Cambridge CB2 8BS, United Kingdom

One Liberty Plaza, 20th Floor, New York, NY 10006, USA

477 Williamstown Road, Port Melbourne, VIC 3207, Australia

314 to 321, 3rd Floor, Plot No.3, Splendor Forum, Jasola District Centre, New Delhi 110025, India

79 Anson Road, #06–04/06, Singapore 079906

Cambridge University Press is part of the University of Cambridge.

It furthers the University's mission by disseminating knowledge in the pursuit of
education, learning and research at the highest international levels of excellence.

www.cambridge.org
Information on this title: www.cambridge.org/9781108496827

First published 2019

Printed in India by Rajkamal Electric Press

A catalogue record for this publication is available from the British Library

Library of Congress Cataloging-in-Publication Data
Names: Sen, Sandeep, author. | Kumar, Amit, 1976- author.
Title: Design and analysis of algorithms / Sandeep Sen, Amit Kumar.
Description: New York, NY, USA : University Printing House, 2019. | Includes
bibliographical references and index.
Identifiers: LCCN 2019002080| ISBN 9781108496827 (hardback : alk. paper) |
ISBN 9781108721998 (paperback : alk. paper)
Subjects: LCSH: Algorithms.
Classification: LCC QA9.58 .S454 2019 | DDC 005.1–dc23
LC record available at https://lccn.loc.gov/2019002080

ISBN 978-1-108-49682-7 Hardback
ISBN 978-1-108-72199-8 Paperback

To the loving memory of my parents, Sisir Sen and Krishna Sen who nourished and inspired my academic pursuits and all my teachers who helped me imbibe the beauty and intricacies of various subjects

– Sandeep Sen

To my parents
– Amit Kumar

Content

List of Figures   xv

List of Tables   xix

Preface   xxi

Acknowledgments   xxv

1 Model and Analysis   1

  1.1 Computing Fibonacci Numbers   1

  1.2 Fast Multiplication   3

  1.3 Model of Computation   4

  1.4 Randomized Algorithms: A Short Introduction   6

    1.4.1 A different flavor of randomized algorithms   8

  1.5 Other Computational Models   10

    1.5.1 External memory model   10

    1.5.2 Parallel model   11

  Further Reading   12

  Exercise Problems   13

2 Basics of Probability and Tail Inequalities   16

  2.1 Basics of Probability Theory   16

  2.2 Tail Inequalities   21

  2.3 Generating Random Numbers   26

    2.3.1 Generating a random variate for an arbitrary distribution   26

2.3.2 Generating random variables from a sequential file   27

2.3.3 Generating a random permutation   29

Further Reading   31

Exercise Problems   31

3 Warm-up Problems   **34**

3.1 Euclid's Algorithm for the Greatest Common Divisor (GCD)   34

3.1.1 Extended Euclid's algorithm   35

3.1.2 Application to cryptography   36

3.2 Finding the kth Smallest Element   37

3.2.1 Choosing a random splitter   38

3.2.2 Median of medians   39

3.3 Sorting Words   41

3.4 Mergeable Heaps   43

3.4.1 Merging binomial heaps   44

3.5 A Simple Semi-dynamic Dictionary   45

3.5.1 Potential method and amortized analysis   46

3.6 Lower Bounds   47

Further Reading   50

Exercise Problems   50

4 Optimization I: Brute Force and Greedy Strategy   **54**

4.1 Heuristic Search Approaches   55

4.1.1 Game trees*   57

4.2 A Framework for Greedy Algorithms   60

4.2.1 Maximum spanning tree   64

4.2.2 Finding minimum weight subset   64

4.2.3 A scheduling problem   65

4.3 Efficient Data Structures for Minimum Spanning Tree Algorithms   66

4.3.1 A simple data structure for Union–Find   68

4.3.2 A faster scheme   69

4.3.3 The slowest growing function?   71

4.3.4 Putting things together   72

4.3.5 Path compression only*   73

4.4 Greedy in Different Ways   74

4.5 Compromising with Greedy   76

4.6 Gradient Descent*   77

    4.6.1 Applications   83

Further Reading   87

Exercise Problems   88

5 Optimization II: Dynamic Programming   **92**

5.1 Knapsack Problem   94

5.2 Context Free Parsing   95

5.3 Longest Monotonic Subsequence   97

5.4 Function Approximation   99

5.5 Viterbi's Algorithm for Maximum Likelihood Estimation   100

5.6 Maximum Weighted Independent Set in a Tree   102

Further Reading   102

Exercise Problems   103

6 Searching   **109**

6.1 Skip-Lists – A Simple Dictionary   110

    6.1.1 Construction of skip-lists   110

    6.1.2 Analysis   111

    6.1.3 Stronger tail estimates   113

6.2 Treaps: Randomized Search Trees   114

6.3 Universal Hashing   117

    6.3.1 Existence of universal hash functions   120

6.4 Perfect Hash Function   121

    6.4.1 Converting expected bound to worst case bound   122

6.5 A log log N Priority Queue*   122

Further Reading   124

Exercise Problems   125

7 Multidimensional Searching and Geometric Algorithms   **128**

7.1 Interval Trees and Range Trees   129

    7.1.1 Two-dimensional range queries   131

7.2 k–d Trees   132

7.3 Priority Search Trees   135

7.4 Planar Convex Hull   137

    7.4.1 Jarvis march   139

    7.4.2 Graham's scan   140

    7.4.3 Sorting and convex hulls   141

7.5 Quickhull Algorithm   142

  7.5.1 Analysis   143

  7.5.2 Expected running time*   145

7.6 Point Location Using Persistent Data Structure   146

7.7 Incremental Construction   149

Further Reading   152

Exercise Problems   153

8 String Matching and Finger Printing   **157**

8.1 Rabin–Karp Fingerprinting   157

8.2 KMP Algorithm   161

  8.2.1 Analysis of the KMP algorithm   165

  8.2.2 Pattern analysis   165

8.3 Tries and Applications   165

Further Reading   168

Exercise Problems   169

9 Fast Fourier Transform and Applications   **171**

9.1 Polynomial Evaluation and Interpolation   171

  9.1.1 Multiplying polynomials   172

9.2 Cooley–Tukey Algorithm   173

9.3 The Butterfly Network   175

9.4 Schonage and Strassen's Fast Multiplication*   176

9.5 Generalized String Matching   179

  9.5.1 Convolution based approach   180

Further Reading   182

Exercise Problems   182

10 Graph Algorithms   **184**

10.1 Depth First Search   184

10.2 Applications of DFS   188

  10.2.1 Strongly connected components (SCC)   188

  10.2.2 Biconnected components   191

10.3 Path Problems   193

  10.3.1 Bellman–Ford SSSP algorithm   194

  10.3.2 Dijkstra's SSSP algorithm   195

  10.3.3 All pair shortest paths algorithm   197

10.4 Computing Spanners for Weighted Graphs   198

10.5 Global Min-cut   201

    10.5.1 The contraction algorithm   202

    10.5.2 Probability of min-cut   203

Further Reading   204

Exercise Problems   205

11 Maximum Flow and Applications   **208**

    11.0.1 Max-Flow Min-Cut   212

    11.0.2 Ford and Fulkerson algorithm   213

    11.0.3 Edmond–Karp augmentation strategy   214

    11.0.4 Monotonicity lemma and bounding the number of iterations   215

11.1 Applications of Max-Flow   216

    11.1.1 Disjoint paths   216

    11.1.2 Bipartite matching   217

    11.1.3 Circulation problems   222

    11.1.4 Project planning   224

Further Reading   226

Exercise Problems   227

12 NP Completeness and Approximation Algorithms   **230**

12.1 Classes and Reducibility   233

12.2 Cook–Levin Theorem   235

12.3 Common NP-Complete Problems   237

12.4 Proving NP Completeness   240

    12.4.1 Vertex cover and related problems   241

    12.4.2 Three coloring problem   242

    12.4.3 Knapsack and related problems   **244**

12.5 Other Important Complexity Classes   247

12.6 Combating Hardness with Approximation   249

    12.6.1 Maximum knapsack problem   251

    12.6.2 Minimum set cover   252

    12.6.3 The metric TSP problem   253

    12.6.4 Three coloring   253

    12.6.5 Max-cut problem   254

Further Reading   254

Exercise Problems   255

13 Dimensionality Reduction*   **258**

13.1 Random Projections and the Johnson–Lindenstrauss Lemma   259

13.2 Gaussian Elimination   262

13.3 Singular Value Decomposition and Applications   264

    13.3.1 Some matrix algebra and the SVD theorem   265

    13.3.2 Low-rank approximations using SVD   267

    13.3.3 Applications of low-rank approximations   269

    13.3.4 Clustering problems   271

    13.3.5 Proof of the SVD theorem   273

Further Reading   275

Exercise Problems   275

14 Parallel Algorithms   **277**

14.1 Models of Parallel Computation   277

14.2 Sorting and Comparison Problems   278

    14.2.1 Finding the maximum   278

    14.2.2 Sorting   282

14.3 Parallel Prefix   287

14.4 Basic Graph Algorithms   291

    14.4.1 List ranking   292

    14.4.2 Connected components   294

14.5 Basic Geometric Algorithms   298

14.6 Relation between Parallel Models   300

    14.6.1 Routing on a mesh   301

Further Reading   303

Exercise Problems   304

15 Memory Hierarchy and Caching   **308**

15.1 Models of Memory Hierarchy   308

15.2 Transposing a Matrix   310

    15.2.1 Matrix multiplication   311

15.3 Sorting in External Memory   313

    15.3.1 Can we improve the algorithm?*   314

15.4 Cache Oblivious Design   316

    15.4.1 Oblivious matrix transpose   317

Further Reading   320

Exercise Problems   321

16 Streaming Data Model   **323**

   16.1 Introduction   323

   16.2 Finding Frequent Elements in a Stream   324

   16.3 Distinct Elements in a Stream   327

   16.4 Frequency Moment Problem and Applications   331

       16.4.1 The median of means trick   334

       16.4.2 The special case of second frequency moment   335

   16.5 Proving Lower Bounds for Streaming Model   337

   Further Reading   339

   Exercise Problems   340

Appendix A Recurrences and Generating Functions   **343**

   A.1 An Iterative Method – Summation   344

   A.2 Linear Recurrence Equations   345

       A.2.1 Homogeneous equations   345

       A.2.2 Inhomogeneous equations   346

   A.3 Generating Functions   346

       A.3.1 Binomial theorem   348

   A.4 Exponential Generating Functions   348

   A.5 Recurrences with Two Variables   349

   Bibliography   351

   Index   363

Figures

1.1   Algorithm for verifying matrix product   9

2.1   Generating a random permutation of n distinct objects   29

3.1   Euclid's algorithm   35

3.2   Algorithm based on median of medians   40

3.3   (a) Recursive construction of binomial tree; (b) Binomial heap of 11
      elements consisting of three binomial trees   44

4.1   Illustration of the induction proof when root node is $\vee$ 59

4.2   Algorithm Gen_Greedy   61

4.3   The matching (a,d) is a maximal independent set, but $(a,b),(c,d)$ is a larger
      maximal independent set   63

4.4   Kruskal's minimum spanning tree algorithm   64

4.5   Successive iterations in Kruskal's greedy algorithm   67

4.6   An example of a Union–Find data structure storing elements $\{1,2,\ldots,12\}$ 69

4.7   An example of a path compression heuristic   70

4.8   Prim's minimum spanning tree algorithm   74

4.9   Boruvka's minimum spanning tree algorithm   76

4.10 A convex function of one variable   78

4.11 Gradient descent algorithm   80

4.12 The convex function is non-differentiable at x 82

4.13 The point P should ideally lie on the intersection of the three circles, but
      there are some measurement errors   83

4.14 A perceptron with inputs $x_1, x_2, \ldots, x_n$ and output determined by the sign of
      $w_0 + w_1 x_1 + \ldots + w_n x_n$ 84

4.15 A plot of the function g as an approximation to the sgn function 85

4.16 The points in P are denoted by dots, and those in N by squares 86

5.1 Recursive Fibonacci sequence algorithm 93

5.2 The recursive unfolding of computing F_6 93

5.3 Table (a) implies that the string *aba* does not belong to the grammar whereas Table (b) shows that *baaba* can be generated from S 96

5.4 In (a), the constant function is an average of the y values which minimizes the sum of squares error. In (b), a 3 step function approximates the 7 point function 99

5.5 For the label *aba* and starting vertex v_1, there are several possible labeled paths like $[v_1, v_3, v_4, v_6], [v_1, v_4, v_5, v_6]$, etc. 101

5.6 In the sequence 13, 5, 8, 12, 9, 14, 15, 2 we have predefined the tree structure but only the first four numbers have been scanned, i.e., 13, 5, 8, 12 104

6.1 The path traversed while searching for the element 87 111

6.2 Diagrams (a) to (d) depict the rotations required to insert the element 20 having priority 58 starting with the treap for the first four elements. Diagram (e) is the final tree after the entire insertion sequence. Diagram (f) shows the schematic for left/right rotations 117

6.3 The shaded leaf nodes correspond to the subset S 123

7.1 The structure of a one-dimensional range search tree where a query interval is split into at most $2\log n$ disjoint canonical (half)-intervals 130

7.2 The rectangle is the union of the slabs represented by the darkened nodes plus an overhanging left segment containing p_6 132

7.3 Rectangular range query used in a k–d tree 133

7.4 Each rectangular subdivision corresponds to a node in the k–d tree and is labeled by the splitting axis – either vertical or horizontal 134

7.5 The query is the semi-infinite upper slab supported by the two bottom points $(0, 4.5)$ and $(10, 4.5)$ 137

7.6 The figure on the left is convex, whereas the one on the right is not convex 137

7.7 Convex hull of points shown as the shaded region 138

7.8 Jarvis March algorithm for convex hull 140

7.9 Merging upper hulls 141

7.10 Left turn(p_m, p_{2j-1}, p_{2j}) is true but $slope(\overline{p_{2j-1}p_{2j}})$ is less than the median slope given by L 144

7.11 The shaded vertical region does not contain any intersection points 147

7.12 An example depicting $\Omega(n^2)$ space complexity for n segments 149

7.13 Path copying technique on adjacent slabs s_5 and s_6 149

7.14 Incremental algorithm for closest pair computation 150

7.15 Maximum number of D-separated points per cell is 4 and the shaded area is the region within which a point can lie with distance less than D from p 152

8.1 Testing equality of two large numbers 159

8.2 Karp–Rabin string matching algorithm 160

8.3 Knuth–Morris–Pratt string matching algorithm 164

8.4 The suffix tree construction corresponding to the string **catca \$**: (i) Suffix **a** starting at position 5, (ii) Suffixes at position 4 and 5, etc. 167

9.1 FFT computation 174

9.2 Computing an eight point FFT using a butterfly network 176

9.3 Matching with wildcards: The pattern is $X = 3\ 2\ 1\ *$ and $r_1 = 6, r_2 = 4, r_3 = 11, r_4 = 0$ all chosen from $[1 \ldots 16]$ corresponding to $p = 17$ 182

10.1 Algorithm for Depth First Search 186

10.2 The pair of numbers associated with each vertex denotes the starting time and finishing time respectively as given by the global counter 187

10.3 The pair of numbers associated with the vertices represent the start and finish time of the DFS procedure 189

10.4 Finding strongly connected components using two DFS 190

10.5 The component graph for the graph on the left is shown on the right 192

10.6 Bellman–Ford single-source shortest path problem 194

10.7 For every vertex, the successive labels over the iterations of the Bellman–Ford algorithm are indicated where i denotes ∞ 196

10.8 Dijkstra's single source shortest path algorithm 196

10.9 An algorithm for weighted 3-spanner 199

10.10 The 3-spanner algorithm – Stages 1 and 2 200

10.11 Stretch bound: (i) Intracluster; (ii) Intercluster 201

10.12 Algorithm for computing t-partition 202

11.1 Greedy algorithm for max-flow: it may not give optimal solution 211

11.2 Example of residual graph 211

11.3 Example of disjoint paths in a graph 217

11.4 Reduction from a matching instance on the left to a max-flow instance on the right 218

11.5 The matching M is shown by dotted edges 220

11.6 Illustration of Hall's theorem 221

11.7 The shaded region consisting of s, A, C, D, E, F represents a min s-t cut of capacity 3 222

11.8 Example of circulation on the left 223

11.9 Figure on the left shows an example of DAG on a set of tasks 225

11.10 Figure for Exercise 11.3. Numbers denote edge capacities 227

12.1 Many-to-one reduction from Π_1 to Π_2 by using a function $f : \mathbb{N} \to \mathbb{N}$ 234

12.2   Graph illustrating the reduction for the 3-CNF formula   242

12.3   Illustration of the reduction proving NP-completeness of the three coloring
       problem.   243

13.1   Gaussian elimination algorithm   263

13.2   A two-dimensional illustration of the SVD subspace V_1 of points
       represented by circular dots, where V_1 denotes the subspace spanned by
       the first column of V 273

14.1   Parallel odd–even transposition sort   282

14.2   Sorting two rows by alternately sorting rows and columns   284

14.3   Shearsort algorithm for a rectangular mesh   285

14.4   Partition sort in parallel   285

14.5   Parallel prefix computation: this procedure computes prefix of $x_a, x_{a+1},$
       $\ldots, x_b$ 287

14.6   Parallel prefix computation using blocking   289

14.7   Recursive unfolding of the prefix circuit with 8 inputs in terms of 4-input
       and 2-input circuits   289

14.8   Parallel list ranking   292

14.9   The star rooted at vertex 5 is connected to all the stars (dotted edges) on the
       right that have higher numbered vertices   296

14.10 Parallel connectivity: We assume that there is a processor assigned to every
       vertex $v \in V$ and to every edge $(u, w) \in E$ 297

14.11 Starting from (r, c), the packet is routed to a random row r' within the same
       column c 303

15.1   The tiling of a $p \times q$ matrix in a row-major layout   310

15.2   Transposing a matrix using minimal transfers   311

15.3   Computing the product $Z = X \cdot Y$ using tiles of size s 312

15.4   Searching a dictionary in external memory   316

15.5   Consider numbers from 1 to 16 arranged according to the Algorithm in
       Fig. 15.4   316

15.6   Algorithm for matrix transpose   317

15.7   Base case: Both A, B fit into cache – no further cache miss   318

15.8   The subsequence $\sigma_{i_1}\sigma_{i_1+1}\ldots\sigma_{i_1+r_1}\sigma_{i_2}$ have $k+1$ distinct elements, whereas
       the subsequence $\sigma_{i_1}\sigma_{i_1+1}\ldots\sigma_{i_1+r_1}$ have k distinct elements   319

16.1   The algorithm A receives input x_t at time t, but has limited space   324

16.2   Boyer–Moore majority voting algorithm   325

16.3   Misra–Gries streaming algorithm for frequent elements   327

16.4   Counting number of distinct elements   328

16.5   Combining reservoir sampling with the estimator for F_k 332

16.6   Estimating F_2 336

Tables

5.1   The dynamic programming table for Knapsack   95

8.1   Finite automaton transition function for the string *aabb* matching   163

8.2   Illustration of matching using KMP failure function f for the pattern *abababca*.   164

12.1   Creating an instance of decision-knapsack from a given instance of 3-SAT   245

14.1   Consecutive snapshots of the list ranking algorithm on 15 elements   293

Preface

This book embodies a distillation of topics that we, as educators, have frequently covered in the past two decades in various postgraduate and undergraduate courses related to *Design and Analysis of Algorithms* in IIT Delhi. The primary audience were the junior level (3rd year) computer science (CS) students and the first semester computer science post-graduate students. This book can also serve the purpose of material for a more advanced level algorithm course where the reader is exposed to alternate and more contemporary computational frameworks that are becoming common and more suitable.

A quick glance through the contents will reveal that about half of the topics are covered by many standard textbooks on algorithms like those by Aho et al. [7], Horowitz et al. [65], Cormen et al. [37], and more recent ones like those by Kleinberg and Tardos [81] and Dasgupta et al. [40]. The first classic textbook in this area, viz., that by Aho et al., introduces the subject with the observation 'The study of algorithms is at the very heart of computer science' and this observation has been reinforced over the past five decades of rapid development of computer science as well as of the more applied field of information technology. Because of its foundational nature, many of the early algorithms discovered about five decades ago continue to be included in every textbook written including this one – for example, algorithms like FFT, quicksort, Dijkstra's shortest paths, etc.

What motivated us to write another book on algorithms are the several important and subtle changes in the understanding of many computational paradigms and the relative importance of techniques emerging out of some spectacular discoveries and changing technologies. As teachers and mentors, it is our responsibility to inculcate the right focus in the younger generation so that they continue to enjoy this intellectually critical activity and contribute to the enhancement of the field of study. As more and more human activities are becoming computer-assisted, it becomes obligatory to emphasize and reinforce the importance of efficient and faster algorithms, which is the core of any automated process. We are often limited and endangered by the instictive use of ill-designed and brute force algorithms, which are often erroneous, leading to fallacious scientific conclusions or incorrect policy decisions. It is therefore important to introduce some formal aspects of algorithm design and analysis into the school curriculum at par with maths and science, and sensitize students about this subject.

Who can use it

The present book is intended for students who have acquired skills in programming as well as basic data structures like arrays, stacks, lists, and even some experience with balanced trees. The authors, with a long experience behind them in teaching this subject, are convinced that algorithm design can be a deceptively hard subject and a gentle exposure is important for, both, understanding and sustaining interest. In IIT Delhi, CS undergraduates do a course in programming followed by a course in data structures with some exposure to basic algorithmic techniques. This book is intended for students having this background and so we have avoided any formal introduction of basic data structures including elementary graph searching methods like BFS/DFS. Instead, the book focusses on a mathematical treatment of the previously acquired knowledge and emphasizes a clean and crisp analysis of any new idea and technique. The CS students in IIT Delhi would have done a course in discrete mathematics and probability before they do this course. The design of efficient algorithms go hand-in-hand with our ability to quickly screen intuitions that lead to poor algorithms – both in terms of efficiency and correctness. We have consciously avoided topics that require long and dry formalism, although we have emphasized rigor at every juncture.

An important direction that we have pursued is based on the significance of adapting algorithm design to the computational environment. Although there has been a long history of research in designing algorithms for real-world models such as parallel and cache-hierarchy models, these have remained in the realms of niche and specialized graduate courses. The tacit assumption in basic textbooks is that we are dealing with uniform cost random access machines (RAMs). It is our firm belief that algorithm design is as much a function of the specific problem as the target model of execution, and failing to recognize this aspect makes the exercise somewhat incomplete and ineffective. Therefore, trying to execute the textbook data structures on a distributed model or Dijkstra's algorithm in a parallel computer would be futile. In summary,

$$Algorithms = ProblemDefinition + Model$$

The last three chapters specifically address three very important environments, namely parallel computing, memory hierarchy, and streaming. They form the core of a course taught in IIT Delhi, *Model Centric Algorithm Design* – some flavor can add diversity to a core course in algorithms. Of course, any addition to a course would imply proportionate exclusion of some other equally important topic – so it is eventually the instructor's choice.

Another recurring theme in the book is the liberal use of randomized techniques in algorithm design. To help students appreciate this aspect, we have described some basic tools and applications in Chapter 2. Even for students who are proficient in the use of probabilistic calculations (we expect all CS majors to have one college level course in probability), may find these applications somewhat non-intuitive and surprising – however, this may also turn into a very versatile and useful tool for anyone who is mathematically minded.

The other major development over the past decade is an increasing popularity of algebraic (particularly spectral) methods for combinatorial problems. This has made the role of conventional continuous mathematics more relevant and important. Reconciling and bridging the two distinct worlds of discrete and continuous methods is a huge challenge to even an experienced researcher, let alone an average student. It is too difficult to address this in a book like ours but we have tried to present some flavor in Chapter 12, which is an introduction to the technique of random projections.

Each chapter is followed by some brief discussion on some historical origins of the problem and pointers to relevant existing literature. The subsections/sections/chapters marked with * are more suitable for the advanced reader and may be skipped by others without loss of continuity.

One of the primary objectives of a course on algorithms is to encourage an appreciation for creativity without sacrificing rigor – this aspect makes algorithm design one of the most challenging and fascinating intellectual pursuit.

Suggested use of the chapters

The material presented in the sixteen chapters can be taught over two semesters at a leisurely pace, for example, in a two sequence course on algorithms. Alternately, for a first course on algorithms (with prior background in basic data structures), the instructor can choose majority portions from Chapters 3 to 11 and parts of Chapter 12. An advanced course can be taught using material from Chapters 12–16. Chapters 14–16 can form the crux of a course on *model centric algorithm design* which can be thought of as a more pragmatic exposure to theory of computation using contemporary frameworks.

Sandeep Sen
Amit Kumar
New Delhi, 2019

Acknowledgments

The authors would like to acknowledge their respective PhD advisors, John Reif and Jon Kleinberg, as their inspiring *gurus* in their journey in the algorithmic world. Sandeep Sen would also like to acknowledge Isaac Scherson, his Masters supervisor, for his motivating support to pursue algorithmic research.

The authors would like to thank many experienced researchers and teachers for their encouraging comments, including Pankaj Agarwal, Gary Miller, Sariel Har Peled, Jeff Vitter, Ravi Kannan, Sachin Maheshwari, Bernard Chazelle, Sartaj Sahni, Arijit Bishnu, and Saurabh Ray. More specifically, the authors would like to acknowledge Kurt Mehlhorn for sharing his notes on Gaussian Elimination in the finite precision model and Surender Baswana for his careful reading of the section on Graph Spanners.

It will also be fitting to acknowledge the numerous students in IIT Delhi, CSE department who have taken our courses in the area of algorithms, whose probing questions and constant prodding have contributed immensely to the shaping of the current contents.

Sandeep Sen would like to acknowledge the patience and support of his wife, Anuradha, who showed exemplary tolerance in sacrificing many hours that was due to her and his son, Aniruddha, for keeping the adrenaline going as a reminder of what the next generation is likely to be interested in. Amit Kumar would like to acknowledge his wife, Sonal, for her unwavering support and patience, and daughters, Aanvi and Anshika, for their love and affection. He would also like to acknowledge his parents for their encouragement and inspiration.

Since typists have long become defunct, it is obligatory to acknowledge the contributions of Donald Knuth for TeX and Leslie Lamport's LaTeX manual for the ease and convenience of writing such textbooks. For the sake of the environment and future of the world, we hope that long-term dissemination of such books will be in the electronic medium.

Model and Analysis

When we make a claim such as *Algorithm A has running time $O(n^2 \log n)$*, we have an underlying computational model where this statement is valid. It may not be true if we change the model. Before we formalize the notion of a *computational model*, let us consider the example of computing Fibonacci numbers.

1.1 Computing Fibonacci Numbers

One of the most popular sequences is the *Fibonacci* sequence defined by

$$F_i = \begin{cases} 0 & i = 0 \\ 1 & i = 1 \\ F_{i-1} + F_{i-2} & \text{otherwise for } i \geq 2 \end{cases}$$

It is left as an exercise problem to show that

$$F_n = \frac{1}{\sqrt{5}}(\phi^n - \phi'^n) \text{ where } \phi = \frac{1 + \sqrt{5}}{2} \quad \phi' = 1 - \phi$$

Clearly, it grows exponentially with n and F_n has $\theta(n)$ bits.

Since the closed form solution for F_n involves the *golden ratio* – an irrational number – we must find a way to compute it efficiently without incurring numerical errors or approximations as it is an integer.

Method 1

By simply using the recursive formula, one can easily argue that the number of operations (primarily additions) involved is proportional to the value of F_n. We just need to unfold the recursion tree where each internal node corresponds to an addition. As we had noted earlier, this leads to an exponential time algorithm and we cannot afford it.

Method 2

Observe that we only need the last two terms of the series to compute the new term. Hence, by applying the principle of *dynamic programming*,[1] we successively compute F_i starting with $F_0 = 0$ and $F_1 = 1$ and use the previously computed terms, F_i and F_{i-1} for $i \geq 2$.

This takes time that is proportional to approximately n additions, where each addition involves adding (increasingly large) numbers. The size of $F\lceil n/2 \rceil$ is about $n/2$ bits; so, the last $n/2$ computations will take $\Omega(n)$ steps [2] culminating in an $O(n^2)$ algorithm.

Since the nth Fibonacci number is at most n bits, it is reasonable to look for a faster algorithm.

Method 3

$$\begin{bmatrix} F_i \\ F_{i-1} \end{bmatrix} = \begin{bmatrix} 1 & 1 \\ 1 & 0 \end{bmatrix} \begin{bmatrix} F_{i-1} \\ F_{i-2} \end{bmatrix}$$

By iterating the aforementioned equation, we obtain

$$\begin{bmatrix} F_n \\ F_{n-1} \end{bmatrix} = \begin{bmatrix} 1 & 1 \\ 1 & 0 \end{bmatrix}^{n-1} \begin{bmatrix} 1 \\ 0 \end{bmatrix}$$

To compute A^n, where A is a square matrix, we recall the following strategy for recursively computing x^n for a real x and positive integer n.

$$\begin{cases} x^{2k} = (x^k)^2 & \text{for even integral powers} \\ x^{2k+1} = x \cdot x^{2k} & \text{for odd integral powers} \end{cases}$$

We can extend this method to compute A^n.

The number of multiplications taken by the aforementioned approach to compute x^n is bounded by $2 \log n$ (the reader can convince oneself by writing a recurrence). However, the actual running time depends on the time taken to multiply two numbers, which in turn depends on their lengths (number of digits). Let us assume that $M(n)$ is the number of (bit-wise) steps to multiply two n bit numbers. The number of steps to implement the aforementioned approach must take into account the lengths of numbers that are being multiplied. The following observations will be useful.

---

[1]The reader who is unfamiliar with this technique may refer to a later chapter, Chapter 5, that discusses it in complete detail.

[2]Adding two k bit numbers takes $\Theta(k)$.

The length of x^k is bounded by $k \cdot |x|$, where $|x|$ is the length of x.

Therefore, the cost of the the squaring of x^k is bounded by $M(k|x|)$. Similarly, the cost of computing $x \times x^{2k}$ can also be bound by $M(2k|x|)$. The overall recurrence for computing x^n can be written as

$$T_B(n) \leq T_B(\lfloor n/2 \rfloor) + M(n|x|)$$

where $T_B(n)$ is the number of bit operations to compute the nth power using the previous recurrence. The solution of the aforementioned recurrence can be written as the following summation (by unfolding)

$$\sum_{i=1}^{\log n} M(2^i|x|)$$

If $M(2i) > 2M(i)$, then this summation can be bounded by $O(M(n|x|))$, that is, the cost of the last squaring operation.

In our case, A is a 2×2 matrix – each squaring operation involves 8 multiplications and 4 additions involving entries of the matrix. Since multiplications are more expensive than additions, let us count the cost of multiplications only. Here, we have to keep track of the lengths of the entries of the matrix. Observe that if the maximum size of an entry is $|x|$, then the maximum size of an entry after squaring is at most $2|x| + 1$ (Why?). The cost of computing A^n is $O(M(n|x|))$, where the maximum length of any entry is $|x|$ (left as an exercise problem). Hence, the running time of computing F_n using Method 3 is dependent on the multiplication algorithm. Well, multiplication is multiplication – what can we do about it? Before that, let us summarize what we know about it. Multiplying two n digit numbers using the add-and-shift method takes $O(n^2)$ steps, where each step involves multiplying two single digits (bits in the case of binary representation), and generating and managing carries. For binary representation, this takes $O(n)$ steps for multiplying with each bit; finally, n shifted summands are added – the whole process takes $O(n^2)$ steps.

Using such a method of multiplication implies that we cannot do better than $\Omega(n^2)$ steps to compute F_n. For any significant (asymptotically better) improvement, we must find a way to multiply faster.

1.2 Fast Multiplication

Problem Given two numbers A and B in binary, we want to compute the product $A \times B$.

Let us assume that the numbers A and B have lengths equal to $n = 2^k$ – this will keep our calculations simpler without affecting the asymptotic analysis.

$$A \times B = (2^{n/2} \cdot A_1 + A_2) \times (2^{n/2} \cdot B_1 + B_2)$$

where A_1 (B_1) is the leading $n/2$ bits of A (B). Likewise, A_2 is the trailing $n/2$ bits of A. We can expand this product as

$$A_1 \times B_1 \cdot 2^{n/2} + (A_1 \times B_2 + A_2 \times B_1) \cdot 2^{n/2} + A_2 \times B_2$$

Observe that multiplication by 2^k can be easily achieved in binary by adding k trailing 0s (likewise, in any radix r, multiplying by r^k can be done by adding trailing zeros). Hence, the product of two n bit numbers can be achieved by recursively computing four products of $n/2$ bit numbers. Unfortunately, this does not improve things (see exercise 1.6).

We can achieve an improvement by reducing it to three recursive calls of multiplying $n/2$ bit numbers by rewriting the coefficient of $2^{n/2}$ as follows

$$A_1 \times B_2 + A_2 \times B_1 = (A_1 + A_2) \times (B_1 + B_2) - (A_1 \times B_1) - (A_2 \times B_2)$$

Although strictly speaking, $A_1 + A_2$ is not $n/2$ bits but at most $n/2 + 1$ bits (Why?), we can still view this as computing three separate products involving $n/2$ bit numbers recursively and subsequently subtracting appropriate terms to get the required products. Subtraction and additions are identical in modulo arithmetic (2's complement), so the cost of subtraction can be bounded by $O(n)$. (What is the maximum size of the numbers involved in subtraction?). This gives us the following recurrence

$$T_B(n) \leq 3 \cdot T_B(n/2) + O(n)$$

where the last term accounts for addition, subtractions, and shifts. It is left as an exercise problem to show that the solution to this recurrence is $O(n^{\log_2 3})$. This running time is roughly $O(n^{1.7})$, which is asymptotically better than n^2 and therefore we have succeeded in designing an algorithm to compute F_n faster than n^2.

It is possible to multiply much faster using a generalization of the aforementioned method in $O(n \log n \log \log n)$ bit operations utilizing Schonage and Strassen's method. However, this method is quite involved as it uses discrete Fourier transform computation over modulo integer rings and has fairly large constants that neutralize the advantage of the asymptotic improvement unless the numbers are a few thousand bits long. It is, however, conceivable that such methods will become more relevant as we may need to multiply large keys for cryptographic/security requirements. We discuss this algorithm in Chapter 9.

1.3 Model of Computation

Although there are a few thousand variations of the computer with different architectures and internal organization, it is best to think about them at the level of the assembly language. Despite architectural variations, the assembly level language support is very similar – the major difference being in the number of registers and the word length of the machine. However, these parameters are also in a restricted range of a factor of two, and hence, asymptotically in the same ballpark. In summary, we can consider any computer as a machine that supports a basic instruction set consisting of arithmetic and logical operations and memory accesses (including indirect addressing). We will avoid cumbersome details of the exact instruction set and assume realistically that any instruction of one machine can be simulated using a constant number of available instructions of another machine. Since analysis of algorithms involves counting the number of operations and not the exact timings (which could differ by an order of magnitude), the aforementioned simplification is justified.

The careful reader would have noticed that during our detailed analysis of Method 3 in the previous section, we were not simply counting the number of arithmetic operations but actually the number of bit-level operations. Therefore, the cost of a multiplication or addition was not unity but proportional to the length of the input. Had we only counted the number of multiplications for computing x^n, it would only be $O(\log n)$. This would indeed be the analysis in a *uniform cost* model, where only the number of arithmetic (also logical) operations are counted and the cost does not depend on the length of the operands. A very common use of this model is for comparison-based problems like sorting, selection, merging, and many data-structure operations. For these problems, we often count only the number of comparisons (not even other arithmetic operations) without bothering about the length of the operands involved. In other words, we implicitly assume $O(1)$ cost for any comparison. This is not considered unreasonable since the size of the numbers involved in sorting does not increase during the course of the algorithm for most of the commonly known sorting problems. On the other hand, consider the following problem of repeated squaring n times starting with 2. The resultant is a number 2^{2^n}, which requires 2^n bits to be represented. It will be very unreasonable to assume that a number that is exponentially long can be written out (or even stored) in $O(n)$ time. Therefore, the uniform cost model will not reflect a realistic setting for this problem.

On the other extreme is the *logarithmic* cost model where the cost of an operation is proportional to the length of the operands. This is very consistent with the physical world and is also similar to the *Turing machine* model which is a favorite of complexity theorists. Our analysis in the previous section is actually done with this model in mind. It is not only the arithmetic operations but also the cost of memory access that is proportional to the length of the address and the operand.

The most commonly used model is something in between. We assume that for an input of size n, any operation involving operands of size $\log n$ [3] takes $O(1)$ steps. This is

---

[3] We can also work with $c \log n$ bits as the asymptotic analysis does not change for a constant c.

justified as follows. All microprocessor chips have specialized *hardware circuits* for arithmetic operations like multiplication, addition, division, etc. that take a fixed number of clock cycles when the operands fit into a word. The reason that $\log n$ is a natural choice for a word is that, even to address an input size n, you require $\log n$ bits of address space. The present high-end microprocessor chips have typically 2–4 GBytes of RAM and about 64 bits word size – clearly 2^{64} exceeds 4 GBytes. We will also use this model, popularly known as *random access machine* (or RAM in short), except for problems that deal with numbers as inputs like multiplication in the previous section where we will invoke the *log cost* model. In the beginning, it is desirable that for any algorithm, we get an estimate of the maximum size of the numbers to ensure that operands do not exceed $\Omega(\log n)$ so that it is safe to use the RAM model.

1.4   Randomized Algorithms: A Short Introduction

The conventional definition of an algorithm demands that an algorithm solves a given instance of a problem *correctly* and *certainly*, that is, for any given instance I, an algorithm $\mathcal{A}$ should return the correct output every time without fail. It emphasizes a deterministic behavior that remains immutable across multiple runs. By exploring beyond this conventional boundary, we have some additional flexibility that provides interesting trade-offs between correctness and efficiency, and also between predictability and efficiency. These are now well-established techniques in algorithm design known as *randomized techniques*. In this section, we provide a brief introduction to these alternate paradigms, and in this textbook, we make liberal use of the technique of randomization which has dominated algorithm design in the past three decades leading to some surprising results as well as simpler alternatives to conventional design techniques.

Consider an array A of n elements such that each element is either colored red or green. We want to output an index i, such that $A[i]$ is green. Without any additional information or structure, we may end up inspecting every element of A to find a green element. Suppose we are told that half the elements are colored green and the remaining red. Even then we may be forced to probe $n/2$ elements of the array before we are assured of finding a green element since the first $n/2$ elements that we probe could be all red. This is irrespective of the distribution of the green elements. Once the adversary knows the probe sequence, it can force the algorithm to make $n/2$ probes.

Let us now assume that all $\binom{n}{n/2}$ choices of green elements are equally likely – in what way can we exploit this? With a little reflection, we see that every element is equally likely to be red or green and therefore, the first element that we probe may be green with probability $= 1/2$. If so, we are done – however, it may not be green with probability $1/2$. Then, we can probe the next location and so on until we find a green element. From our earlier argument, we may have to probe at most $n/2$ locations before we succeed. But there is a crucial difference – it is very *unlikely* that in a random placement of green elements, all the first $n/2$ elements are red. Let us make this more precise.

If the first $m < n/2$ elements are red, it implies that all the green elements got squeezed in the $n - m$ locations. If all placements are equally likely, then the probability of this scenario is

$$\frac{\binom{n-m}{n/2}}{\binom{n}{n/2}} = \frac{(n-m)! \cdot (n/2)!}{n! \cdot (n/2-m)!} = \frac{(n-m)(n-m-1)\cdots(n/2-m+1)}{n(n-1)\cdots(n/2+1)}$$

It is easy to check that this probability is at most $e^{-m/2}$. Therefore, the expected number of probes is at most

$$\sum_{m \geq 0} (m+1) \cdot e^{-m/2} = O(1)$$

In the previous discussion, the calculations were based on the assumption of random placement of green elements. Can we extend it to the general scenario where no such assumption is required? This turns out to be surprisingly simple and obvious once the reader realizes it. The key to this is – instead of probing the array in a pre-determined sequence $A[1], A[2], \ldots$, we probe using a random sequence, say $j_1, j_2, \ldots, j_n$, where $j_1, \ldots, j_n$ is a permutation of $\{1, \ldots, n\}$.

How does this change things ? Since $n/2$ locations are green, a random probe will yield a green element with probability $1/2$. If it is not green, then the subsequent random probes (limited to the unprobed locations) will have even higher probability of the location having a green element. This is a simple consequence of conditional probability given that all the previous probes yielded red elements. To formalize, let X be a random variable that denotes the number of probes made to find the first green element. Then,

$\Pr[X = k] =$ The probability that the initial $k-1$ probes are red and the k-th probe is green

$$\leq 1/2^k$$

The reader must verify the correctness of this expression. The expression can also be modified to yield

$$\Pr[X \geq k] \leq \sum_{i=k}^{i=n/2} 1/2^i \leq 1/2^{k-1},$$

and the expected number of probes is at most $O(1)$.

This implies that the number of probes not only decreases exponentially with k but is *independent of the placement of the green elements*, that is, the worst-case scenario is over all possible input arrays. Instead of relying on the randomness of the placement (which is not in our control), the algorithm itself uses a random probe sequence matching the same phenomenon. This is the essence of a *randomized* algorithm. In this case, the final result is always correct, that is, a green element is output but the running time (number of probes) is a random variable and there is a trade-off between the number of probes k and the probability of termination within k probes.

If the somewhat hypothetical problem of finding a green element from a set of elements has not been convincing in terms of its utility, here is a classical application of the aforementioned solution. Recall the quicksort sorting algorithm. In quicksort, we partition a given set of n numbers around a pivot. It is known that the efficiency of the algorithm depends primarily on the relative sizes of the partition – the more balanced they are in size, the better. Ideally, one would like the pivot to be the median element so that both sides of the partition are small. Finding the median element is a problem in itself; however, any element around the median is almost equally effective, say an element with rank[4] between $[\frac{n}{4}, \frac{3n}{4}]$ will also lead to a balanced partitioning. These $n/2$ elements can be thought of as the green elements and so we can apply our prior technique. There is a slight catch – how do we know that the element is green or red? For this, we need to actually compute the rank of the probed element, which takes $n-1$ comparisons but this is acceptable since the partitioning step in quicksort takes n steps and will subsume this. However, this is not a complete analysis of quicksort which is a recursive algorithm; we require more care that will be discussed in a later chapter dealing with selections.

1.4.1   A different flavor of randomized algorithms

Consider a slight twist on the problem of computing the product of two $n \times n$ matrices $C = A \times B$. We are actually given A, B, C and we have to verify if C is indeed the product of the two matrices A and B. We may be tempted to actually compute $A \times B$ and verify it element by element with C. In other words, let $D = A \times B$ and we check if $C - D = \mathbb{O}^n$, where the right-hand side is an $n \times n$ matrix whose elements are identically 0.

This is a straightforward and simple algorithm, except that we will pay the price for computing the product which is not really necessary for the problem. Using elementary method for computing matrix products, we will need about $O(n^3)$ multiplications and additions[5], whereas an ideal algorithm could be $O(n^2)$ steps, which is the size of the input. To further simplify the problem and reduce dependency on the size of each element, let us consider Boolean matrices and review addition modulo 2. Examine the algorithm described in Figure 1.1. It computes three matrix vector products – BX, $A(BX)$, and CX–incurring a total of $3n^2$ operations which matches the input size and therefore, is optimal.

---

[4]The rank of x is the number of elements in the set smaller than x.

[5]There are sophisticated and complex methods to reduce the number of multiplications below n^3 but they are still much more than n^2.

Procedure Verifying matrix product(A, B, C)

---

1 Input: A, B, C are $n \times n$ matrices over GF(2);
2 Output: If $A \cdot B = C$ then Yes else No;
3 Choose a random 0–1 vector X;
4 **if** $A \cdot (B \cdot X) = C \cdot X$ **then**
5 | Return **YES**;
6 **else**
7 | Return **NO**

Figure 1.1 *Algorithm for verifying matrix product*

Observation If $A(BX) \neq CX$, then $AB \neq C$.

However, the converse, that is, $A(BX) = C \implies AB = C$ is not easy to see. On the contrary, consider the following example, which raises serious concerns.

Example 1.1 $A = \begin{bmatrix} 1 & 1 \\ 1 & 0 \end{bmatrix}$ $B = \begin{bmatrix} 0 & 1 \\ 1 & 0 \end{bmatrix}$ $C = \begin{bmatrix} 1 & 0 \\ 0 & 1 \end{bmatrix}$ $AB = \begin{bmatrix} 1 & 1 \\ 1 & 0 \end{bmatrix}$

$$X = \begin{bmatrix} 1 \\ 0 \end{bmatrix} \quad ABX = \begin{bmatrix} 1 \\ 0 \end{bmatrix} \quad CX = \begin{bmatrix} 1 \\ 0 \end{bmatrix}$$

$$X' = \begin{bmatrix} 0 \\ 1 \end{bmatrix} \quad ABX' = \begin{bmatrix} 1 \\ 0 \end{bmatrix} \quad CX' = \begin{bmatrix} 0 \\ 1 \end{bmatrix}$$

Clearly, the algorithm is not correct if we choose the first vector. Instead of giving up on this approach, let us get a better understanding of the behavior of this simple algorithm.

Claim 1.1 *For an arbitrary vector (non-zero) Y and a random vector X, the probability that the dot product $X \cdot Y = 0$ is less than 1/2.*

There must be at least one $Y_i \neq 0$ – choose that X_i last; with probability $1/2$, it will be non-zero. For the overall behavior of the algorithm, we can claim the following.

Claim 1.2 *If $A(BX) \neq CX$, then $AB \neq C$, that is, the algorithm is always correct if it answers NO. When the algorithm answers YES, then $\Pr[AB = C] \geq 1/2$.*

If $AB \neq C$, then in $AB - C$, at least one of the rows is non-zero and from the previous claim, the dot product of a non-zero vector with a random vector is non-zero with probability $1/2$. It also follows that by repeating this test and choosing independently another random vector when it returns YES, we can improve the probability of success and our confidence in the result. If the algorithm returns k consecutive YES, then $\Pr[AB \neq C] \leq \frac{1}{2^k}$.

The reader may have noted that the two given examples of randomized algorithms have distinct properties. In the first example, the answer is always correct but the running time has a probability distribution. In the latter, the running time is fixed, but the answer may be incorrect with some probability. The former is known as *Las Vegas* and the latter is referred to as *Monte Carlo* randomized algorithm. Although in this particular example, the Monte Carlo algorithm exhibits asymmetric behavior (it can be incorrect only when the answer is YES), it need not be so.

1.5   Other Computational Models

There is clear trade-off between the simplicity and the fidelity achieved by an abstract model. One of the obvious (and sometimes serious) drawbacks of the RAM model is the assumption of unbounded number of registers since the memory access cost is uniform. In reality, there is a memory hierarchy comprising registers, several levels of cache, main memory, and finally the disks. We incur a higher access cost as we go from registers toward disks and for technological reasons, the size of the faster memory is limited. There could be a disparity of 10^5 between the fastest and the slowest memory which makes the RAM model somewhat suspect for larger input sizes. This has been redressed by the *external memory model*.

1.5.1   External memory model

In this model, the primary concern is the number of disk accesses. Given the rather high cost of a disk access compared to any CPU operation, this model actually ignores all other costs and counts only the number of disk accesses. The disk is accessed as contiguous memory locations called *blocks*. The blocks have a fixed size B and the simplest model is parameterized by B and the size of the faster memory M. In this two-level model, the algorithms are only charged for transferring a block between the internal and external memory; all other computations are free. The cost of sorting n elements is $O\left(\frac{n}{B} \log_{M/B} \frac{n}{B}\right)$ disk accesses and this is also optimal. To see this, we can analyze M/B-way merge sort in this model. Note that one block from each of the M/B sorted streams can fit into the main memory. Using appropriate data structures, we can generate the next B elements of the output and we can write an entire block to the output stream. Hence, the overall number of I-Os per phase is $O(n/B)$ since each block is read and written exactly once. The algorithm makes $O(\frac{n/B}{M/B})$ passes, yielding the required bound.

There are further refinements to this model that parameterizes multiple levels and also accounts for internal computation. As the model becomes more complicated, designing algorithms also becomes more challenging and often more laborious. We discuss algorithm design and analysis in this model and many variations in Chapter 15.

1.5.2 Parallel model

The basic idea of parallel computing is extremely intuitive and a fundamentally intellectual pursuit. At the most intuitive level, it symbolizes what can be achieved by cooperation among individuals in terms of expediting an activity. It is not about division of labor (or specialization), but actually assuming similar capabilities. Engaging more laborers clearly speeds up construction; similarly, using more than one processor is likely to speed-up computation. Ideally, by using p processors, we would like to obtain a p-fold speed-up over the conventional algorithms; however, the principle of decreasing marginal utility shows up. One of the intuitive reasons for this is that with more processors (as with more individuals), the communication requirements tend to dominate after a while. But more surprisingly, there are algorithmic constraints that pose serious limitations to our objective of obtaining proportional speed-up.

This is best demonstrated in the model called *PRAM* (or parallel random access machine) which is the analog of RAM. Here p processors are connected to a shared memory and the communication happens through reading and writing in a globally shared memory. It is left to the algorithm designer to avoid read and write conflicts. It is further assumed that all operations are synchronized globally and that there is no cost of synchronization. In this model, there is no extra overhead for communication as it is charged in the same way as a local memory access. Even in this model, it has been shown that it is not always possible to obtain ideal speed-up. As an example, consider the elementary problem of finding the minimum of n elements. It has been proved that with n processors, the time (parallel time) is at least $\Omega(\log\log n)$. For certain problems, like *depth first search* of graphs, it is known that even if we use any polynomial number of processors, we cannot obtain polylogarithmic time! So, clearly not all problems can be parallelized effectively.

A more realistic parallel model is the *interconnection network* model that has an underlying communication network, usually a regular topology like a two-dimensional mesh, hypercube, etc. These can be embedded into VLSI (very large scale integration) chips and be scaled according to our needs. To implement a parallel algorithm, we have to design efficient schemes for data routing.

A very common model of parallel computation is a hardware circuit comprising basic logic gates. The signals are transmitted in parallel through different paths and the output is a function of the input. The *size* of the circuit is the number of gates and the (parallel) time is usually measured in terms of the maximum path length from any input gate to the output gate (each gate contributes to a unit delay). Those familiar with circuits for addition and comparison can analyze them in this framework. The carry–save adder is a low-depth circuit that adds two n-bit numbers in about $O(\log n)$ steps, which is much faster than a sequential circuit that adds one bit at a time taking n steps.

An example Given numbers $x_1, x_2, \ldots, x_n$, consider the problem of computing the terms $S_i = \sum_{j=1}^{i} x_j$ for all $1 \leq i \leq n$. Each term corresponds to a partial sum. It is trivial to compute all the partial sums in $O(n)$ steps. Computing S_i for each i can be done in parallel using a binary tree of depth $\lceil \log i \rceil$, where the inputs are given at the leaf nodes and each internal node corresponds to a summation operation. All the summations at the same level can be done simultaneously and the final answer is available at the root. Doing this computation independently for each S_i is wasteful since $S_{i+1} = S_i + x_{i+1}$ that will be about $O(n^2)$ additions compared to the sequential complexity of $O(n)$.

Instead we use the following idea. Add every odd–even pair of inputs into a single value $y_{i/2} = x_{i-1} + x_i$, for every even i (assume n is a power of two). Now compute the partial sums $S_1', S_2', \ldots, S_{n/2}'$ recursively. Note that $S_j' = \sum_{k=1}^{2j} x_k = S_{2j}$, that is, half the terms can be computed this way. To obtain S_{2j+1}, $0 \leq j \leq n/2 - 1$, add x_{2j+1} to S_j'. This can also be done simultaneously for all terms.

This recursive description can be unfolded to yield a parallel circuit for the computation. The algorithm can be generalized for any arbitrary associative operation and is known as *parallel prefix* or *scan* operation. Using an appropriately defined composition function for a semi-adder (adding two bits given a carry), we can construct the carry–save adder circuit. In Chapter 14, we formally introduce parallel computation models and discuss parallel algorithm design techniques for many basic problems including parallel prefix.

One of the most fascinating developments is the quantum model, which is inherently parallel but also fundamentally different from the previous models. A breakthrough result in recent years is a polynomial time algorithm [134] for factorization, which forms the basis of many cryptographic protocols in the conventional model. The interested reader may learn the basics of quantum computing from introductory textbooks like the one by Nielsen and Chuang [111].

Biological computing models is a very active area of research where scientists are trying to assemble a machine out of DNA strands. It has potentially many advantages over silicon-based devices and is inherently parallel. Adleman [2] was one of the earliest researchers to construct a prototype to demonstrate its potential.

Further Reading

The dependence between algorithm designs and computation models is often not highlighted enough. One of the earliest textbooks on algorithm design [7] had addressed this very comprehensively by establishing precise connections between random access machine (RAM) and random access stored program (RASP) as well as between the uniform and the logarithmic cost models. However, over the last two decades, researchers have shown how to exploit word models to improve algorithms based on comparison models – see Fredman and Willard [54] – that breaches the $\Omega(n \log n)$ lower bound for comparison sorting. Shamir [132] had shown that factorization of a given integer can be done in $O(\log n)$ arithmetic steps if very large integers can be allowed as operands. A very esoteric field of theoretical computer science is *complexity theory*, where precise relations are characterized between various computational models [13, 114].

Fibonacci sequence is one of the most popular recurrences in computer science and also quite useful in applications such as Fibonacci search (see Knuth's work [83]) and Fibonacci heaps [53]. The divide-and-conquer algorithm for multiplication is known as Karatsuba's algorithm (described by Knuth [82]). Algorithms for multiplication and division attracted early attention [17] and continues to be a tantalizing issue, as is it is indeed asymptotically harder than addition.

Randomized algorithm and probabilistic techniques opened up an entirely new dimension in algorithm design which is both elegant and powerful. Starting with the primality testing algorithms [95, 103, 136], it provided researchers with many surprising alternatives that changed the perspective of computer science. Readers are encouraged to refer to the textbook by Motwani and Raghavan [106] for a very comprehensive application of such methods.

In the later chapters of this book, we provide a more detailed introduction to alternate models of algorithm design such as parallel, external memory, and streaming models. An experienced algorithm designer is expected to find the right match between an algorithm and a model for any specific problem.

Exercise Problems

1.1 Solve the following recurrence equations given $T(1) = O(1)$

   (a) $T(n) = T(n/2) + bn \log n$

   (b) $T(n) = aT(n-1) + bn^c$

1.2 Show that

$$F_n = \frac{1}{\sqrt{5}}(\phi^n - \phi'^n) \quad \phi = \frac{1 + \sqrt{5}}{2} \quad \phi' = 1 - \phi$$

where F_n is the nth Fibonacci number. Use the recurrence for F_n and solve it using the generating function technique.

Prove that

$$F_n = 1 + \sum_{i=0}^{n-2} F_i$$

1.3 An AVL tree is a balanced binary search tree that satisfies the following invariant. At every internal node (including the root node), the heights of the left subtree and the right subtree can differ by at most one.

Convert this invariant into an appropriate recurrence to show that an AVL tree with n nodes has height bounded by $O(\log n)$.

1.4 Show that the solution to the recurrence $X(1) = 1$ and

$$X(n) = \sum_{i=1}^{n} X(i)X(n-i) \text{ for } n > 1$$

is $X(n+1) = \frac{1}{n+1}\binom{2n}{n}$.

1.5 Show that the cost of computing A^n is $O(M(n|x|))$, where A is a 2×2 matrix and the largest element is x. Here $|.|$ denotes the size of a number.

1.6 (i) Show that the recurrence $T(n) = 4T(n/2) + O(n)$ has a solution $T(n) = \Omega(n^2)$.

(ii) The improved recurrence for multiplying two n-bit numbers is given by

$$T_B(n) \leq 3 \cdot T_B(n/2) + O(n)$$

With an appropriate terminating condition, show that the solution for the bit complexity $T_B(n)$ is $O(n^{\log_2 3})$.

(iii) Extend the idea of doing a two-way divide-and-conquer algorithm to multiply two n-bit numbers, to a four-way division by saving the number of lower order multiplications. Is there any advantage in extending this further ?

1.7 Given two polynomials $P_A(n) = a_{n-1}x^{n-1} + a_{n-2}x^{n-2} + \ldots + a_0$ and $P_B(n) = b_{n-1}x^{n-1} + b_{n-2}x^{n-2} + \ldots + b_0$, design a subquadratic ($o(n^2)$) time algorithm to multiply the two polynomials. You can assume that the coefficients a_i and b_i are $O(\log n)$ bits and can be multiplied in $O(1)$ steps.

1.8

$$\text{Let } fact(n) = \begin{cases} \binom{n}{2} \cdot (n/2)!^2 & \text{if } n \text{ is even} \\ n \cdot (n-1)! & n \text{ is otherwise} \end{cases}$$

This equation is similar to the recurrence for fast computation of x^n. Can you make use of it to compute a fast algorithm for computing factorials?

1.9 Let $p(x_1, x_2, \ldots, x_n)$ be a multivariate polynomial in n variables and degree d over a field $\mathbb{F}$ such that $p()$ is not identically 0. Let $\mathbb{I} \subseteq \mathbb{F}$ be a finite subset. Then the number of elements $Y \in \mathbb{I}^n$ such that $p(Y) = 0$ is bounded by $|\mathbb{I}|^{n-1} \cdot d$. Note that Y is an n tuple.

(i) Prove this using induction on n and the fundamental theorem of algebra.

(ii) Give an alternate proof of the matrix product verification $C = A \cdot B$ using this result.

Hint: What is the degree and the field size in this case?

1.10 Comparison model In problems related to selection and sorting, the natural and intuitive algorithms are based on comparing pairs of elements. For example, the minimum among a given set of n elements can be found in exactly $n-1$ comparisons.

(i) Show that no algorithm can correctly find the minimum in fewer than $n-1$ comparisons.

(ii) Show that $3\frac{n}{2}$ comparisons are sufficient and necessary to find the minimum and the maximum (both) of n elements.

1.11 Lop-sided search A company has manufactured shock-proof watches and it wants to test the strength of the watches before it publicizes its warranty that will be phrased as "Even if it drops from the Xth floor, it will not be damaged." To determine what X is, we have to conduct real experiments by dropping the watch from different floors. If it breaks when we drop it from, say, 10th floor, then X is clearly less than 10. However, in this experiment, we have lost a watch.

(i) If the watch can withstand a fall from the Xth floor but not the X+1th, what is the minimum number of trials we have to conduct if we are allowed to destroy at most one watch?

(ii) Let a pair $T_n = (k, m)$ denote that, to determine that the watch can withstand a fall from the nth floor (but not $n+1$), we can do this by breaking at most k watches and m trials. The previous problem alluded to $k = 1$.

Determine $(2, m)$ where you need to express m as a function of n.

(iii) For any constant integer k, determine m by writing an appropriate recurrence.

Contrast your results with binary search.

1.12 Given a positive integer n, design an efficient algorithm to find all the primes $\leq n$. Recall that the number of primes is $\Theta(n/\log n)$, so the running time should be close to this.

Hint: Use the sieve technique where we begin with a size n array and progressively cancel all the multiples of primes. The algorithm should not visit a location too many times if we want to achieve a running time close to $O(n)$, which is the time required to initialize the array.

1.13 In the field $\mathbb{F}_2$, there are two elements, namely 0 and 1, and addition and multiplication are performed modulo 2. Let y be a non-zero vector of length n. We choose a random vector x of length n by choosing each coordinate independently to be 0 or 1 uniformly at random. Prove that the probability that the dot product $(x \cdot y)$ mod 2 is 0 is exactly $1/2$.

Basics of Probability and Tail Inequalities

Randomized algorithms use random coin tosses to guide the progress of the algorithm. Although the actual performance of the algorithm may depend on the outcomes of these coin tosses, it turns out that one can often show that with reasonable probability, the algorithm has the desired properties. This model can dramatically improve the power of an algorithm. We will give examples where this ability can lead to very simple algorithms; in fact, sometimes, randomization turns out to be necessary. In this chapter, we begin with the basics of probability theory. We relate the notion of a random variable with the analysis of a randomized algorithm – often, the running time of a randomized algorithm will be a random variable. We will then describe techniques for bounding the probability of a random variable exceeding certain values, thereby bounding the running time.

Note Since randomized techniques have been extensively used as a basic tool, this chapter lays down some of the foundations of such applications for readers who are not familiar with this methodology. For others, this chapter can be used as reference as and when required.

2.1 Basics of Probability Theory

In this section, we provide a brief review of the axiomatic approach to probability theory. We will deal with the discrete case only. We begin with the notion of a sample space, often denoted by Ω. It can be thought of as the set of outcomes (or elementary events) in an experiment. For example, if we are rolling a dice, then Ω can be defined as the set of 6 possible outcomes. In an abstract setting, we will define Ω to be any set (which will be finite or countably infinite). To see an example where Ω can be infinite, consider the following experiment: we keep tossing a coin till we see a heads. Here the set of possible outcomes are infinite – for any integer $i \geq 0$, there is an outcome consisting of i tails followed by a heads. Given a sample space Ω, a *probability measure* Pr assigns a non-negative real value p_ω to each elementary event $\omega \in \Omega$. The probability measure Pr should satisfy the following condition:

$$\sum_{\omega \in \Omega} p_\omega = 1. \tag{2.1.1}$$

A *probability space* consists of a sample space Ω with a *probability measure* associated with the elementary events. In other words, a probability space is specified by a pair (Ω, Pr) of sample space and probability measure. Observe that the actual probability assigned to each elementary event (or outcome) is part of the axiomatic definition of a probability space. Often one uses prior knowledge about the experiment to come up with such a probability measure. For example, if we assume that a dice is fair, then we could assign equal probability, that is, $1/6$ to all the 6 outcomes. However, if we suspect that the dice is biased, we could assign different probabilities to different outcomes.

Example 2.1 *Suppose we are tossing 2 coins. In this case, the sample space is $\{HH, HT, TH, TT\}$. If we think all 4 outcomes are equally likely, then we could assign probability 1/4 to each of these 4 outcomes. However, assigning probability $0.3, 0.5, 0.1, 0.1$ to these 4 outcomes also results in a probability space.*

We now define the notion of an *event*. An event is a subset of Ω. The probability of an event E is defined as $\sum_{\omega \in E} p_\omega$, that is, the total sum of probabilities of all the outcomes in E.

Example 2.2 *Consider the experiment of throwing a dice, that is, $\Omega = \{1, 2, 3, 4, 5, 6\}$, and suppose the probabilities of these outcomes (in this sequence) are $0.1, 0.2, 0.3, 0.2, 0.1, 0.1$. Then, $\{2, 4, 6\}$ is an event (which can also be defined as the event that the outcome is an even number) whose probability is $0.2 + 0.2 + 0.1 = 0.5$.*

The following properties follow immediately from the definition of the probability of an event (proof deferred to exercises):

1. For all $A \subset \Omega$, $0 \leq \text{Pr}[A] \leq 1$

2. $\text{Pr}[\Omega] = 1$

3. For mutually disjoint events $E_1, E_2, \dots$, $\text{Pr}[\cup_i E_i] = \sum_i \text{Pr}[E_i]$

The principle of inclusion–exclusion also has its counterpart in the probabilistic world, namely

Lemma 2.1

$$\Pr[\cup_i E_i] = \sum_i \Pr[E_i] - \sum_{i<j} \Pr[E_i \cap E_j] + \sum_{i<j<k} \Pr[E_i \cap E_j \cap E_k] \ldots$$

Example 2.3 *Suppose we pick a number uniformly at random from 1 to 1000. We would like to calculate the probability that it is divisible by either 3 or 5. We can use the principle of inclusion–exclusion to calculate this. Let E be the event that it is divisible by either 3 or 5. Let E_1 be the event that it is divisible by 3 and E_2 be the event that it is divisible by 5. Clearly, $E = E_1 \cup E_2$. By the inclusion–exclusion principle*

$$\Pr[E] = \Pr[E_1] + \Pr[E_2] - \Pr[E_1 \cap E_2].$$

Clearly E_1 happens if we pick a multiple of 3. The number of multiples of 3 in the range $[1, 1000]$ is $\lfloor 1000/3 \rfloor = 333$, and so, $\Pr[E_1] = \frac{333}{1000}$. Similarly, $\Pr[E_2] = \frac{200}{1000}$. It remains to compute $\Pr[E_1 \cap E_2]$. But note that this is exactly the probability that the number is divisible by 15, and so, it is equal to $\frac{\lfloor 1000/15 \rfloor}{1000} = \frac{66}{1000}$. Thus, the desired probability is $467/1000$.

Definition 2.1 *The* conditional probability *of E_1 given E_2 is denoted by $\Pr[E_1|E_2]$ and is given by*

$$\frac{\Pr[E_1 \cap E_2]}{\Pr[E_2]}$$

assuming $\Pr[E_2] > 0$.

Definition 2.2 *A collection of events $\{E_i | i \in I\}$ is* independent *if for all subsets $S \subset I$*

$$\Pr[\cap_{i \in S} E_i] = \Pi_{i \in S} \Pr[E_i]$$

Remark E_1 and E_2 are independent if $\Pr[E_1|E_2] = \Pr[E_1]$.

The notion of independence often has an intuitive meaning – if two events depend on experiments which do not share any random bits respectively, then they would be independent. However, the converse may not be true, and so the only way to verify if two events are independent is to check the aforementioned condition.

Example 2.4 *Suppose we throw two dice. Let E_1 be the event that the sum of the two numbers is an even number. It is easy to check that $\Pr[E_1] = 1/2$. Let E_2 be the event that the first die has outcome "1". Clearly, $\Pr[E_2] = 1/6$. It is also clear that $\Pr[E_1 \cap E_2]$ is $1/12$ – indeed, for $E_1 \cap E_2$ to occur, the second die can have only 3 outcomes. Since $\Pr[E_1 \cap E_2] = \Pr[E_1] \cdot \Pr[E_2]$, these two events are independent.*

We now come to the notion of a *random variable*.

Definition 2.3 *A* random variable *(r.v.) X is a real-valued function over the sample space, $X : \Omega \to \mathbb{R}$.*

In other words, a random variable assigns a real value to each outcome of an experiment.

Example 2.5 *Consider the probability space defined by the throw of a fair die. Let X be function which is 1 if the outcome is an even number and 2 if the outcome is an odd number. Then, X is a random variable. Now consider the probability space defined by the throw of two fair dice (where each of the 36 outcomes are equally likely). Let X be a function which is equal to the sum of the values of the two dice. Then, X is also a random variable which takes values in the range* $\{2, \ldots, 12\}$.

With each random variable X, we can associate several events. For example, given a real x, we can define the event $[X \geq x]$ as the set $\{\omega \in \Omega : X(\omega) \geq x\}$. One can similarly define the events $[X = x], [X < x]$, and in fact, $[X \in S]$ for any subset S of real numbers.[1] The probability associated with the event $[X \leq x]$ (respectively, $[X < x]$) is known as the *cumulative density function*, cdf (respectively, *probability density function* or pdf); it helps us characterize the behavior of the random variable X. As in the case of events, one can also define the notion of independence for random variables. Two random variables X and Y are said to be independent if for all x and y in the range of X and Y respectively

$$\Pr[X = x, Y = y] = \Pr[X = x] \cdot \Pr[Y = y].$$

It is easy to check from this definition that if X and Y are independent random variables, then

$$\Pr[X = x | Y = y] = \Pr[X = x].$$

As in the case of events, we say that a set of random variables $X_1, \ldots, X_n$ are mutually independent if for all reals $x_1, \ldots, x_n$, where x_i lies in the range of X_i, for all $i = 1, \ldots, n$,

$$\Pr[X_1 = x_1, X_2 = x_2, \ldots, X_n = x_n] = \prod_{i=1}^{n} \Pr[X_i = x_i].$$

The *expectation* of an r.v. X, whose range lies in a (countable) set R, is denoted by $\mathbb{E}[X] = \sum_{x \in R} x \cdot \Pr[X = x]$. The expectation can be thought of as the typical value of X if we conduct the corresponding experiment. One can formalize this intuition – the law of large numbers states that if we repeat the same experiment many times, then the average value of X is very close to $\mathbb{E}[X]$ (and gets arbitrarily close as the number of experiments goes to infinity).

A very useful property of expectation, called the *linearity property*, can be stated as follows.

Lemma 2.2 *If X and Y are random variables, then*

$$\mathbb{E}[X + Y] = \mathbb{E}[X] + \mathbb{E}[Y]$$

Remark Note that X and Y do not have to be independent!

---

[1] We are only considering the case when X can be countably many different values.

Proof: Let us consider R to be the union of ranges of X and Y – we will assume that R is countable, though the result holds in general as well. We can also assume that both X and Y have range R (if $r \in R$ is not in the range of X, we can add it to the range of X with the provision that $\Pr[X = r] = 0$). Then,

$$\mathbb{E}[X + Y] = \sum_{r_1 \in R, r_2 \in R} (r_1 + r_2) \Pr[X = r_1, Y = r_2].$$

We proceed as follows:

$$\sum_{r_1 \in R, r_2 \in R} (r_1 + r_2) \Pr[X = r_1, Y = r_2] = \sum_{r_1 \in R, r_2 \in R} r_1 \cdot \Pr[X = r_1, Y = r_2]$$

$$+ \sum_{r_1 \in R, r_2 \in R} r_2 \Pr[X = r_1, Y = r_2]. \qquad (2.1.2)$$

If X and Y were independent, we could have just written $\Pr[X = r_1, Y = r_2]$ as $\Pr[X = r_1] \cdot \Pr[Y = r_2]$, and the result would follow trivially.

Now observe that $\sum_{r_1 \in R, r_2 \in R} r_1 \cdot \Pr[X = r_1, Y = r_2]$ can be written as $\sum_{r_1 \in R_1} r_1 \cdot \sum_{r_2 \in R_2} \Pr[X = r_1, Y = r_2]$. But we can see that $\sum_{r_2 \in R_2} \Pr[X = r_1, Y = r_2]$ is just $\Pr[X = x_1]$, and so $\sum_{r_1 \in R_1} r_1 \cdot \sum_{r_2 \in R_2} \Pr[X = r_1, Y = r_2]$ is the same as $\mathbb{E}[X]$. One can similarly show that the other term in the RHS of Eq. (2.1.2) is equal to $\mathbb{E}[Y]$. □

The linearity of the expectation property has many surprising applications, and can often be used to simplify many intricate calculations.

Example 2.6 *Suppose we have n letters meant for n different people (with their names written on the respective letters). Suppose we randomly distribute the letters to the n people (more formally, we assign the first letter to a person chosen uniformly at random, the next letter to a uniformly chosen person from the remaining $n - 1$ persons, and so on). Let X be the number of persons who receive the letter meant for them. What is the expectation of X? We can use the definition of X to calculate this quantity, but the reader should check that even the expression of $\Pr[X = r]$ is non-trivial, and then, adding up all such expressions (weighted by the corresponding probability) is a long calculation. We can instead use linearity of expectation to compute $\mathbb{E}[X]$ in a very simple manner as follows. For each person i, we define a random variable X_i, which takes only two values – 0 or 1 [2]. We set X_i to 1 if this person receives the correct letter, otherwise to 0. It is easy to check that $X = \sum_{i=1}^n X_i$, and so, by linearity of expectation, $\mathbb{E}[X] = \sum_i \mathbb{E}[X_i]$. It is now easy to compute $\mathbb{E}[X_i]$. Indeed, it is equal to $0 \cdot \Pr[X_i = 0] + 1 \cdot \Pr[X_i = 1] = \Pr[X_i = 1]$. Now, $\Pr[X_i = 1]$ is $1/n$ because this person receives each of the n letters with equal probability. Therefore, $\mathbb{E}[X] = 1$.*

Lemma 2.3 *For independent random variables X, Y,*

$$\mathbb{E}[X \cdot Y] = \mathbb{E}[X] \cdot \mathbb{E}[Y]$$

---

[2] These are called indicator random variables and often simplify calculations in many situations.

Proof:

$$\mathbb{E}[XY] = \sum_i \sum_j x_i \cdot y_j P(x_i, y_j) \text{ where } P \text{ denotes joint distribution,}$$

$$= \sum_i \sum_j x_i \cdot y_j p_X(x_i) \cdot p_Y(y_j) \text{ from independence of } X, Y$$

$$= \sum_i x_i p_X(x_i) \sum_j y_j p_Y(y_j)$$

$$= \mathbb{E}[X] \cdot \mathbb{E}[Y] \qquad\qquad \square$$

As in the case of events, we can also define conditional expectation of a random variable given the value of another random variable. Let X and Y be two random variables. Then, the *conditional expectation* of X given $[Y = y]$ is defined as

$$\mathbb{E}[X|Y = y] = \sum_x \mathrm{Pr}\, x \cdot [X = x|Y = y]$$

The *theorem of total expectation* that can be proved easily states that

$$\mathbb{E}[X] = \sum_y \mathbb{E}[X|Y = y]$$

2.2 Tail Inequalities

In many applications, especially in the analysis of randomized algorithms, we would like to bound the running time of our algorithm (or the value taken by some other random variable). Although one can compute the expectation of a random variable, it may not give any useful information about how likely the random variable is going to be close to its expectation. For example, consider a random variable which is uniformly distributed in the interval $[0, n]$, for some large number n. Its expectation is $n/2$, but the probability that it lies in the interval $[n/2(1 - \delta), n/2(1 + \delta)]$ is only 2δ, where δ is a small constant. We will see examples of other random variables where this probability will be very close to 1. Therefore, to say something more meaningful about a random variable, one needs to look beyond its expectation. The law of large numbers states that if we take many independent trials of a random variable, then the average value taken by the random variable over these trials converges (almost certainly) to the expectation. However, it does not say anything about how fast this convergence happens, or how likely the random variable is going to be close to its expectation if we perform this experiment only once.

In this section, we give various inequalities which bound the probability that a random variable deviates from its expectation by a large amount. The foremost such inequality is Markov's inequality, which just uses the expectation of a random variable. As mentioned earlier, it may not yield very strong bounds, but it is the best one can say when we do not have any other information about the random variable.

As a running example, we will use a modification of the experiment considered in the previous chapter. We are given an array A of size m (which is even). Half of the elements in A are colored red and the rest are colored green. We perform the following experiment n times independently: pick a random element of A, and check its color. Define X as a random variable which counts the number of times we picked a green element. It is easy to show, using linearity of expectation, that $\mathbb{E}[X]$ is $n/2$. We would now be interested in tail inequalities which bound the probability that X deviates from its mean.

Markov's inequality Let X be a non-negative random variable. Then,

$$\Pr[X \geq k\mathbb{E}[X]] \leq \frac{1}{k} \tag{2.2.3}$$

This result is really an 'averaging' argument (for example, in any class consisting of n students, at most half the students can get twice the average marks). The proof of this result also follows easily. Let R be the range of $X \geq 0$.

$$\mathbb{E}[X] = \sum_{r \in R} r \cdot \Pr[X = r] \geq \sum_{r \in R : r \geq k\mathbb{E}[X]} r \cdot \Pr[X = r] \geq k\mathbb{E}[X] \cdot \sum_{r \in R : r \geq k\mathbb{E}[X]} \Pr[X = r]$$

$$= k\mathbb{E}[X] \Pr[X \geq k\mathbb{E}[X]]$$

Canceling $\mathbb{E}[X]$ on both sides yields Markov's inequality. Unfortunately, there is no symmetric result which bounds the probability of events $[X < k\mathbb{E}[X]]$, where $k < 1$. To see why Markov's inequality cannot yield a two-sided bound, consider the following example.

Example 2.7 *Let X be a random variable which takes two values – 0 with proability $(1 - 1/n)$, and n^2 with probability $1/n$ (think of n as a large number). Then, $\mathbb{E}[X]$ is n. However, $\Pr[X < n/2]$ is $1 - 1/n$, which is very close to 1.*

We now apply this inequality on our running example.

Example 2.8 *In the example of array A with elements colored red or green, we know that $\mathbb{E}[X] = n/2$. Therefore, we see that $\Pr[X > 3n/4] \leq 1/4$.*

Note that we get a very weak bound on the probability that $[X \geq 3n/4]$ in Example 2.7. Ideally, one would think that the probability of this event would go down as we increase n (and indeed, this is true). However, Markov's inequality is not strong enough to prove this. The reason for this is that one can easily design random variables X whose expectation is $n/2$ but the probability of going above $3n/4$ is at most $2/3$. The extra information, that X is a sum of several independent random variables, is not exploited by Markov's inequality. Moreover, notice that we cannot say anything about the probability of the event $[X \leq n/4]$

using Markov's inequality. We now show that there are inequalities which can exploit facts about higher moments of X, and give stronger bounds.

The notion of *expectation of a random variable* can be extended to functions $f(X)$ of random variable X in the following natural way (we can think of $Y := f(X)$ as a new random variable)

$$E[f(X)] = \sum_{r \in R} \Pr[X = r] \cdot f(r)$$

The variance of a random variable is given by $\mathbb{E}[X^2] - \mathbb{E}[X]^2$. Consider the random variable X in Example 2.7. Its variance is equal to

$$\mathbb{E}[X^2] - \mathbb{E}[X]^2 = n^3 - n^2$$

Let us now compute the variance of the random variable in our running example. We first show that if X_1 and X_2 are two independent random variables, then variance of $X_1 + X_2$ is sum of the variance of the two random variables. The variance of $X_1 + X_2$ is given by

$$
\begin{aligned}
\mathbb{E}[(X_1 + X_2)^2] - \mathbb{E}[X_1 + X_2]^2 &= \mathbb{E}[X_1^2] + \mathbb{E}[X_2^2] + 2\mathbb{E}[X_1 X_2] - \mathbb{E}[X_1]^2 - \mathbb{E}[X_2]^2 - 2\mathbb{E}[X_1]\mathbb{E}[X_2] \\
&= \mathbb{E}[X_1^2] - \mathbb{E}[X_1]^2 + \mathbb{E}[X_2^2] - \mathbb{E}[X_2]^2
\end{aligned}
$$

because $\mathbb{E}[X_1 X_2] = \mathbb{E}[X_1]\mathbb{E}[X_2]$ (we use independence of these two random variables here). The same observation extends by induction to the sum of several random variables. Let us apply this observation to our running example. Let X_i be the random variable which is 1 if we pick a green element on the ith trial, 0 otherwise. Variance of X_i is $\mathbb{E}[X_i^2] - \mathbb{E}[X_i]^2$. Since X_i is a 0–1 random variable, $\mathbb{E}[X_i^2] = \mathbb{E}[X_i]$, and so, its variance is $1/2 - 1/4 = 1/4$. Let X denote the total number of green elements seen. Hence, $X = \sum_{i=1}^{n} X_i$ and its variance is $n/4$.

If we have bounds on the variance of a random variable, then the following gives a stronger tail bound

Chebychev's inequality

$$\Pr[|X - \mathbb{E}[X]| \geq t] \leq \frac{\sigma}{t^2} \tag{2.2.4}$$

where σ is the variance of X. The proof of this inequality follows from applying Markov's inequality on the random variable $Y := (X - \mathbb{E}[X])^2$. Observe that this is a two-sided inequality – not only does it bound the probability that X goes much above its mean, but also the probability of X going much below its mean.

Example 2.9 *We now apply this inequality to our running example. We get*

$$\Pr[X \geq 3n/4] \leq \Pr[|X - \mathbb{E}[X] \geq n/4|] \leq \frac{n/4}{9n^2/16} = \frac{4}{9n}$$

Thus, this probability goes to 0 as n goes to infinity.

We see in Example 2.9 that Chebychev's inequality gives a much stronger bound than Markov's inequality. In fact, it is possible to get much stronger bounds. Chebychev just uses bounds on the second moment of X. With knowledge of higher moments, we can give tighter bounds on the probability that X deviates from its mean by a large amount. If $X = \sum_i^n X_i$ is the sum of n mutually independent random variables where each X_i is a Bernoulli random variable (i.e., takes values 0 or 1 only), then

Chernoff bounds gives

$$\Pr[X \geq (1+\delta)\mu] \leq \frac{e^{\delta\mu}}{(1+\delta)^{(1+\delta)\mu}} \tag{2.2.5}$$

where δ is any positive parameter and μ denotes $\mathbb{E}[X]$. The analogous bound for deviations below the mean is as follows:

$$\Pr[X \leq (1-\delta)\mu] \leq \frac{e^{\delta\mu}}{(1+\delta)^{(1+\delta)\mu}} \tag{2.2.6}$$

where δ lies between 0 and 1.

Before we get into the proof of these bounds, we state more usable versions which often suffice in practice. It is easy to check that for any $\delta > 0$, $\ln(1+\delta) > \frac{2\delta}{2+\delta}$. Therefore,

$$\delta - (1+\delta)\ln(1+\delta) \leq -\frac{\delta^2}{2+\delta}$$

Taking exponents on both sides, we see that

$$\frac{e^{\delta\mu}}{(1+\delta)^{(1+\delta)\mu}} \leq e^{-\frac{\delta^2\mu}{2+\delta}}$$

Thus, we get the following:

- For $0 \leq \delta \leq 1$,

$$\Pr[X \geq (1+\delta)\mu] \leq e^{-\delta^2\mu/3} \tag{2.2.7}$$

  and

$$\Pr[X \leq (1-\delta)\mu] \leq e^{-\delta^2\mu/3} \tag{2.2.8}$$

- For $\delta > 2$,

$$\Pr[X \geq (1+\delta)\mu] \leq e^{-\delta\mu/2} \tag{2.2.9}$$

$$\mathrm{Prob}(X \geq m) \leq \left(\frac{np}{m}\right)^m e^{m-np} \tag{2.2.10}$$

We now give a proof of the Chernoff bound, Eq. (2.2.5). The proof for Eq. (2.2.6) is analogous.

$$\Pr[X \geq (1+\delta)\mu] = \Pr[e^{\lambda X} \geq e^{\lambda(1+\delta)\mu}] \leq \frac{\mathbb{E}[e^{\lambda X}]}{e^{\lambda(1+\delta)\mu}}$$

where λ is a positive parameter that we shall fix later, and the last inequality follows from Markov's inequality. Notice that $\mathbb{E}[e^{\lambda X}] = \mathbb{E}[\prod_{i=1}^{n} e^{\lambda X_i}] = \prod_{i=1}^{n} \mathbb{E}[e^{\lambda X_i}]$ because $X_1, \ldots, X_n$ are mutually independent. Let p_i denote the probability with which X_i takes the value 1. Then, $\mathbb{E}[e^{\lambda X_i}] = (1-p_i) + p_i \cdot e^{\lambda} = 1 + p_i(e^{\lambda}-1) \leq e^{p_i(e^{\lambda}-1)}$, because $1+x \leq e^x$ for any positive x. Since $\mu = \sum_{i=1}^{n} p_i$, we get,

$$\Pr[X \geq (1+\delta)\mu] \leq \frac{e^{\mu(e^{\lambda}-1)}}{e^{\lambda(1+\delta)\mu}}$$

Now we choose $\lambda > 0$ to minimize the right-hand side, that is, to minimize $e^{\lambda} - \lambda(1+\delta)$. It is easy to check that this is minimized at $\lambda = \ln(1+\delta)$. Substituting this value of λ in the RHS of the aforementioned inequality gives us the Chernoff bound, Eq. (2.2.5).

Example 2.10 *We now apply Chernoff bound to our running example. Here $\mu = n/2$. Using $\delta = 1/2$ in Eq. (2.2.7), we get*

$$\Pr[X \geq 3n/4] \leq e^{-n/12}$$

Note that for large values of n, this is a much sharper bound than the one obtained using Chebychev's inequality.

Example 2.11 *(Balls in bins) Suppose we throw n balls into n bins, where each ball is thrown independently and uniformly at random into one of the bins. Let Y_i denote the number of balls which fall in bin i. We are interested in the random variable $Y := \max_{i=1}^{n} Y_i$, that is, the maximum number of balls which fall in a bin. We will use Chernoff bound to show that Y is $O(\ln n)$ with high probability. Let us first consider a fixed bin i and show that Y_i is $O(\ln n)$ with high probability. For a ball j, let X_j be the indicator random variable which is 1 if ball j falls in bin i, 0 otherwise. Clearly, $\Pr[X_j = 1]$ is $1/n$. Now, $Y_i = \sum_{j=1}^{n} X_j$, and so, $\mathbb{E}[Y_i] = 1$. Since $X_1, \ldots, X_n$ are independent Bernoulli random variables, we can apply Eq. (2.2.9) with $\delta = 4 \ln n$ to get*

$$\Pr[Y_i \geq 4\ln n + 1] \leq e^{-2\ln n} = 1/n^2.$$

Now we use union bound to get

$$\Pr[Y \geq 4\ln n + 1] \leq \sum_{i=1}^{n} \Pr[Y_i \geq 4\ln n + 1] \leq 1/n.$$

Thus, with probability at least $1 - 1/n$, no bin gets more than $4\ln n + 1$ balls.

It turns out that one can get a sharper bound if we use Eq. (2.2.5) directly. It is left as an exercise to show that Y is $O(\ln n/\ln\ln n)$ with high probability.

Example 2.12 *Suppose we toss a fair coin n times independently. What is the absolute value of the difference between the number of Heads and the number of Tails? Using Chernoff bounds, one can show that this random variable is very likely to be $O(\sqrt{n})$. To see this, let X_i be the indicator random variable which is 1 if the outcome of the ith coin toss is Heads, 0 otherwise. Then the random variable $X = \sum_{i=1}^{n} X_i$ counts the number of Heads which are seen during this experiment. Clearly, $\mu := \mathbb{E}[X] = n/2$. Using $\delta = 3/\sqrt{n}$ in Eq. (2.2.7) and in Eq. (2.2.8), we see that $\Pr[|X - n/2| \geq \sqrt{n}]$ is at most e^{-3}, which is about 0.05.*

2.3   Generating Random Numbers

The performance of any randomized algorithm is closely dependent on the underlying random number generator (RNG) in terms of efficiency. A common underlying assumption is the availability of an RNG that generates a number uniformly in some range $[0,1]$ in unit time or alternately $\log N$ independent random *bits* in the discrete case for the interval $[0,\ldots,N]$. This primitive is available in all standard programming languages – we will refer to this RNG as $\mathcal{U}$. We will need to adapt this to various scenarios that we describe in the following subsections.

2.3.1   Generating a random variate for an arbitrary distribution

We consider a discrete distribution $\mathcal{D}$, which is specified by distribution function $f(s)$, $s = 1,\ldots,N$. We would like to generate a random variate according to $\mathcal{D}$. The distribution $\mathcal{D}$ can be thought of as generating a random variable X with weight $w_i = f(i)$, where $\sum_i w_i = 1$. A natural way to sample from such a distribution is as follows. We can divide the interval $[0,1]$ into consecutive subintervals $I_1, I_2, \ldots$ such that I_j has length w_j. Now, using the RNG $\mathcal{U}$, we sample a random point in the interval $[0,1]$. If it falls in the interval I_j, we output j. It is easy to see that the probability that this random variable takes value j is exactly $f(j)$.

As stated earlier, this process can take $O(N)$ time because we need to figure out the interval in which the randomly chosen point lies. We can make this more efficient by using binary search. More formally, let $F(j)$ denote $\sum_{i=1}^{j} f(i)$ – it is also called the *cumulative distribution function* (CDF) of $\mathcal{D}$. Clearly, the sequence $F(1), F(2), \ldots, F(N) = 1$ forms a monotonically non-decreasing sequence. Given a number x in the range $[0,1]$, we can use binary search to find the index j such that x lies between $F(j)$ and $F(j+1)$. Therefore, we can sample from the distribution in $O(\log N)$ time.

This idea of dividing the unit interval into discrete segments does not work for a continuous distribution (for example, the normal distribution). However, we can still use a simple extension of the previous idea. A continuous distribution is specified by a CDF $F()$, where $F(s)$ is supposed to indicate the probability of taking a value less than or equal to s. We assume that $F()$ is continuous (note that $F(-\infty) = 0$ and $F(+\infty) = 1$). In order to sample from this distribution, we again sample a value x uniformly from $[0,1]$ using $\mathcal{U}$. Let s be a value such that $F(s) = x$ (we are assuming we can compute F^{-1}; in the discrete case, we were using a binary search procedure instead). We output the value s. It is again easy to check that this random variable has a distribution given by $\mathcal{D}$.

2.3.2 Generating random variables from a sequential file

Suppose a file contains N records from which we would like to sample a subset of n records uniformly at random. There are several approaches to this basic problem:

- *Sampling with replacement* We can use $\mathcal{U}$ to repeatedly sample an element from the file. This could lead to *duplicates*.

- *Sampling without replacement* We can use the previous method to choose the next sample but we will reject duplicates. The result is a uniform sample but the efficiency may suffer. In particular, the expected number of times we need to invoke the RNG for the kth sample is $\frac{N}{N-k}$ (see exercises).

- *Sampling in a sequential order* Here we want to pick the samples $S_1, S_2, \ldots, S_n$ in an increasing order from the file, that is, $S_i \in [1 \ldots N]$ and $S_i < S_{i+1}$. This has applications to processes where we can scan the records exactly once and retracing is not possible.

  Suppose we have selected $S_1, \ldots, S_m$ so far, and scanned the first t elements. Conditioned on these events, we select the next element (as S_{m+1}) with probability $\frac{n-m}{N-t}$. Again, we implement this process by choosing a random value x in the range $[0,1]$ using $\mathcal{U}$ and then checking if x happens to be more or less than $\frac{n-m}{N-t}$.

In order to show that this random sampling procedure is correct, let us calculate the probability that this process selects elements $s_1, \ldots, s_n$, where $1 \le s_1 \le s_2 \le \ldots \le s_n \le N$. Let us condition on the fact that $S_1 = s_1, \ldots, S_m = s_m$. What is the probability that $S_{m+1} = s_{m+1}$? For this to happen, we must not select any of the elements in $s_m + 1, \ldots, s_{m+1} - 1$, and then select s_{m+1}. The probability of such an event is exactly

$$\frac{n-m}{N-s_{m+1}} \cdot \prod_{t=s_m+1}^{s_{m+1}-1} \left(1 - \frac{n-m}{N-t}\right)$$

Taking the product of this expression for $m = 1, \ldots, n$, we see that the probability of selecting $s_1, \ldots, s_n$ is exactly $\frac{1}{\binom{N}{n}}$.

Although the aforementioned procedure works, $\mathcal{U}$ is called N times. The following is a more efficient process which calls $\mathcal{U}$ fewer number of times. It is easy to check that the distribution of $S_{i+1} - S_i$ is given by (see exercises)

$$F(s) = 1 - \frac{\binom{(N-t-s)}{(n-m)}}{\binom{(N-t)}{(n-m)}} \quad s \in [t+1, N] \tag{2.3.11}$$

Thus, we can sample random variables from the distribution $S_1, S_2 - S_1, \ldots, S_n - S_{n-1}$, and then select the corresponding elements.

- *Sampling in a sequential order from an arbitrarily large file*: This case is the same as earlier except that we do not know the value of N. This is the typical scenario in a streaming algorithm (see Chapter 15).

In this case, we always maintain the following invariant:

Among the i records that we have scanned so far, we have a sample of n elements chosen uniformly at random from these i elements.

Note that the invariant makes sense only when $i \geq n$ because the n samples are required to be distinct. Further, when $i = n$, the first n records must be chosen in the sample. Now assume that this invariant holds for some $i \geq n$. Let $S_{n,i}$ denote the random sample of n elements at this point of time. When we scan the next record (which may not happen if the file has ended), we want to restore this invariant for the $i + 1$ records. Clearly the $i + 1$th record needs to be in the sample with some probability, say p_{i+1} and if picked, one of the previous sampled records must be replaced.

Note that $p_{i+1} = \frac{n}{i+1}$. This follows from the fact that there are $\binom{i+1}{n}$ ways of selecting n samples from the first $i + 1$ elements, and exactly $\binom{i}{n-1}$ of these contain $i + 1$. Therefore,

$$p_{i+1} = \frac{\binom{i}{n-1}}{\binom{i+1}{n}} = \frac{n}{i+1}$$

If the (i+1)th record is indeed chosen, we drop one of the previously chosen n samples with equal probability. To see this, notice that the invariant guarantees that the set $S_{n,i}$ is a uniformly chosen sample of n elements. We claim that dropping one of the samples uniformly at random gives us $S_{n-1,i}$, that is, a uniform $n - 1$ sample. The probability that a specific subset of $n - 1$ elements, say S^* is chosen is the probability that $S^* \cup \{x\}$ was chosen, ($x \notin S^*$), and x was dropped. You can verify that

$$\frac{1}{n} \cdot (i - n + 1) \cdot \frac{1}{\binom{i}{n}} = \frac{1}{\binom{i}{n-1}}$$

where the term $(i - n + 1)$ represents the number of choices of x. The RHS is the uniform probability of an $n - 1$ sample. Thus, the sampling algorithm is as follows: when we consider record $i + 1$, we select it in the sample with probability $\frac{n}{i+1}$ – if it gets selected, we drop one of the earlier chosen samples with uniform probability.

2.3.3 Generating a random permutation

Many randomized algorithms rely on the properties of random permutation to yield good expected bounds. Some algorithms like Hoare's quicksort or randomized incremental construction actually start from the assumption of an initial random order. However, the input may not have this property; in which case, the onus is on the algorithm to generate a random permutation. Broadly speaking, any such algorithm must have access to random numbers and also ensure that all the permutations of the input objects are equally likely outcomes.

We describe the algorithm in Figure 2.1. The algorithm runs in n iterations; in the ith iteration, it assigns x_i to a random location in the permutation. It places the ordered elements (according to the random permutation) in an array A. Note that the size of A is slightly larger than n, and so, some positions in A will remain empty at the end. Still, we can read the permutation from A by scanning it from left to right.

Procedure Random permutation($\{x_1, x_2, \ldots, x_n\}$)

1 *Input* : Objects $\{x_1, x_2, \ldots, x_n\}$;
2 *Output*: A random permutation $\Pi = \{x_{\sigma(1)}, x_{\sigma(2)}, \ldots, x_{\sigma(n)}\}$;
3 Initialize an array of size $m(> n)$ as *unmarked* ;
4 **for** $i = 1$ *to* n **do**
5   **while** $A[j]$ *is marked* **do**
6   Generate a random number $j \in_{\mathcal{U}} [1, m]$;
7      $A[j] \leftarrow i$;
8   *mark* $A[j]$;
9 Compress the marked locations in $A[1, n]$ and Return A where
   $\sigma(A[j]) = j$;

Figure 2.1   *Generating a random permutation of n distinct objects*

In the array A, the algorithm marks the locations which are occupied. The main loop tries to assign x_i to a random location among the unmarked (unoccupied) locations in the array A. For this, it keeps trying until it finds a free position. We need to prove the following

(i) After termination, all permutations are equally likely.

(ii) The expected number of executions of the loop is not too large – preferably linear in n.

(iii) Returning the n elements in contiguous locations takes m steps.

To balance (ii) and (iii), we have to choose m somewhat carefully. We make some simple observations

Claim 2.1 *If the number of unmarked locations in A is t, then each of the t locations is chosen with equal likelihood.*

This follows from a simple application of conditional probability, conditioned on a location being unmarked. Consider any fixed set N of distinct n locations. Conditioned on assigning the elements $x_1, x_2, \ldots, x_n$ to N, all permutations of $x_1, x_2, \ldots, x_n$ are equally likely. Again this follows from the observation that, after $x_1, x_2, \ldots, x_i$ are assigned, x_{i+1} is uniformly distributed among the unoccupied $n - i$ locations. Since this holds for any choice of N, the unconditional distribution of the permutations is also the same.

The number of iterations depend on the number of unsuccessful attempts to find an unassigned location. The probability of finding an unassigned location after i assignments is $\frac{m-i}{m} = 1 - \frac{i}{m}$. Since the locations are chosen independently, the expected number of iterations to find a free location for x_{i+1} is $\frac{m}{m-i}$ and from the linearity of expectation, the total expected number of iterations is

$$\sum_{i=0}^{n-1} \frac{m}{m-i} = m \left(\frac{1}{m} + \frac{1}{m-1}, \ldots, \frac{1}{m-n+1} \right) \tag{2.3.12}$$

For $m = n$, this is $O(n \log n)$, whereas for $m = 2n$, this becomes $O(n)$. Since the probabilities are independent, we can obtain concentration bounds for deviation from the expected bounds using Chernoff–Hoeffding bounds as follows.

What is the probability that the number of iterations exceed $3n$ for $m = 2n$? This is equivalent to finding fewer than n assignments in $3n$ iterations. Let $p_i = \frac{2n-i}{2n}$, then for $i \leq n$, $p_i \geq 1/2$, where p_i is the probability of finding a free location for x_i. Let us define 0–1 random variables X_i such that $X_i = 1$ if the ith iteration is successful, that is, we find an unmarked location. To terminate, we need n unmarked locations. From our previous observation, $\Pr[X_i = 1] \geq 1/2$. Hence, $\mathbb{E}[\sum_{i=1}^{3n} X_i] \geq 3n/2$. Let $X = \sum_i X_i$ be the number of successes in $3n/2$ iterations. Then, X is a sum of independent Bernoulli random variables and a straightforward application of Chernoff bounds (Eq. (2.2.8) shows that

$$\Pr[X < n] = \Pr[X < (1 - 1/3)\mathbb{E}[X]] \leq \exp \left(-\frac{3n}{36} \right)$$

which is inverse exponential.

Claim 2.2 *A random permutation of n distinct objects can be generated in $O(n)$ time and $O(n)$ space with high probability.*

The reader would have noted that as m grows larger, the probability of encountering a marked location decreases. Therefore, it is worth estimating for what value of m, there will be exactly n iterations with high probability, that is, no reassignment will be necessary. This could be useful in online applications where we need to generate random permutations. Using equation (2.2.10), we can bound the probability that the number of random assignments in a location exceeds 1 as

$$\left(\frac{n}{2m}\right)^2 e^{2-n/m} \leq O(n^2/m^2)$$

Note that the expected number of assignments in a fixed location $\mu = \frac{n}{m}$. From union bound, the probability that any of the m locations has more than 1 assignment is bound by $O(\frac{n^2}{m})$. Hence, by choosing $m = \Omega(n^2)$, with probability $1 - O(\frac{n^2}{m})$, the number of iterations is n, that is, there is no reassignment required.

Further Reading

There are several excellent textbooks on introductory probability theory and randomized algorithms [105, 106, 126]. Most of the topics covered in this chapter are classical, and are covered in these texts in more detail. Chernoff bounds are among the most powerful tail inequalities when we are dealing with independent random variables. There are similar bounds which sometimes give better results depending on the parameters involved, for example, Hoeffding's bound. Maintaining a random sample during a streaming algorithm is a common subroutine used in many streaming algorithms (see e.g., Chapter 16). The idea that picking n elements out of an array of size $2n$ or more results in small repetitions is often used in many other applications, for example, hashing (see Chapter 6).

Exercise Problems

2.1 Consider the experiment of tossing a fair coin till two heads or two tails appear in succession.

   (i) Describe the sample space.
   (ii) What is the probability that the experiment ends with an even number of tosses?
   (iii) What is the expected number of tosses?

2.2 A chocolate company is offering a prize for anyone who can collect pictures of n different cricketers, where each wrap has one picture. Assuming that each chocolate can have any of the pictures with equal probability, what is the expected number of chocolates one must buy to get all the n different pictures?

2.3 There are n letters which have corresponding n envelopes. If the letters are put blindly in the envelopes, show that the probability that none of the letters goes into the right envelope tends to $\frac{1}{e}$ as n tends to infinity.

2.4 Imagine that you are lost in a new city where you come across a crossroad. Only one of them leads you to your destination in 1 hour. The others bring you back to the same point after 2, 3, and 4 hours respectively. Assuming that you choose each of the roads with equal probability, what is the expected time needed to arrive at your destination?

2.5 A gambler uses the following strategy. The first time he bets Rs. 100 – if he wins, he quits. Otherwise, he bets Rs. 200 and quits regardless of the result. What is the probability that he goes back a winner assuming that he has probability 1/2 of winning each of the bets. What is the generalization of his strategy?

2.6 **Gabbar Singh problem** Given that there are 3 consecutive blanks and three consecutive loaded chambers in a pistol, and you start firing the pistol from a random chamber, calculate the following probabilities.

(i) The first shot is a blank. (ii) The second shot is also a blank given that the first shot was a blank. (iii) The third shot is a blank given that the first two were blanks.

2.7 In the balls in bins example (Example 2.11), show that the maximum number of balls in any bin is $O(\ln n / \ln \ln n)$ with high probability.

2.8 Suppose we throw m balls independently and uniformly at random in n bins. Show that if $m \geq n \ln n$, then the maximum number of balls received by any bin is $O(m/n)$ with high probability.

2.9 Three prisoners are informed by the jailer that one of them will be acquitted without divulging the identity. One of the prisoners requests the jailer to divulge the identity of one of the other prisoner who will not be acquitted. The jailer reasons that since at least one of the remaining two will not be acquitted, he would not be divulging the secret and reveals the identity. However this makes the prisoner very happy. Can you explain this?

2.10 For random variables X, Y, show that

(i) $\mathbb{E}[X \cdot Y] = \mathbb{E}[Y \times \mathbb{E}[X|Y]]$

(ii) $\mathbb{E}[\mathbb{E}[X|Y]] = \mathbb{E}[X]$

(iii) $\mathbb{E}[\phi_1(X_1) \cdot \phi_2(X_2)] = \mathbb{E}[\phi_1(X_1)] \cdot \mathbb{E}[\phi_2(X_2)]$ for functions ϕ_1, ϕ_2 of random variables.

2.11 Give an example to show that even if $\mathbb{E}[X \cdot Y] = \mathbb{E}[X] \cdot \mathbb{E}[Y]$, the random variables X, Y may not be independent.

Hint: Consider X and some appropriate function of X.

2.12 Let $Y = \sum_{i=1}^{n} X_i$, where X_is are identically distributed random variables with expectation μ. If n is a non-negative integral random variable, then Y is known as *random sum*. Show that $\mathbb{E}[Y] = \mu \cdot \mathbb{E}[n]$.

2.13 Let Y be a random variable that denotes the number of times a fair die must be rolled till we obtain a six. Assume that the outcomes are independent of each other. How many times do we have to roll the die to obtain k successes?

Let X be a random variable that denotes this, then

(i) Compute $\mathbb{E}[X]$.

(ii) Show that $\Pr[X \geq 10k] \leq \frac{1}{2^k}$ using Chernoff bounds.

The distribution of Y is known as geometric distribution and X is known as negative binomial distribution.

2.14 For a discrete random variable X, e^{Xs} is known as the *moment generating function*. Let $M(s) = \mathbb{E}[e^{sX}]$. Show that

$\mathbb{E}[X^k] = \frac{d^k M}{ds^k}|_{s=0}, k = 1, 2, \ldots$. This is a useful formulation for computing the kth moment of a random variable.

Hint: Write down the series for e^{sX}.

2.15 Let $G(n, p)$ be a graph on n vertices where we add an edge between every pair of vertices independently with probability p. Let X denote the number of edges in the graph $G(n, p)$. What is the expectation of X? What is the variance of X?

2.16 Let $G(n, p)$ be as stated in Exercise 2.16. A triangle in this graph is a set of three vertices $\{u, v, w\}$ (note that it is an unordered triplet) such that we have edges between all the three pairs of vertices. Let X denote the number of triangles in $G(n, p)$. What are the expectation and the variance of X?

2.17 Consider the algorithm for sampling from a continuous distribution in Section 2.3.1. Prove that the random variable has the desired distribution.

2.18 Consider the problem of uniformly sampling n distinct elements from a file containing N elements. Suppose we have already sampled a set S of k elements. For the next element, we keep on selecting a uniform sample from the file till we get an element which is not in S. What is the expected number of times we need to sample from the file?

2.19 Consider the problem of sampling in a sequential order. Prove that the distribution of $S_i - S_{i-1}$ is given by the expression in Eq. (2.3.11).

3

CHAPTER

Warm-up Problems

In this chapter, we discuss some basic algorithmic problems. Each of these problems requires a new technique and its analysis depends on the underlying computational model. These are analyzed using basic techniques which should be familiar to the reader.

3.1 Euclid's Algorithm for the Greatest Common Divisor (GCD)

Euclid's algorithm for computing the greatest common divisor (gcd) of two positive integers is allegedly the earliest known algorithm in a true sense. It is based on two very simple observations. Given two positive integers a, b, their gcd satisfies

$$\gcd(a,b) = \gcd(a, a+b)$$

$$\gcd(a,b) = b \text{ if } b \text{ divides } a$$

The reader is encouraged to prove this rigorously. The aforementioned equations also imply that $\gcd(a,b) = \gcd(a-b,b)$ for $b < a$ and repeated application of this fact implies that $\gcd(a,b) = \gcd(a \mod b, b)$, where mod denotes the remainder operation. Hence, we have essentially derived Euclid's algorithm, described formally in Figure 3.1.

Let us now analyze the running time of Euclid's algorithm in the bit computational model (i.e., we count the number of bit operations needed for the computation). Since it depends on integer division, which is a topic in its own right, let us compute the number of iterations of Euclid's algorithm in the worst case.

Procedure Algorithm Euclid-GCD(a,b)

---

1 **Input**: Positive integers a,b such that $b \leq a$;
2 **Output** GCD of a,b;
3 Let $c = a \mod b$;
4 **if** $c = 0$ **then**
5 $\quad\mid\quad$ return b
6 **else**
7 $\quad\mid\quad$ return Euclid-GCD(b,c)

Figure 3.1   *Euclid's algorithm*

Observation 3.1 *The number* $a \mod b \leq \frac{a}{2}$, *that is the size of* $a \mod b$ *is strictly less than* $|a|$.

This is a simple case analysis based on $b \leq \frac{a}{2}$ and $b > \frac{a}{2}$. As a consequence of this observation, it follows that the number of iterations of Euclid's algorithm is bounded by $|a|$, or equivalently $O(\log a)$. This bound is actually tight. Hence, by using the long division method to compute mod, the running time is bounded by $O(n^3)$, where $n = |a| + |b|$.

3.1.1   Extended Euclid's algorithm

If you consider the numbers defined by the linear combinations of a,b, namely, $\{xa + yb|, x,y \text{ are integers}\}$, it is known that

$$\gcd(a,b) = \min\{xa + yb|xa + yb > 0\}$$

To prove this, let $\ell = \min\{xa + yb|xa + yb > 0\}$. Clearly, $\gcd(a,b)$ divides ℓ and hence, $\gcd(a,b) \leq \ell$. We now prove that ℓ divides a (also b). Let us assume by contradiction that $a = \ell q + r$, where $\ell > r > 0$. Now $r = a - \ell q = (1 - xq)a - (yq)b$ contradicting the minimality of ℓ.   □

For some applications, we are interested in computing x and y corresponding to $\gcd(a,b)$. We can compute them recursively along with Euclid's algorithm.

Claim 3.1 *Let* (x',y') *correspond to* $\gcd(b,a \mod b)$, *that is,* $\gcd(b,a \mod b) = x' \cdot b + y' \cdot (a \mod b)$. *Then,* $\gcd(a,b) = y' \cdot a + (x' - q)b$, *where q is the quotient of the integer division of a by b.*

The proof is left as an exercise problem.

One immediate application of the extended Euclid's algorithm is for computing the inverse in a multiplicative prime field F_q^*, where q is prime. $F_q^* = \{1,2,\ldots,(q-1)\}$, where the multiplication is performed modulo q. It is known[1] that for every number $x \in F_q^*$, there exists $y \in F_q^*$ such that $x \cdot y \equiv 1 \mod q$, which is also called the inverse of x. To compute the inverse of a, we can use the extended Euclid algorithm to find s, t such that $sa + tq = 1$ since a is relatively prime to q. By taking remainder modulo q, we see that $s \mod q$ is the required inverse. This result can be extended to $Z_N^* = \{x | x$ is relatively prime to $N\}$. First show that Z_N^* is closed under multiplication modulo N, that is, $a, b \in Z_N^*$ $a \cdot b \mod N \in Z_N^*$. The proof is left as an exercise problem.

3.1.2 Application to cryptography

The RSA (Rivest–Shamir–Adleman) public cryptosystem is one of the most important discoveries in computer science. Suppose we want to send an encrypted message to a receiver. One idea would be that the sender and the receiver share a secret key among themselves and then use this secret to encrypt and decrypt messages. However, this solution is not scalable – how would they share the secret key? If this involves communication over an insecure channel, then we are back to the same problem. A more elegant solution was proposed by Rivest, Shamir, and Adleman in 1977, and is now known as the RSA public key cryptosystem. Here the receiver generates two keys – one is called the *public* key which is known to everyone, and the other is called the *private* key which is known to the receiver only. Anyone who wants to send a message to the receiver will encrypt the message using the public key, but one can decrypt the message only if she knows the private key!

The RSA cryptosystem works on the following principle. We (i.e., the receiver) first pick two large prime numbers p and q, and let $n = p \cdot q$. Think of the message as an integer m such that $0 \leq m < n$. Let $\phi(n) = (p-1) \cdot (q-1)$, which is known as Eulier's totient function, and is the private key. Let d, e be integers such that $e \cdot d \equiv 1 \mod \phi(n)$, where e, d are co-primes to $\phi(n)$. The public key is e. The encryption of message m is done by computing $m^e \mod n$ and decryption is done by computing $(m^e)^d \mod n$. Note that $(m^e)^d \equiv m^{k\phi(n)+1} \mod p \equiv m \mod p$. Similarly, $(m^e)^d \equiv m \mod q$ implying that $(m^e)^d \equiv m \mod (pq) \equiv m \mod n$. The last step follows from the Chinese remainder theorem (see Exercise 8.1 for a formal statement of the theorem). The underlying assumption is that given n, e, it is hard to compute d.

The two main algorithmic steps here are computing exponentiation, which can be done using the divide and conquer technique described in Chapter 1, and computing the multiplicative inverse which can be done using the extended Euclid's algorithm.

---

[1] since it forms a group

3.2 Finding the kth Smallest Element

Problem Given a set S of n elements, and an integer k, $1 \leq k \leq n$, find an element $x \in S$ such that the rank of x is k. The rank of an element in a set S is k if $x = x_k$ in the sorted sequence $x_1, x_2, \ldots, x_n$ of the elements in S. We will denote the rank of x in S by $R(x, S)$.

Since S can be a multi-set, the position of x in the sorted sequence is not uniquely defined. We can however make the elements unique by (hypothetically) appending extra bits to them. For example, if S is an array, we can append $\log n$ trailing bits equal to the index of each element in the array. So the ith element x_i can be thought of as a pair (x_i, i). This makes all the elements in S unique. The case $k = 1$ ($k = n$) corresponds to finding the minimum (maximum) element.

We can easily reduce the selection problem to that of sorting. First we sort S and then report the kth element of the sorted sequence. But this also implies that we cannot circumvent the lower bound of $\Omega(n \log n)$ for comparison-based sorting. If we want a faster algorithm, we cannot afford to sort. For instance, when $k = 1$ or $k = n$, we can easily select the minimum (maximum) element using $n - 1$ comparisons. The basic idea for a faster selection algorithm is based on the following observation.

Given an element $x \in S$, we can answer the following query in $n - 1$ comparisons:

> Is x the kth element or is x larger than the kth element or is x smaller than the kth element?

This is easily done by comparing x with all elements in $S - \{x\}$ and finding the rank of x. Using an arbitrary element x as a filter, we can subsequently confine our search for the kth element to either

(i) $S_> = \{y \in S - \{x\} | y > x\}$ if $R(x, S) < k$ or

(ii) $S_< = \{y \in S - \{x\} | y < x\}$ if $R(x, S) > k$

In the fortuitous situation, $R(x, S) = k$, x is the required element. In case (i), we must find the k'th element in $S_>$, where $k' = k - R(x, S)$.

Suppose $T(n)$ is the worst case running time for selecting the kth element for any k; then, we can write the following recurrence

$$T(n) \leq \max\{T(|S_<|), T(|S_>|)\} + O(n)$$

A quick inspection tells us that if we can ensure $\max\{|S_<|, |S_>|\} \leq \varepsilon n$ for some $1/2 \leq \varepsilon < \frac{n-1}{n}$ (for all recursive calls as well), $T(n)$ is bounded by $O(\frac{1}{1-\varepsilon} \cdot n)$. So it could vary between $\Omega(n)$ and $O(n^2)$ – where a better running time is achieved by ensuring a smaller value of ε.

An element x used to divide the set is often called a *splitter* or a *pivot*. We will now discuss methods to select a good *splitter*. From our previous discussion, we would like to select a splitter that has a rank in the range $[\varepsilon \cdot n, (1 - \varepsilon) \cdot n]$ for a *fixed* fraction ε. Typically, ε will be chosen as $1/4$.

3.2.1 Choosing a random splitter

Let us analyze the situation where the splitter is chosen uniformly at random from S, that is, any of the n elements is equally likely to be chosen as the splitter. This can be done using standard routines for random number generation in the range $(1, 2, \ldots, n)$. A central observation is as follows:

For a randomly chosen element $r \in S$, the probability

$$\Pr\{n/4 \leq R(r, S) \leq 3n/4\} \geq 1/2$$

It is easy to verify, in linear time, if the rank $R(r, S)$ falls in this range; if it does not, then we choose another element *independently* at random. This process is repeated till we find a splitter in the aforementioned range – let us call such a splitter a *good* splitter.

How many times do we need to repeat the process?

To answer this, we have to take a slightly different view. One can argue easily that there is no guarantee that we will terminate after some fixed number of trials, while it is also intuitively clear that it is extremely unlikely that we need to repeat this more than say 10 times. The probability of failing 9 consecutive times, when the success probability of picking a good splitter is $\geq 1/2$ independently is $\leq \frac{1}{2^9}$. More precisely, the *expected* [2] number of trials is bounded by 2. Hence, in (expected) two trials, we will find a good splitter that reduces the size of the problem to at most $\frac{3}{4}n$. This argument can be repeated for the recursive calls, namely, the expected number of splitter selection (and verification of its rank) is 2. If n_i is the size of the problem after i recursive calls with $n_0 = n$, then the expected number of comparisons done during the ith recursive call is $2n_i$. The total expected number of comparisons X after t calls can be written as $X_0 + X_1 + \ldots + X_t$, where t is sufficiently large such that the problem size $n_t \leq C$ for some constant C (we can choose other stopping criteria) and X_i is the number of comparisons done at stage i. By taking expectation on both sides

$$E[X] = E[X_1 + X_2 + \ldots + X_t] = E[X_1] + E[X_2] + \ldots + E[X_t]$$

From the previous discussion, $E[X_i] = 2n_i$; moreover $n_i \leq \frac{3}{4}n_{i-1}$. Therefore, the expected number of comparisons is bounded by $8n$.

Let us analyze the original recursive algorithm, where we choose a random splitter and proceed with the relevant subproblem. Let $\bar{T}(n)$ be the expected time for selection of the kth ranked element (for any k). Since each element is equally likely to be the splitter, we can do a case analysis based on the rank of the random splitter x compared to k.

Case rank(x) < k: The subproblem size is $n - \text{rank}(x)$ for each of the $k - 1$ possibilities for x.

---

[2] Please refer to Chapter 2 for a quick recap of basic measures of discrete probability.

Case rank(x) > k: The subproblem size is rank(x) − 1 for each of the $n - k - 1$ possibilities for *x*.

As each individual case has probability $\frac{1}{n}$, we can write the recurrence as

$$\bar{T}(n) = \frac{1}{n} \sum_{i=n-1}^{k-1} \bar{T}(i) + \frac{1}{n} \sum_{j=n-1}^{n-(k-1)} \bar{T}(j) + O(n) \tag{3.2.1}$$

We then verify that this recurrence has the worst case behavior for $k = n/2$ assuming that $\bar{T}(i)$ grows monotonically with *i*. Then, we have to find the solution of

$$\bar{T}(n) = \frac{2}{n} \sum_{i=n/2}^{n-1} \bar{T}(i) + c'n$$

which can be verified as $\bar{T}(n) = cn$ for $c > 4c'$ by induction as follows

$$\begin{aligned} \bar{T}(n) &= \frac{2}{n} \sum_{i=0}^{n/2-1} [c(n/2+i)] + c'n \\ &\leq c[\frac{n}{2} + \frac{n}{4}] + cn/4 = cn \end{aligned} \tag{3.2.2}$$

3.2.2 Median of medians

The aforementioned algorithm required random number generation. What if we are required to give a deterministic algorithm? It turns out that one can still come up with a linear time algorithm for selection of the *k*th smallest element, but it is more involved. This is typical of many randomized algorithms – getting rid of randomness used by an algorithm often leads to more complicated algorithms.

Consider the deterministic algorithm given in Figure 3.2 for a selection that finds the *k*th ranked element in a given (unordered) set *S* of *n* elements.

The algorithm begins by successively forming groups of 5 consecutive elements. It finds the median of each of these groups (of size 5) in constant time (it could just sort them as it would take constant time only). Among these set of *n*/5 medians, it picks their median element (i.e., median of medians) as the splitter. The reader may note that this algorithm is very similar to the previous strategy, except for the choice of *M* – which was randomly selected. Let us first estimate how many elements are guaranteed to be smaller than *M* since from our previous observation, we would like to prune at least a constant fraction of *S* going into the recursive calls.

Without loss of generality, assume all elements are distinct; this implies that there are about *n*/10 medians[3] that are smaller than *M*. For each such median, there are 3 elements

---

[3] Strictly speaking, we should be using the floor function but we are avoiding the extra symbols and it does not affect the analysis.

that are smaller than M, giving a total of at least $n/10 \cdot 3 = 3n/10$ elements smaller than M. Likewise, we can argue that there are at least $3n/10$ elements larger than M. Therefore, we can conclude that $3n/10 \leq R(M,S) \leq 7n/10$, which satisfies the requirement of a good splitter.

Procedure Algorithm MoMSelect(S,k)

1 **Input** A set S of n elements. ;
2 **Output** The kth ranked element of S. ;
3 Partition S into 5-element groups arbitrarily – denote the groups by
   $G_1, \ldots, G_t$, where $t = \lceil n/5 \rceil$. ;
4 For $i = 1, \ldots t$ let m_i denote the median of G_i (rank 3 element) and
   let $S' = \{m_1, m_2, \ldots, m_t\}$. ;
5 Let M be the median of S' and $m = R(M,S)$, i.e., the rank of M. ;
6 Let $S_< = \{x \in S | x < M\}$ and $S_> = S - S_<$;
7 **if** $m = k$ **then**
8 $\quad$ return M
9 **else**
10 $\quad$ **if** $k < m$ **then**
11 $\quad\quad$ MoMSelect $(S_<, k)$
12 $\quad$ **else**
13 $\quad\quad$ MoMSelect $(S_>, m - r)$

Figure 3.2 *Algorithm based on median of medians*

The next question is how to find M which is the median of medians. Each m_i can be determined in $O(1)$ time because we are dealing with groups of size 5. However, finding the median of $n/5$ elements is like going back to square one! But it is $n/5$ elements instead of n and therefore, we can apply a recursive strategy, that is, the line 5 in the algorithm would be

$$M = \text{MoMSelect}\left(S', \frac{|S'|}{2}\right).$$

We can write a recurrence for running time as follows

$$T(n) \leq T\left(\frac{7n}{10}\right) + T\left(\frac{n}{5}\right) + O(n)$$

where the second recursive call is to find the median of medians (for finding a good splitter). After we find the splitter (by recursively applying the same algorithm), we use it to reduce the original problem size to at most $\frac{7n}{10}$. Note that for this to be a linear time algorithm, it is not enough that the splitter satisfies the requirements of a good splitter. It is easy to check that the solution to a recurrence of the form

$$T(n) \leq T(\alpha n) + T(\beta n) + O(n)$$

is linear if $\alpha + \beta < 1$. The proof is left as an exercise. Since $7n/10 + n/5 = 9n/10$, it follows that the aforementioned algorithm is a linear time algorithm.

3.3   Sorting Words

Problem Given n words $w_1, w_2, \ldots, w_n$ of lengths $l_1, l_2, \ldots, l_n$ respectively, arrange the words in lexicographic order. A word is an ordered sequence of characters from a given alphabet Σ.

Recall that lexicographic ordering refers to the dictionary ordering. Let $N = \sum_i l_i$, that is, the cumulative length of all words. A single word may be very long and we cannot assume that it fits into a single word of the computer. So, we cannot use straightforward comparison sorting. Let us recall some basic results about integer sorting.

Claim 3.2 *n integers in the range $[1..m]$ can be sorted in $O(n+m)$ steps.*

We maintain an array of m lists for bucket b_i, where $1 \leq i \leq m$. For multiple integers in the same bucket, we form a linked list. After this, the lists are output by scanning all the (non-empty) lists in sequence. The first phase takes $O(n)$ steps and the second phase requires a scan of all the m buckets. Note that we could speed-up the procedure by skipping the empty buckets but we do not maintain that information.

A sorting algorithm is considered *stable* if the relative order of input elements having identical values is preserved in the sorted output. Clearly, the previous sorting, often called *bucket* sorting is inherently stable since the linked lists containing all the same-valued elements are built in order of the input elements.

Claim 3.3 *Using stable sorting, n integers in the range $[1..m^k]$ can be sorted in $O(k(n+m))$ steps.*

We think of each integer as a k-tuple in radix m. We now apply the $O(n+m)$ time stable sorting algorithm for each of the digits, starting from the least significant digit first. We leave it as an exercise to show that after we have applied this algorithm on all the digits, the integers will be arranged in sorted sequence. This algorithm, also called *radix sort*, is an ideal algorithm for sorting numbers with a small number of digits. When we consider the algorithm for sorting words, the running time will be $O(L(n + |\Sigma|))$, where $L = \max\{l_1, l_2, \ldots, l_n\}$. This is not satisfactory since $L \cdot n$ can be much larger than N (size of input).

The reason that the aforementioned method is potentially inefficient is that many words may be much shorter than L and hence, by considering them to be length L words (by hypothetical trailing blanks), we are increasing the input size asymptotically. If we consider radix sort as a possible solution, the words would have to be left-aligned, that is, all words should begin from the same position. For example, consider the English words $\{cave, bat, at\}$. Since the largest string has four letters, a straightforward application of radix sort will have four rounds as follows. Note that we have used the character _ to represent blank which is the hypothetical lowest rank character.

b	a	t	_		a	t	_	_		b	a	t	_		a	t	_	_
a	t	_	_		b	a	t	_		c	a	v	e		b	a	t	_
c	a	v	e		c	a	v	e		a	t	_	_		c	a	v	e

To make radix sort efficient and to avoid redundant comparison (of blanks), we should not consider a word until the radix sort reaches the right boundary of the word. The radix sort will take a maximum of L rounds and a word of length l will start participating from the $L-l+1$ iteration. This can be easily achieved. A bigger challenge is to reduce the range of sorting in each iteration depending on which symbols of the alphabet participate.

Given a word $w_i = a_{i,1}a_{i,2}\ldots a_{i,l_i}$, where $a_{i,j} \in \Sigma$, we form the following pairs – $(1, a_{i,1}), (2, a_{i,2}), \ldots$. There are N such pairs from the n words and we can think of them as length two strings, where the first symbol is from the range $[1 \ldots L]$ and the second symbol is from Σ. We can sort them using radix sort in two rounds in time proportional to $O(N+L+|\Sigma|)$, which is $O(N+|\Sigma|)$ since $N > L$. From the sorted pairs, we know exactly which symbols appear in a given position (between 1 and L) – let there be m_i words that have non-blank symbols in position i. When considering position i in the radix sort algorithm, we will like to sort these m_i words only (according to their ith digit), because the remaining words will have a blank here, and so, will appear before all these m_i words.

Continuing with the previous example, we obtain the following pairs

$cave : (1, c), (2, a), (3, v), (4, e)$
$bat : (1, b), (2, a)(3, t)$
$at : (1, a), (2, t)$

The sorted order of the pairs is given by the following:

$(1, a), (1, b), (1, c), (2, a), (2, a), (2, t), (3, t), (3, v), (4, e)$

Each pair maintains a pointer to the original word so that given a set of pairs, we can recover the set of words which correspond to these pairs. Now we go back to sorting the given words using radix sort where we will use the information available from the sorted pairs. We start with $i = L$ and decrease it till we reach $i = 0$. For each value of i, let W_i denote the words which have at least m_i symbols. We maintain the invariant that after we have seen digits $i+1$ till L, we have the sorted sequence of words W_{i+1} according to these digits. As we change i, we also maintain a pointer in the sequence of sorted pairs which points to the first pair for which the first symbol is i. In iteration i, we apply stable sort to pairs which have i as their first symbol (for example, if $i = 2$ in the given example, we will apply stable sort to the pairs $(2, a), (2, a), (2, t)$). We need to clarify what 'stable' means here. Note that we have an ordering of words from the previous iteration – this ordering arranges the words in W_{i+1} (and any word which is not in W_{i+1} appears before the words in W_{i+1}). While applying the stable sorting algorithm on the ith digit of all words in W_i, we maintain the invariant mentioned earlier. To maintain this invariant, we allocate an array of size m_i, where we place the pointers to the words in W_i. We must also take care of the new words that start participating in the radix sort – once a word participates, it will participate in all future rounds. (Where should the new words be placed within its symbol group?)

In the given example, $m_4 = 1$, $m_3 = 2$, $m_2 = 3$, and $m_1 = 4$. After two rounds, the table has two elements, viz., *bat* and *cave*. When *at* enters, it must be implicitly before any of the existing strings, since it has blanks as the trailing characters.

$$
\begin{array}{|llll|}
\hline
b & a & t & \\
c & a & v & e \\
\hline
\end{array}
\qquad
\begin{array}{|llll|}
\hline
a & t & & \\
b & a & t & \\
c & a & v & e \\
\hline
\end{array}
$$

The analysis of this algorithm can be done by looking at the cost of each radix sort which is proportional to $\sum_{i=1}^{L} O(m_i)$ which can be bounded by N. Therefore, overall running time of the algorithm is the sum of sorting the pairs and the radix sort. This is given by $O(N + |\Sigma|)$. If $|\Sigma| < N$, then the optimal running time is given by $O(N)$.

3.4 Mergeable Heaps

Heaps[4] are one of the most common implementation of priority queues and are known to support the operations *min, delete-min, insert, delete* in logarithmic time. A complete binary tree (often implemented as an array) is one of the simplest ways to represent a heap. In many situations, we are interested in an additional operation, namely, combining two heaps into a single heap. A binary tree does not support fast (polylogarithmic) merging and is not suitable for this purpose – instead we use *binomial trees*.

A binomial tree B_i of *order i* is recursively defined as follows:

- B_0 is a single node.

- For $i \geq 0$, B_{i+1} is constructed from two B_is by making the root node of one B_i a left child of the other B_i.

The following properties for B_i can be proved using induction (left as an exercise).

---

[4] We are assuming min heaps.

Claim 3.4   *(i) The number of nodes in B_i equals 2^i.*

  (ii) The height of B_k is k (by definition B_0 has height 0).

  (iii) There are exactly $\binom{i}{k}$ nodes at depth k for $k = 0, 1, \ldots$

  (iv) The children of B_i are roots of $B_{i-1}, B_{i-2}, \ldots, B_0$.

A **binomial heap** is an ordered set of binomial trees such that for any i there is at most one B_i.

Let us refer to this property as the *unique-order* property. We actually maintain a list of the root nodes in increasing order of their degrees.

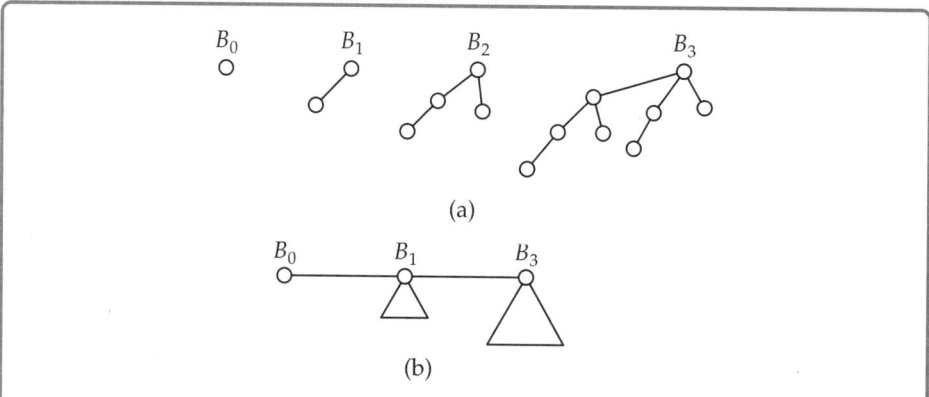

Figure 3.3   *(a) Recursive construction of binomial tree; (b) Binomial heap of 11 elements consisting of three binomial trees*

We can think of this property as a binary representation of a number where the ith bit from right is 0 or 1; in the latter case, its contribution is 2^i (for LSB $i = 0$). Figure 3.3 illustrates a Binomial heap contructed using the Binomial trees. From this analog, a binomial heap on n elements has $\log n$ binomial trees. Therefore, finding the minimum element can be done in $O(\log n)$ comparisons by finding the minimum of the $\log n$ roots.

3.4.1 Merging binomial heaps

Merging two binomial heaps amounts to merging the root lists and restoring the unique-order property. First, we merge the two root lists of size at most $\log n$ into one list so that the trees of the same degree are consecutive in this list (this is similar to the merging procedure in merge sort, and we can do this in time proportional to the total length of these two lists). Subsequently, we walk along the lists, combining two trees of the same degree whenever we find them – they must be consecutive. In other words, whenever we see two B_i trees, we combine them into one B_{i+1} tree by making the root of one tree, a child of the root of the other tree (the choice of the root which becomes the child of the other root depends on which of these is storing the larger value, so that the heap property is preserved). Note that to preserve the property that the list maintains the trees in sorted order of degree, it is best to scan the list from highest degree first. In this way, whenever we replace two B_i trees by one B_{i+1} tree, this property would be preserved. It is possible that a B_{i+1} tree appeared before the two B_i trees in this sequence. In this case, we will have to merge the two B_{i+1} trees as well into a B_{i+2} tree, and so on. Combining two binomial trees takes $O(1)$ time, so the running time is proportional to the number of times we combine.

Claim 3.5 *Two binomial heaps can be combined in $O(\log n)$ steps where the total number of nodes in the two trees is n.*

Every time we combine two trees, the number of binomial trees decreases by one, so there can be at most $2 \log n$ times when we combine trees.

Remark The reader may compare this with the method for summing two numbers in binary representation. This procedure can be used to implement the operation *delete-min* – the details are left as an exercise.

Inserting a new element is easy – add a node to the root list and merge. Deletion takes a little thought. Let us first consider an operation *decrease-key*. This happens when a key value of a node x decreases. Clearly, the min-heap property of the parent node, *parent*(x) may not hold. But this can be restored by exchanging the node x with its parent. The operation may have to be repeated at the parent node. This continues until the value of x is greater than its current parent or x does not have a parent, that is, it is the root node. The cost is the height of a binomial tree which is $O(\log n)$.

To delete a node, we decrease the key value to $-\infty$, so that it becomes the root node. Now, it is equivalent to the operation *delete-min* which is left as an exercise problem.

3.5 A Simple Semi-dynamic Dictionary

Balanced binary search trees like AVL (Adelson-Velskii and Landis) trees, red–black trees, etc., support both search and updates in the worst case $O(\log n)$ comparisons for n keys. These trees inherently use dynamic structures like pointers which actually slow down memory access. Arrays are inherently superior since they support direct memory access; however, they are not amenable to inserts and deletes.

Consider the following scheme for storing n elements in multiple arrays $A_0, A_1, \ldots, A_k$ such that A_i has length 2^i. Each A_i that exists contains 2^i elements in sorted order – there is no ordering between different arrays. Only those A_i exists for which the ith bit b_i in the binary representation of n is non-zero (recall that this representation is unique). Therefore, $\sum_i b_i \cdot |A_i| = n$ and maximum number of occupied arrays is $\log n$.

For searching, we do binary search in all the arrays that takes $O(\log^2 n)$ steps ($O(\log n)$ steps for each array). To insert, we compare the binary representations of n and $n+1$. There is a unique smallest suffix (of the binary representation of n) that changes from $11 \ldots 1$ to $100 \ldots 0$, that is, n is $w011 \ldots 1$ and $n+1$ is $w100 \ldots 0$. Consequently, all the elements of those A_i for which the ith bit becomes 0 are *merged* into an array that corresponds to the bit that becomes 1 (and is also large enough to hold all elements including the newly inserted element). It is left as an exercise (Problem 3.6) to show how these lists can be merged into a single list using $O(2^j)$ comparisons, where A_j is the final sorted list (and contains 2^j elements).

Clearly this could be much larger than $O(\log n)$, but notice that A_j will continue to exist for the next 2^j insertions and therefore, averaging over the total number of insertions gives us a reasonable cost. As an illustration, consider a binary counter and let us associate the cost of incrementing the counter as the number of bits that undergo changes. Observe that at most $\log n$ bits change during a single increment but mostly it is much less. Overall, as the counter is incremented from 0 to $n-1$, bit b_i changes at most $n/2^i$ times, $1 \le i$. So roughly there are $O(n)$ bits that change, implying $O(1)$ changes on the average.

In the case of analyzing insertion in arrays, by analogy, the total number of operations needed to carry out the sequence of merging that terminates at A_j is $\sum_{s=1}^{j-1} O(2^s)$, which is $O(2^j)$. Therefore, the total number of operations over the course of inserting n elements can be bounded by $\sum_{j=1}^{\log n} O(n/2^j \cdot 2^j)$, which is $O(n \log n)$. In other words, the average cost of insertion is $O(\log n)$ that matches the tree-based schemes.

To extend this analysis more formally, we introduce the notion of potential-based *amortized* analysis.

3.5.1   Potential method and amortized analysis

To accurately analyze the performance of an algorithm, let us denote $\Phi(i)$ as a function that captures the *state* of an algorithm or its associated data structure at any stage i. We define *amortized* work done at step i of an algorithm as $w_i + \Delta_i$, where w_i is the actual number of steps[5], as $\Delta_i = \Phi(i) - \Phi(i-1)$, which is referred to as the difference in potential. Note that the total work done by an algorithm over t steps is $W = \sum_{i=1}^{i=t} w_i$. On the other hand, the total amortized work is

$$\sum_{i=1}^{t} (w_i + \Delta_i) = W + \Phi(t) - \Phi(0)$$

If $\Phi(t) - \Phi(0) \ge 0$, amortized work is an upperbound on the actual work.

---

[5] This may be hard to analyze.

Example 3.1 *For the counter problem, we define the potential function of the counter as the number of 1s of the present value. Then, the amortized cost for a sequence of 1s changing to 0 is 0 plus the cost of a 0 changing to 1 resulting in $O(1)$ amortized cost.*

Example 3.2 *A stack supports push, pop, and empty stack operations. Define $\Phi()$ as the number of elements in the stack. If we begin from an empty stack, $\Phi(0) = 0$. For a sequence of push, pop, and empty stack operations, we can analyze the amortized cost. Amortized cost of push is 2, for pop it is 0, and for empty stack it is negative. Therefore, the bound on amortized cost is $O(1)$ and the cost of n operations is $O(n)$. Note that the worst case cost of an empty stack operation can be very high.*

Let us try to define an appropriate potential function for the search data structure analysis. We define the potential of an element in array A_i as $c(\log n - i)$ for some suitable constant c. This implies that the cost of insertion is $c \log n$ for each new element. This could lead to a sequence of merges and from our previous observation, the merging can be done in $O(2^j)$ steps if it involves j arrays. For concreteness, let us assume that it is $\alpha 2^j$ for some constant α. Since the elements are moving to a higher numbered list, the potential is actually decreasing by the number of levels each element is moving up. The decrease in potential can be bounded by

$$\sum_{i=0}^{i=j-1} c2^i(j-i) = \sum_{i=0}^{j-1} \frac{c \cdot i \cdot 2^{j-1}}{2^i} \leq c'2^j \text{ for some appropriate } c'$$

By balancing out α and c, the aforementioned relation can be bounded by $O(1)$. Therefore, the total amortized cost of inserting an element can be bounded by $O(\log n)$ which is the initial potential of an element at level 0.

3.6 Lower Bounds

Although designing a faster algorithm gives us a lot of satisfaction and joy, the icing on the cake is to show that our algorithm is the best possible by proving a matching lower bound. A lower bound is with reference to a computational model with certain capabilities and limitations that constrain the behavior of any algorithm solving a specific problem. The lower bound is also with reference to a problem for which we have to design an algorithm in the given computational model.

The most common example is that of showing a $\Omega(n \log n)$ lower bound in the comparison model. In earlier sections, we have shown that this can be circumvented by using hashing-based algorithms that avoid pairwise comparisons.

Let us highlight the arguments for the $\Omega(n \log n)$ lower bound. First, we abstract every algorithm as a binary tree where the root corresponds to the input sequence and the leaves correspond to each of the $n!$ output permutations. It is left as an exercise problem to the reader to justify this observation. Each internal node correspond to a specific pair that is being compared by the algorithm and the two children correspond to the relations $>, \leq$. For a specific input, the execution of the algorithm corresponds to a path in this tree and the number of nodes in this path is the number of comparisons done. Note that in this model, we are not charged for any operations other than comparisons, in particular, for any processing done to infer the (partial) ordering between successive comparisons. For example, if two comparisons yield that $a < b$ and $b < c$, then the information $a < c$ can be deduced without additional comparisons and one is not charged for this.

Then, we can invoke the following classical result.

Lemma 3.1 *For any binary tree having N leaves, the average length of a root–leaf path is at least* $\Omega(\log N)$.

A formal proof is left to the reader as an exercise problem. As a corollary, the average (and therefore the worst case) number of comparisons used for sorting is $\Omega(\log(n!))$, which is $\Omega(n \log n)$ from Stirling's approximations.

If all input permutations are equally likely, then $\Omega(n \log n)$ is also a lower bound on the average complexity of sorting that is attained by quicksort.

The following elegant result connects the average complexity to the expected complexity of a randomized algorithm.

Theorem 3.1 *Let $\mathcal{A}$ be a randomized algorithm for a given problem and let $\mathbb{E}_{\mathcal{A}}(I)$ denote the expected running time of $\mathcal{A}$ for input I. Let $T_{\mathcal{D}}(A)$ denote the average running time of a deterministic algorithm A over all inputs chosen from distribution $\mathcal{D}$. Then*

$$\max_{I} \mathbb{E}_{\mathcal{A}}(I) \geq \min_{A} T_{\mathcal{D}}(A)$$

Proof: If we fix the random bits for a randomized algorithm, then its behavior is completely deterministic. Let us denote the family of algorithms by $\mathcal{A}^s$ when the choice of the random string is s. The lower bound of the average behavior of $\mathcal{A}^s$ for inputs with distribution $\mathcal{D}$ is given by $T_{\mathcal{D}}(\mathcal{A}^s)$. The average of the *expected* running time of the randomized algorithm $\mathcal{A}$ on inputs having distribution $\mathcal{D}$ can be written as

$$\sum_{I \in \mathcal{D}} \sum_{s} \Pr(s) \cdot T(\mathcal{A}^s(I)) = \sum_{s} \Pr(s) \sum_{I \in \mathcal{D}} T(\mathcal{A}^s(I)) \text{ by interchanging the summation}$$

Since every $\mathcal{A}^s$ is a deterministic algorithm, their average running time over inputs having distribution $\mathcal{D}$ is at least $T_{\mathcal{D}}^*$, where $T_{\mathcal{D}}^* = \min_A T_{\mathcal{D}}(A)$. So the RHS is at least $\sum_s \Pr(s) T_{\mathcal{D}}^* \geq T_{\mathcal{D}}^*$ as $\sum_s \Pr(s) = 1$. This implies that for at least one input I^*, the expected running time must exceed the average value $T_{\mathcal{D}}^*$ that proves the result about the worst case expected bound $\mathbb{E}_{\mathcal{A}}(I^*)$. $\qquad \square$

In the context of sorting, we can conclude that quicksort is optimal even among the family of randomized sorting algorithms.

We now consider a more basic problem that helps us to get an alternate proof for sorting.

Element distinctness (ED): Given a set S of n elements, we want to determine if for all pairs of elements $x, y \in S$, $x \neq y$.

This is a decision problem, that is, the output is *YES/NO*. For example, the answer is YES for the input $[5, 23, 9, 45.2, 38]$ and NO for the set $[43.2, 25, 64, 25, 34.7]$. We can use sorting to solve this problem easily since all elements with equal values will be in consecutive locations of the sorted output. Therefore, the ED problem is *reducible* to the problem of sorting. We will discuss the notion of *reducibility* more formally in Chapter 12.

Therefore, any upper bound on sorting is an upper bound on ED and any lower bound on ED will apply to sorting. To see this, suppose there is a $o(n \log n)$ algorithm for sorting; then, we can obtain an algorithm for ED by first sorting followed by a linear time scan to find duplicate elements. This will give us an $o(n \log n)$ algorithm for ED.

Given this relationship, the nature of the domain set S is crucial for the time complexity of ED. For example, if $S = [1, 2, \ldots, n^2]$, then ED can be solved in $O(n)$ time using radix sort. Here, we will focus on the complexity of ED in the comparison model where the interesting question is if it is easier to compute than sorting.

Consider the input $[x_1, x_2, \ldots, x_n]$ as a point p in the Euclidean space $\mathbb{R}^n$. Consider the hyperplanes in $\mathbb{R}^n$ corresponding to the equations $x_i = x_j$ $i \neq j$. A point p is classified as YES iff it is NOT incident on any of the hyperplanes.

Claim 3.6 *The hyperplanes partitions the space into $n!$ disconnected[6] regions.*

Clearly any two distinct permutations π_1, π_2 must be separated by such a hyperplane since there is at least one pair of elements whose ordering are different across π_1, π_2. So the number of such regions exceed $n!$. For the equality the reader is encouraged to complete the proof. Any algorithm for ED can be represented as a comparison tree where each internal node corresponds to a comparison of the kind $x_k \leq x_\ell$ and the two children corresponds to the two possibilities. The algorithm navigates through this hypothetical tree and reaches a leaf node where the given input is classified as YES/NO. Consider any path in the comparison tree – this corresponds to the intersection of inequalities $x_k \leq x_\ell$ that are half-plane and hence convex. Since the intersection of convex regions is convex, any node in the path corresponds to a connected convex region C in $\mathbb{R}^n$. Therefore, this tree must have at least $n!$ leaf nodes as the region corresponding to a leaf node must be completely contained within one of the $n!$ partitions described in the previous claim.

---

[6] It implies that any path between two regions must intersect at least one of the hyperplanes.

Note that the segment joining two points p_1, p_2 that reach the same leaf node with a YES label completely lies in the convex region corresponding to this leaf node. That is, all the input points represented by this segment will yield a YES answer. If p_1, p_2 correspond to distinct permutations then the segment joining p_1, p_2 will intersect their separating hyperplane at some point $q \in \mathbb{R}^n$ which corresponds to a NO answer. Therefore, the algorithm will classify q incorrectly on reaching the leaf node. The same argument can be extended easily even when we allow stronger primitives than comparisons like general linear inequalities $\sum_i^n a_i x_i \leq 0$. So we can conclude the following

Theorem 3.2 *The lower bound for the element-distinctness problem in a linear decision tree model is $\Omega(n \log n)$ for an input $[x_1, x_2, \ldots, x_n] \in \mathbb{R}^n$.*

Further Reading

The RSA algorithm [125] forms the basis of many cryptographic protocols. It is based on the (conjectured) hardness of a number theoretic problem. There exist public key cryptographic systems based on other hardness assumptions [46]. The randomized selection algorithm is conceptually simpler than the deterministic 'median of medians' algorithm [22] and is described in Floyd and Rivest [49]. This is typical of many problems – andomization often simplifies the algorithm. The nuts and bolts problem [85] is another such example. The binomial heap was invented by Vuileman [151]. A related and theoretically somewhat superior data structure called Fibonacci heaps was first reported by Fredman and Tarjan [53]. The relationship between the worst case randomized time complexity and the average case time complexity was shown by Yao [155].

Exercise Problems

3.1 Construct an input for which the number of iterations in Euclid's algorithm is $\Theta(n)$, where n is the sum of the sizes of the input numbers.

3.2 Prove the following claim about the extended Euclid's algorithm.

Let (x', y') be the integral multipliers corresponding to $\gcd(b, a \mod b)$, that is, $\gcd(b, a \mod b) = x' \cdot b + y' \cdot (a \mod b)$. Then show that $\gcd(a, b) = y' \cdot a + (x' - q)b$, where q is the quotient of the integer division of a by b.

3.3 Extend the algorithm for computing inverse modulo N for a non-prime number N, where $Z_N^* = \{x | x$ is relatively prime to $N\}$. First show that Z_N^* is closed under the multiplication modulo N, that is, $a, b \in Z_N^*$ $a \cdot b \mod N \in Z_N^*$.

3.4 Analyze the complexity of the encryption and decryption of the RSA cryptosystem including finding the inverse pairs e, d as described in Section 3.1.2.

3.5 Prove that the recurrence given by Eq. (3.2.1) for the recursive selection algorithm attains its worst case behavior for $k = n/2$.

Hint: Compare the recurrence expressions for $k = n/2$ versus $k \neq n/2$.

3.6 Given a set S of n numbers, $x_1, x_2, \ldots, x_n$, and an integer k, $1 \leq k \leq n$, design an algorithm to find $y_1, y_2, \ldots, y_{k-1}$ ($y_i \in S$ and $y_i \leq y_{i+1}$) such that they induce k partitions of S of roughly equal size. Namely, let $S_i = \{x_j | y_{i-1} \leq x_j \leq y_i\}$ be the ith partition and assume $y_0 = -\infty$ and $y_k = \infty$. The number of elements in S_i should be $\lfloor n/k \rfloor$ or $\lfloor n/k \rfloor + 1$.

Note: If $k = 2$, then it suffices to find the median.

3.7 By using an appropriate terminating condition, show that $T(n) \in O(n)$ for the deterministic algorithm based on median of medians.

(a) Try to minimize the leading constant by adjusting the size of the group.

(b) What is the space complexity of this algorithm?

3.8 An element is *common* if it occurs more than $n/4$ times in a given set of n elements. Design an $O(n)$ algorithm to find a *common* element if one exists.

3.9 For n distinct elements $x_1, x_2, \ldots, x_n$ with positive weights $w_1, w_2, \ldots, w_n$ such that $\sum_i w_i = 1$, the *weighted median* is the element x_k satisfying

$$\sum_{i | x_i < x_k} w_i \leq 1/2 \quad \sum_{i | x_i \geq x_k, i \neq k} w_i \leq 1/2$$

Describe an $O(n)$ algorithm to find such an element. Note that if $w_i = 1/n$, then x_k is the (ordinary) median.

3.10 Given two sorted arrays A and B of sizes m and n respectively, design an algorithm to find the median in $O(\text{polylog}(m+n))$.

(You can do this in exactly $O(\log(m+n))$ steps).

Can you generalize it to m sorted arrays?

3.11 **Multi-set sorting** Given n elements among which there are only h distinct values, show that you can sort in $O(n \log h)$ comparisons.

Further show that if there are n_α elements with value α, where $\sum_\alpha n_\alpha = n$, then we can sort in time

$$O(\sum_\alpha n_\alpha \cdot \log(\frac{n}{n_\alpha} + 1))$$

3.12 **Sorting in linear time** Consider an input S of n real numbers α_i $1 \leq i \leq n$ that are independently and uniformly chosen at random from the interval $[0, 1]$. We use the following algorithm to sort S.

(i) Hash $x_i \in [0,1]$ to the location $A(\lceil x_i \cdot n \rceil)$ where A is an array of length n. If there is more than one element, create a list in the corresponding location.

(ii) Sort each of the lists using some simple algorithm like selection sort. If $A(i)$ has n_i elements, this will take $O(n_i^2)$ comparisons.

(iii) Concatenate the sorted chains to output the sorted set of elements. Show that the expected running time of this algorithm is $O(n)$. The reader may want to reflect on why this is possible even though the average case lower bound for sorting is $\Omega(n \log n)$ comparisons.

3.13 The *mode M* of a set $S = \{x_1, x_2, \ldots, x_n\}$ is the value that occurs most frequently (in case of ties, break them arbitrarily). For example, among $\{1.3, 3.8, 1.3, 6.7, 1.3, 6.7\}$, the mode is 1.3. If the mode has frequency $m(\leq n)$, then design an $O\left(n \log \left(\frac{n}{m} + 1\right)\right)$ algorithm to find the mode – note that m is not known initially.

3.14 Instead of the conventional two-way merge sort, show how to implement a k-way ($k \geq 2$) merge sort using appropriate data structure in $O(n \log n)$ comparisons. Note that k is not necessarily fixed (but can be a function of n).

3.15 * We want to sort n integers in the range $0 \ldots 2^{b-1}$ (b bits each) using the following approach. Let us assume that b is a power of 2. We divide each integer into two $b/2$ bit numbers – say x_i has two parts x_i' and x_i'', where x_i' is the more significant part. We view the more significant bits as buckets and create lists of $b/2$ bit numbers by associating the lower significant $b/2$ bit numbers with the bucket with the more significant bits. Namely, x_i'' is put into the list corresponding to x_i'. To merge the list, we now add the $b/2$ bit numbers corresponding to the non-empty buckets to the earlier list (to distinguish, we can mark them). We can now sort the list of $b/2$ bit integers recursively and output the merged list by scanning the sorted elements. Note that this list can have more than n numbers since we also added the buckets. Suggest a method to avoid this blow up (since it is not good for recursion) and analyze this algorithm.

Hint: You may want to aim for an $O(n \log b)$ performance in a model that has word size of b bits. What is the space used?

3.16 **Odd–even merge sort** Consider the following (recursive) algorithm for merging two sorted sequences S_1 and S_2. Let $S_{i,j}$ denote the jth element in the sorted sequence S_i and for $i = 1, 2$, let

$$S_i^E = \{S_{i,2}, S_{i,4}, S_{i,6} \ldots\} \text{ (all the even numbered elements)}$$

$$S_i^O = \{S_{i,1}, S_{i,3}, S_{i,5} \ldots\} \text{ (all the odd numbered elements)}$$

The algorithm (recursively) merges S_1^E with S_2^E and S_1^O with S_2^O. Denote the two merged sequences by S^E and S^O. Intersperse the two sequences starting with the smallest element of S^O. (Interspersing $a_1, a_2, a_3 \ldots$ with $b_1, b_2, b_3 \ldots$ produces $a_1, b_1, a_2, b_2 \ldots$).

For example, if we have $S_1 = [2, 6, 10, 11]$ and $S_2 = [4, 7, 9, 15]$, then after merging the odd numbered elements, we get $S^O = [2, 4, 9, 10]$ and similarly, $S^E = [6, 7, 11, 15]$. After interspersing, we obtain $[2, 6, 4, 7, 9, 11, 10, 15]$.

(i) Prove that the smallest element of S^O is the smallest element in the entire set.

(ii) If the interspersed sequence is $\alpha_1, \alpha_2, \alpha_3, \ldots, \alpha_{2i}, \alpha_{2i+1}, \ldots$, show that we can obtain a sorted sequence by comparing the pairs $\alpha_{2i}, \alpha_{2i+1}$ independently. Hence, we need another $n/2$ comparisons to complete the merging.

(iii) How will you use odd–even merge to design a sorting algorithm and what is the running time?

3.17 Show that the delete-min operation a binomial heap can be implemented in $O(\log n)$ steps using merging.

3.18 Starting from an empty tree, show that the amortized incremental cost of building a binomial heap by successive insertions is $O(1)$. (No other updates are allowed). Compare this with the cost of building a binary heap.

3.19 You are given k sorted lists $S_0, S_1, \ldots, S_{k-1}$, where S_i contains 2^i elements. Design an efficient algorithm to merge all the given lists into a single sorted list in $O(\sum_{i=0}^{i=k-1} |S_i|)$ steps.

3.20 We have a set of n nuts and n bolts such that there are n unique pairs of nuts and bolts. There are no measuring gauge available and the only way that we can test a nut against a bolt is to try and see if it exactly fits or the nut is oversized or the bolt is oversized. Design a strategy that minimizes the number of trials comparing a nut and a bolt.

Note that two nuts or two bolts cannot be compared against each other directly.

3.21 Given an array $A = x_1 x_2 \ldots x_n$, the *smallest nearest value* corresponding to x_i is defined as $V_i = \min_{j > i} \{j | x_j < x_i\}$. It is undefined if all elements to the right of x_i are larger than x_i. The *all smallest nearest value* problem (ANSV) is to compute for all i for a given array A. For example, for the array $[5, 3, 7, 1, 8]$, $V_1 = 2, V_2 = 4, V_3 = 4, V_4 = U, V_5 = U$. Here U means undefined.

Design a linear time algorithm for the ANSV problem.

3.22 Prove Lemma 3.1. Generalize the result to k-ary trees for any $2 \le k < n$.

Complete the $\Omega(n \log n)$ proof for sorting by an argument that the comparison tree for any sorting algorithm must have at least $\Omega(n!)$ leaf nodes. Justify why this result does not contradict the earlier exercise problem 3.11 on the upper bound for multi-set sorting.

3.23 The recurrence for the running time of randomized quicksort is given by

$$T(n) \le T(|S_<|) + T(|S_>|) + O(n)$$

using the same notations as in the partition procedure in Section 3.2. Derive a bound on the expected running time by suitably modifying the analysis of the selection problem.

4

CHAPTER

Optimization I: Brute Force and Greedy Strategy

Optimization problems are used to model many real-life problems. Therefore, solving these problems is one of the most important goals of algorithm design. A general optimization problem can be defined by specifying a set of constraints that defines a subset in some underlying space (like the Euclidean space $\mathbb{R}^n$) called the *feasible* subset and an objective function that we are trying to maximize or minimize, as the case may be, over the feasible set. The difficulty of solving such problems typically depends on how 'complex' the feasible set and the objective function are. For example, a very important class of optimization problems is *linear programming*. Here the feasible subset is specified by a set of linear inequalities (in the Euclidean space); the objective function is also linear. A more general class of optimization problems is *convex programming*, where the feasible set is a convex subset of a Euclidean space and the objective function is also *convex*. Convex programs (and hence, linear programs) have a nice property that any local optimum is also a global optimum for the objective function. There are a variety of techniques for solving such problems – all of them try to approach a local optimum (which we know would be a global optimum as well). These notions are discussed in greater detail in a later section in this chapter. The more general problem, the so-called *non-convex programs*, where the objective function and the feasible subset could be arbitrary can be very challenging to solve. In particular, *discrete optimization* problems, where the feasible subset could be a (large) discrete subset of points falls under this category.

In this chapter, we first discuss some of the most intuitive approaches for solving such problems. We begin with heuristic search approaches, which try to search for an optimal solution by exploring the feasible subset in some principled manner. Subsequently, we introduce the idea of designing algorithms based on the greedy heuristic.

4.1   Heuristic Search Approaches

In heuristic search, we explore the search space in a structured manner. Observe that in general, the size of the feasible set (also called the set of feasible solutions) can be infinite. Even if we consider some discrete approximations to the feasible set (or if the feasible set itself is discrete), the set of feasible solutions can be exponentially large. In such settings, we cannot hope to look at every point in the feasible set. Heuristic search approaches circumvent this problem by pruning out parts of the search space where we are sure that the optimal solution does not lie. These approaches are widely used in practice, and are often considered a general purpose technique for many difficult optimization problems.

We illustrate the ideas behind this technique by considering the *0–1 knapsack* problem. The 0–1 knapsack problem is defined as follows. The input consists of a parameter C, which is the capacity of a knapsack, n objects of volumes $\{w_1, w_2, \ldots, w_n\}$, and profits $\{p_1, p_2, \ldots, p_n\}$. The objective is to choose a subset of these n objects that fits into the knapsack (i.e., the total volume of these objects should be at most C) such that the total profit of these objects is maximized.

We can frame this problem as a discrete optimization problem. For each object i, we define a variable x_i, which could be either 0 or 1. It should be 1 if the solution selects object i in the knapsack, 0 otherwise. Note that the feasible subset in this optimization problem is a subset of $\{0,1\}^n$. The knapsack problem can also be formally stated as follows:

$$\text{Maximize} \sum_{i=0}^{n} x_i \cdot p_i \text{ subject to } \sum_{i=0}^{n} x_i \cdot w_i \leq C, \text{ and } (x_1, \ldots, x_n) \in \{0,1\}^n$$

Note that the constraint $x_i \in \{0,1\}$ is not linear, otherwise we could use linear programming. A simplistic approach to solving this problem would be to enumerate all subsets of the n objects, and select the one that satisfies the constraints and maximizes the profits. Any solution that satisfies the knapsack capacity constraint is called a *feasible* solution. The obvious problem with this strategy is the running time which is at least 2^n corresponding to the power-set of n objects. Instead of thinking of the search space as the set of all subsets of objects, we now think of it in a more structured manner. We can imagine that the solution space is generated by a binary tree where we start from the root with an empty set and then move left or right according to the selection of the first object (i.e., value of the variable x_1). At the second level, we again associate the left and right branches with the choice of x_2. Thus, each node in the tree corresponds to a *partial solution*

– if it is at depth j from the root, then the values of variables $x_1, \ldots, x_j$ are known at j. In this way, the 2^n leaf nodes correspond to each possible subset of the power-set which corresponds to an n length 0–1 vector. For example, the vector $000 \ldots 01$ corresponds to the subset that only contains the object n.

Thus, the simplistic approach just means that we look at every leaf node in this tree, and see whether the objects chosen by the solution at the leaf node fit in the knapsack. Among all such leaves, we pick the best solution. This is just a re-statement of the brute force strategy of looking at all possible 2^n different solutions. However, we can devise clever ways of reducing the search space. For example, suppose we traverse the tree (in a top–down manner), and reach a node v. This node corresponds to a partial solution, and assume that the objects which have been picked in this partial solution have total volume more than the knapsack size. At this moment, we know that there is no use exploring the sub-tree below v, because the partial solution corresponding to v itself does not fit in the knapsack.

One can devise more intricate strategies for pruning the search space. A very high level idea would be the following. We maintain a parameter T, which denotes the profit of the best solution that we have obtained thus far while traversing the tree. For each node v, let $P(v)$ denote the partial solution (objects chosen) till v that we want to extend by reaching the set of leaves in the sub-tree rooted at v. For each node v in the tree, we maintain two values, $L(v)$ and $U(v)$, which are supposed to be lower and upper bounds on the best solution among all leaves such that the extended solution lies in the range $[P(v) + L(v), P(v) + U(v)]$. When our algorithm reaches a node v, and if $T > P(v) + U(v)$, it need not explore the sub-tree below v at all. However, if $T < P(v) + L(v)$, then we are guaranteed to improve the current best solution and the algorithm must explore the sub-tree. Note that the bounds $L(v)$ and $U(v)$ may not be fixed – the algorithm updates them as it proceeds.

Consider a node v at level j in the tree, implying that $P(v)$ corresponds to a partial solution, where we have decided which objects to choose among $1, 2, \ldots, j$. Now suppose the partial solution fits in the knapsack; it occupies weight $W(v)$ and has profit $P(v)$. For $U(v)$, let ρ denote the maximum density of an object among $j+1, \ldots, n$, where the density of an object is the ratio of its profit to its weight. We observe that $U(v)$ can be set to $(C - W(v)) \cdot \rho$. Indeed, the objects chosen in the partial solution for v already occupy $W(v)$ space, and so, we can only add $C - W(v)$ more weight to it. Any such object added after v would contribute at most ρ units of profit per unit weight.

Example 4.1 *Let the capacity of the knapsack be 15; the weights and profits are as follows:*

Profits	10	10	12	18
Weight	2	4	6	9

We will use the strategy described earlier for setting $L(v)$ and $U(v)$. Observe that the densities of the objects are 5, 2.5, 2, and 2 respectively. Initially, our estimate $T = 0$. For the root node v, $L(v)$

is 0 and $U(v)$ is $5 \times 15 = 75$. Consider the left child of the root node which corresponds to including the first object. For this node, call it w, the remaining capacity of the knapsack is 13, and T becomes 10 (because there is a solution of value 10 – just take this object). By proceeding this way, we obtain $T = 38$ for the set of objects $\{1, 2, 4\}$. Exploring further, we will come to a stage where we would have included object 1 and decided against including object 2. Call this node u. Hence, $L(u) = 10$, and residual capacity is 13. Should we explore the sub-tree regarding $\{3, 4\}$? Since the densities of both these objects are 2, we get $U(u) = 2 \times 13 + 10 = 36 < T = 38$. So we need not search this sub-tree. By continuing in this fashion, we may be able to prune large portions of the search tree. However, it is not possible to obtain any provable improvements.

This method of pruning a search is called *branch and bound* and although it is clear that it is advantageous to use the strategy, there may not be any provable savings in the worst case.

4.1.1 Game trees*

Game trees[1] represent a game between two players who alternately make moves trying to win the game. For example, consider the game "tic-tac-toe". This game is played on a 3×3 board with 9 squares. Let us call the two players A and B. Initially, all 9 squares are empty. The two players make a move alternately – player A writes the symbol $\times$ on one of the empty squares, whereas the player B writes the symbol $\bigcirc$ on one of the empty squares. The player who first gets three of her symbols along a straight line in the board (diagonal, vertical, or horizontal) wins.

The set of all strategies in this game can be represented by a giant tree, where each node in the tree corresponds to a configuration of the board. A configuration of the board is obtained by labeling the squares on the board which can be realized during a game. Note that certain labelings of the board are not configurations. For example, suppose we label 4 of the squares as $\times$ and 1 square as $\bigcirc$. We can never reach such a labeling because players take turns. In this tree, the root node corresponds to the configuration where all squares are empty. Further, player A makes a move – she has 9 choices. Therefore, the root node has 9 children, each corresponding to writing the $\times$ symbol on one of the 9 squares. Consider a child v of the root (which has exactly one $\times$ symbol). At v, it is B's turn to move. Player B has 8 choices, and so this node will have 8 children. As we can see, this tree has 9 levels, with odd levels (we denote the top level consisting of the root as level 1) corresponding to player A's turn and even levels for player B. Further, a node at which we have three symbols of the same kind lying on a straight line corresponds to a win situation for one of the players. Such a node will be a leaf node in the tree. Similarly, a node where all the squares have been labeled will be a leaf node (and may correspond to a scenario where no one wins).

---

[1] The reader is expected to be familiar with the notion of conditional expectation.

For the sake of simplicity, let us consider a two-player game where one of the players always wins when the game stops (at a leaf node). For reasons which will become clear soon, we shall call the nodes corresponding to player A's turn (i.e., those at odd levels) as 'OR' nodes denoted by $\vee$; similarly, we shall call the nodes at even levels as 'AND' nodes denoted by $\wedge$. Such a tree is often called an *AND–OR* tree. Let 1 represent a win for player A and 0 represent a loss for player A. These numbers are flipped for player B. The leaf nodes correspond to the final state of the game and are labeled 1 or 0 corresponding to win or loss for player A. We describe a rule for labeling each internal node of the tree (in a bottom-up fashion). A $\vee$ node has value 1 if one of the children is 1, and 0 otherwise, and so it is like the Boolean function *OR*. A $\wedge$ node behaves like the Boolean AND function – it is 0 if one of the children is 0, and 1 if all of the children are 1. The interpretation of this 0–1 assignment to nodes is as follows. A node is labeled 1 if there is a winning strategy for player A irrespective of how player B plays; whereas, a label of 0 indicates that no matter how A plays, there is always a winning strategy for player B. The player A at the root can choose any of the branches that leads to a win. However, at the next level, she is at the mercy of the player B – only when all branches for B lead to a win for A, will player A have a winning strategy; otherwise, the player B can inflict a loss on player A. Note that in such situations, we are not banking on mistakes made by either player; we are only concerned with guaranteed winning strategies.

For concreteness, we will consider game trees where each internal node has two children. Hence, the evaluation of this game tree works as follows. Each leaf node is labeled 0 or 1 and an internal node is labeled $\wedge$ or $\vee$ – these will compute the Boolean function of the value of the two child nodes. The value of the game tree is the value available at the root node. This value indicates which player has a winning strategy – note that one of the two players will always have a winning strategy, because it will be labeled 0 or 1. Consider a game tree of depth $2k$ – it has $2^{2k} = 4^k$ nodes. Thus, it seems that it will take about $O(4^k)$ time to evaluate such a game tree. We now show that with clever use of randomness, one can reduce the *expected* time to $O(3^k)$ evaluations, which is a considerable saving since the exponent changes.

The basic idea can be explained with the help of a single level $\wedge$ tree. Suppose we are evaluating the $\wedge$ node at the root; assume that it evaluates to 0. Therefore, at least one of the two leaf children happens to be 0. If we happened to look at this child before the other one, we need not evaluate the other child. Apriori, it is difficult to tell which of the two leaf children is 0 – but if we choose a child randomly, then the *expected* number of lookups is

$$\Pr[\text{ first child is } 0] \cdot 1 + \Pr[\text{ first child is not } 0] \cdot 2 = \frac{1}{2} \cdot 1 + \frac{1}{2} \cdot 2 = \frac{3}{2}$$

a saving of 4/3 factor over the naive strategy of probing both the children. This is a conditional expectation that the $\wedge$ node is 0. Note that in the case where both children are 0, the expectation is 1; so we are considering the worst case scenario. For the other case, when the $\wedge$ node evaluates to 1, there is no saving by this strategy. We still have to probe both the children. However, any interesting game tree will have at least two levels, one $\wedge$ and the other $\vee$. Then we can see that for an $\wedge$ node to be 1, both the child $\vee$ nodes must be 1. Now for these $\vee$ nodes, we can use the aforementioned strategy to save the number of probes. In essence, we are applying the branch and bound method to this problem, and we obtain a provable improvement by evaluating the two children at a node in a random order.

Now consider the general case of a tree with depth $2k$ (i.e., 4^k leaf nodes) and alternating $\wedge$ and $\vee$ nodes, each type having k levels. We will show that the expected number of leaf nodes visited is 3^k by induction on k. The base case (for $k=1$) is left to the reader as an exercise problem. Assume that the statement is true for trees of depth $2(k-1)$ $k \geq 2$. We will use $N(v)$ to denote the number of leaf nodes evaluated of a sub-tree rooted at v and $\mathbb{E}[N(v)]$ to denote its expected value.

Now consider such a tree of depth $2k$. There are two cases depending on whether the root is labeled $\vee$ or $\wedge$. Let us consider the case where the root has label $\vee$; hence, its two children, say y and z, are labeled $\wedge$ (Fig. 4.1). The children of y and z are $\vee$ nodes with $2(k-1)$ depth.

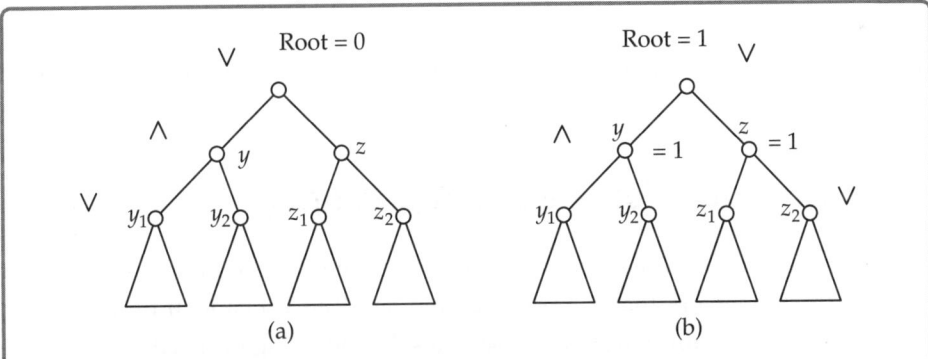

Figure 4.1 *Illustration of the induction proof when root node is $\vee$. The two cases (i) and (ii) correspond to when the root node equals 0 and 1.*

We will consider the two cases.

(i) The root evaluates to 0: Since the root is an $\vee$ node, both y and z must evaluate to 0. Since these are $\wedge$ nodes, it must be the case that at least one child of y is 0 (and similarly for z). It now follows from the earlier argument that with probability $1/2$, we will end up evaluating the leaf nodes of only one of the children of y (and similarly for z). Using the induction hypothesis for the children y_1, y_2 of y, we obtain that the expected number of evaluations for the sub-tree below y satisfies

$$\mathbb{E}[N(y)] = \frac{1}{2} \cdot \mathbb{E}[N(y)|\text{ one children evaluated}] + \frac{1}{2}\mathbb{E}[N(y)|\text{ both child evaluated}]$$

$= 1/2 \cdot 3^{k-1} + 1/2 \cdot 2 \cdot 3^{k-1} = 3^k/2$. We obtain an identical expression for the expected number of evaluations below z, and therefore, the total expected number of evaluations is 3^k.

(ii) The root evaluates to 1: At least one of the $\wedge$ nodes y, z must be 1. Assume without loss of generality that the node y evaluates to 1. With probability $1/2$, we will probe y first; then, we need not look at z. To evaluate y, we will have to look at both the children of y, which are at depth $2(k-1)$. Applying induction hypothesis on children of y, we see that the expected number of evaluations for the sub-tree below y is $2 \cdot 3^{k-1}$. We obtain the same expression for the sub-tree below z. Therefore, the expected number of evaluations is $1/2 \cdot 2 \cdot 3^{k-1} + 1/2 \cdot 4 \cdot 3^{k-1} = 3^k$, where the first term corresponds to the event that we pick y first (and so do not evaluate z at all), and the second term corresponds to the event that we pick z first, and so may evaluate both y and z.

In summary, for an $\vee$ root node, regardless of the output, the expected number of evaluations is bounded by 3^k. We can express this in terms of the total number of leaves. Note that if N denotes the number of leaves, then $N = 4^k$, and so, the expected number of evaluations is $N^{\log_4 3} = N^\alpha$, where $\alpha < 0.8$. The case when the root is an AND node is left as an exercise.

4.2   A Framework for Greedy Algorithms

There are very few algorithmic techniques for which the underlying theory is as precise and clean as the framework that is presented here. Let S be a set and M be a subset[2] of 2^S. Then, (S, M) is called a **subset system** if it satisfies the following property.

> For all subsets $T \in M$, if $T' \subset T$, then $T' \in M$.

Note that the empty subset $\phi \in M$. The family of subsets M is often referred to as *independent* subsets and one may think of M as the *feasible* subsets.

Example 4.2 *Let $G = (V, E)$ be an undirected graph. Consider the subset system (E, M), where M consists of all subsets of E which form a forest (recall that a set of edges form a forest if they do not induce a cycle). It is easy to see that this satisfies the property for a subset system.*

Given a subset system, we can define a natural optimization problem as follows. For any weight function $w : S \to \mathbb{R}^+$, we would like to find a subset in M for which the

---

[2] M is a family of subsets of S.

cumulative weight of the elements is *maximum* among all choices of subsets from M. We refer to such a subset as an *optimal* subset. Note that this is a non-trivial problem because the size of M could be exponential in S, and we may only have an implicit description of M (as in the earlier example[3]). In such a case, we cannot afford to look at every subset in M and evaluate the total weight of elements in it. An intuitive strategy to find such a subset is the greedy approach explained in Fig. 4.2.

Procedure GenGreedy(S,M)

1 **Input** $S = \{e_1, e_2, \ldots, e_n\}$ in decreasing order of weights;
2 $T = \phi$.
3 **for** $i = 1$ *to* n **do**
4   **if** $T \cup \{e_i\} \in M$ **then**
5       $T \leftarrow T \cup \{e_i\}$

6 Output T as the solution

Figure 4.2   *Algorithm Gen_Greedy*

The running time of the algorithm is dependent mainly on the test for *independence* which depends on the specific problem. Even if M is not given explicitly, we assume that an implicit characterization of M can be used to perform the test. In the example of forests in a graph, we just need to check if the set T contains a cycle or not.

What seems more important is the following question – Is T the maximum weight subset in M? This is answered by the following result.

Theorem 4.1 *The following are equivalent.*

1. *Algorithm Gen_Greedy outputs the optimal subset for any choice of the weight function.*

2. **Exchange property**
   For any pair of subsets $S_1, S_2 \in M$, where $|S_1| < |S_2|$, there exists an element $e \in S_2 - S_1$ such that $S_1 \cup \{e\} \in M$.

3. **Rank property**
   For any $A \subset S$, all maximal independent subsets of A have the same cardinality. A subset T of A is a maximal independent subset if $T \in M$, but $T \cup \{e\} \notin M$ for any $e \in (A - T)$. This is also called the rank of the subset system.

A subset system satisfying any of the aforementioned three conditions is called a *matroid*. The theorem can be used to establish properties 2 or 3 to justify that a greedy

---

[3] The number of spanning trees of a complete graph is n^{n-2} from Cayley's formula.

approach works for the problem. On the contrary, if we prove that one of the properties does not hold (by a suitable counterexample), the greedy approach may not return an optimal subset.

Proof: We will prove it by the following cyclic implications:
Property 1 $\implies$ Property 2, Property 2 $\implies$ Property 3, Property 3 $\implies$ Property 1.

Property 1 implies Property 2 We prove the contrapositive. Suppose Property 2 does not hold for some subsets S_1 and S_2. That is, we cannot add any element from $S_2 - S_1$ to S_1 and keep it independent. We will show that Property 1 does not hold. Let p denote S_1 (and hence, $|S_2| \geq p + 1$). We now define a weight function on the elements of S such that the greedy algorithm fails to output an optimal subset. We define the weight function on the elements of S as follows:

$$w(e) = \begin{cases} p+2 & \text{if } e \in S_1 \\ p+1 & \text{if } e \in S_2 - S_1 \\ 0 & \text{otherwise} \end{cases}$$

The greedy algorithm will pick up all elements from S_1 and then it will not be able to choose any element from $S_2 - S_1$. Therefore, the solution given by the greedy algorithm has weight $(p+2)|S_1| = (p+2) \cdot p$. Now consider the solution consisting of elements of S_2. The total weight of elements in S_2 is $(p+1)|S_2 - S_1| + (p+2)|S_1| > (p+2) \cdot p$. Thus, the greedy algorithm does not output an optimal subset, that is, Property 1 does not hold.

Property 2 implies Property 3 Let S_1 and S_2 be two maximal independent subsets of A, and suppose, for the sake of contradiction, that $|S_1| < |S_2|$. Then, Property 2 implies that we can add an element $e \in S_2 - S_1$ to S_1 and keep it independent. However, this contradicts the assumption that S_1 is maximal. Therefore, the two sets must have the same size.

Property 3 implies Property 1 Again we will prove the contrapositive. Suppose Property 1 does not hold, that is, there is a choice of weights $w(e)$ such that the greedy algorithm does not output an optimal subset. Let $e_1, e_2, \ldots, e_n$ be the edges chosen by the greedy algorithm in decreasing order of their weights. Call this set E_1. Further, let $e_1', e_2', \ldots, e_m'$ be the edges of an optimal solution in decreasing order of their weights – call this set E_2. First observe that the solution E_1 is maximal – indeed, if we can add an element e to the greedy solution and keep it independent, then the greedy algorithm should have added e to the set T (as described in Procedure Gen_Greedy). It follows from Property 3 that $m = n$.

Since the weight of the greedy solution is not maximum, there must be a $j \leq m$ such that $w(e_j) < w(e_j')$. Otherwise, the fact that $m = n$ implies that the weight of E_1 is at least that of E_2. Let $A = \{e \in S | w(e) \geq w(e_j')\}$ be the set of elements whose weight is at least $w(e_j')$. The subset $\{e_1, e_2, \ldots, e_{j-1}\}$ is maximal with respect to A (Why?). All the elements in

$\{e_1', e_2', \ldots, e_j'\}$ form an independent subset of A that has greater cardinality. This shows that Property 3 does not hold. □

Many natural problems can be modeled by subset systems. We describe some well-known examples of matroids and the corresponding maximum weight independent set problem in this section.

Example 4.3 Half-matching problem *Given a directed graph with non-negative edge weights, we would like to find out the maximum weighted subset of edges such that the in-degree of any node is at most 1. Let us see how to phrase this as a maximum weight independent set problem in a matroid.*

The definition of the subset system should be clear – the set S is the set of edges in the directed graph and M is the family of all subsets of edges E' such that no vertex has an in-degree more than 1 in the sub-graph induced by E'. Let us now show that this subset system is a matroid. We prove Property 2. Consider two subsets S_p and S_{p+1} with p and $p+1$ edges respectively. Let V_p be the set of vertices which form the head of the edges in S_p, that is, $V_p = \{u : \exists e = (v,u) \in S_p\}$. By the definition of an independent set, note that $|V_p| = |E_p|$. Define V_{p+1} similarly. Since $|V_{p+1}| > |V_p|$, there is a vertex $u \in V_{p+1} - V_p$. Consider the edge e in E_{p+1} whose head is u. Clearly, $e \notin E_p$ and adding e to E_p will preserve the independence of this set. Therefore, this subset system is a matroid.

Example 4.4 Maximum weight bipartite matching *We now give an example of an important subset system which is not a matroid. Let G be a bipartite graph where edges have weights. A matching in G is a subset of edges which do not share a common vertex. The maximum weight matching problem seeks to find a matching for which the total weight of the edges in it is maximum. As earlier, we can define a subset system corresponding to matchings in G. We define a subset system (S, M), where S is the set of edges in G and M consists of all subsets of edges which form a matching. However, this subset system is not a matroid.*

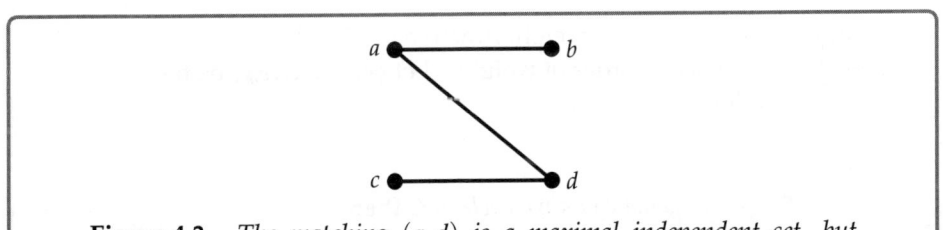

Figure 4.3   *The matching (a,d) is a maximal independent set, but $(a,b), (c,d)$ is a larger maximal independent set.*

To see why, consider a simple bipartite 'zig-zag' graph (shown in Fig. 4.3). There are two maximal independent sets here – one with cardinality 2 and the other having only 1 edge. Therefore, Property 3 is violated. In fact, algorithms for finding maximum weight matchings turn out to be much more complex than simple greedy strategies.

4.2.1 Maximum spanning tree

In the maximum spanning tree problem, we are given an undirected graph $G = (V, E)$ and a weight function $w : E \to \mathbb{R}$. Assuming that the graph is connected, we would like to find a spanning tree for which the total weight of edges in it is maximum. It is natural to define a subset system here – (S, M), where S is the set of edges, and M consists of those subsets of S which form a forest (i.e., do not contain a cycle). For any $A \subset E$, let $V(A)$ denote the vertex set of the edges in A. Note that a maximum independent subset here will be a spanning tree. We know that every spanning tree in a connected graph has $n - 1$ edges, where n is the number of vertices and equals $n - k$ if there are k connected components. Hence, the maximal subset of A has rank $V(A) - k$ if there are k connected components in the subgraph induced by the edges in A. Therefore, it follows from Property 3 that this set system is a matroid.

4.2.2 Finding minimum weight subset

The greedy algorithm for matroids finds an independent set of maximum total weight. Is it possible to extend the algorithm to finding the minimum weighted maximal independent subset, for example, minimum spanning trees (MST)? The well-known Kruskal's algorithm (see Figure 4.4) seems identical to the greedy framework except that it chooses the minimum weight element at each stage. Do we need to develop an analogous theory for minimization? Fortunately, we can simply use a reduction to the maximization problem. Replacing the weight of each element by its negation does not work because the greedy algorithm requires that all weights be non-negative.

Procedure Kruskal(G, w)

1 **Input** Graph $G = V, E$, a weight function $w : E \Rightarrow \mathbb{R}$;
2 Sort E in increasing order of weights. Let $\{e_1, e_2, \ldots, e_m\}$ be the sorted order;
3 $T = \Phi$;
4 **for** $i = 1$ *to* m **do**
5 $\quad$ **if** $T \cup \{e_i\}$ *does not contain a cycle in G* **then**
6 $\quad\quad T \leftarrow T \cup \{e_i\}$
7 Output T as MST of G.

Figure 4.4   *Kruskal's minimum spanning tree algorithm*

Suppose the maximum weight of any element in S is $g = \max_{x \in S}\{w(x)\}$. We define another related weight function $w'(x) = g - w(x), \forall x \in S$. Thus, $w'(x) \geq 0$. Suppose we now run the Gen_Greedy algorithm with the weight function w'. This produces a maximum weight independent subset with respect to the weight function w'. Let this subset be $\{y_1, y_2, \ldots, y_n\}$ in decreasing order of weights $w'(y_i)$, where n equals the size of any maximal independent set in the matroid, that is, its rank (Property 3).

$$\sum_{i=1}^{n} w'(y_i) = \sum_{i=1}^{n} (g - w(y_i)) = ng - \sum_{i} w(y_i)$$

This implies that $\sum_i w(y_i)$ must be the minimum among all maximal independent subsets (else we improve the maximum under w'). Moreover, $y_1, y_2, \ldots, y_n$ must be in increasing order of weights under w. This means that if we run the Gen_Greedy algorithm by picking the smallest feasible element at each step, we obtain the minimum weighted independent subset. Kruskal's algorithm is a special case of this fact.

The crucial argument in this reduction is based on the *rank property* of matroids that enabled us to express the weight of the minimum subset in terms of subtraction from the fixed term ng, where n is the rank of the matroid. If n was not fixed for all maximal independent subsets, the argument would fail.

4.2.3   A scheduling problem

We now give another application of the greedy algorithm. Unlike other examples, the construction of the corresponding matroid for this application is not immediately obvious. Consider the following scheduling problem. We are given a set of jobs $J_1, J_2, \ldots, J_n$. Each job J_i has a deadline d_i for completion and a corresponding penalty p_i if it exceeds its deadline. There is one machine on which each of them needs to be processed (in some order) and each job takes unit amount of time to complete. Our goal is to process the jobs in an order such that the the total penalty incurred by the jobs that are not completed before their deadlines is minimized. Equivalently, we want to maximize the penalty of the jobs that get completed before their deadlines.

In order to apply the greedy algorithm, we define a subset system (S, M), where S is the set of all jobs. A set A of jobs is *independent* if there exists a schedule to complete all jobs in A without incurring any penalty, that is, all jobs in A can be completed within their deadlines. We prove that this set system is a matroid by showing that it has Property 2. Recall that Property 2 states that given any two independent sets A, B with $|B| > |A|$, there exists a job $J \in B - A$ such that $\{J\} \cup A$ is independent. We prove this by induction on $|A|$. If $|A| = 0$, this is trivial. Now assume Property 2 holds whenever $|A| = m - 1$. Pick two independent sets A, B, with $|A| = m < n = |B|$.

Consider a feasible schedule F_A for A, that is, an ordering of jobs in A such that each job finishes before its deadline. Note that the ith job in this ordering finishes at time i because all jobs are of unit size. Let this ordering be $A_1, A_2, \ldots, A_m$ (note that $A_1, \ldots, A_m$ are the jobs in the set A). Similarly, consider a similar schedule F_B for B, and let the ordering be $B_1, B_2, \ldots, B_n$.

Note that the deadline of B_n is at least n (because it finishes before its deadline in F_B). If $B_n \notin A$, then we can add B_n to A and schedule it as the last job – this job will finish at time $m+1$, whereas it was finishing at time n in schedule B. Since $m+1 \leq n$, this job will finish before its deadline. Hence, assume $B_n \in A$. Now form sets A' and B' by removing B_n from A and B respectively. By induction hypothesis, there is a job $J \in B' - A'$ such that $A' \cup \{J\}$ is independent. Let A'' denote $A' \cup \{J\}$ – we know that there is an ordering $F_{A''}$ of jobs in A'' such that every job finishes by its deadline. Now we claim that $A'' \cup \{B_n\}$, which is the same as stating that $A \cup \{J\}$ is also independent. Indeed, consider the following schedule – first process the jobs in A'' according to $F_{A''}$, and then process B_n. Since $|A''| = m$, note that B_n will finish at time $m+1 \leq n$, which is before its deadline. Thus, we see that $A \cup \{J\}$ is independent. Since $J \in B - A$, Property 2 follows.

Now that our subset system is a matroid, we can use the greedy algorithm to solve the maximization problem. The only remaining detail is how to verify if a set of jobs is independent. For this, we just need to order the jobs in increasing order by their deadlines and check if this ordering satisfies all the deadlines (see Exercises).

4.3 Efficient Data Structures for Minimum Spanning Tree Algorithms

In this section, we re-visit the greedy algorithm for the minimum spanning tree in (connected) undirected graphs. This algorithm (Fig. 4.4) is also known as Kruskal's algorithm and was discovered much before the matroid theory was developed. We present it again without the matroid notation. We first sort the edges in increasing order of weights. The algorithm maintains a set T of edges which will eventually form the desired spanning tree. It considers edges in this order, and adds an edge e to T only if adding e to T does not create a cycle (i.e., if the set remains independent in the matroid sense).

The key to an efficient implementation is the *cycle test*, that is, how do we quickly determine if adding an edge induces a cycle in T? We can view Kruskal's algorithm as a process that starts with a forest of singleton vertices and gradually connects the graph by adding edges to the set T and hence, grow the trees. In fact, at any point of time, the set T will be a forest. An edge e will be added to T only if the endpoints of e do not lie in the same connected component (i.e., tree) of T. Adding such an edge will create a cycle (Fig. 4.5). Conversely, if the endpoints of such an edge lie in different trees of T, then we can add e to T without creating a cycle. When we add such an edge to T, two connected components of T merge into one connected component.

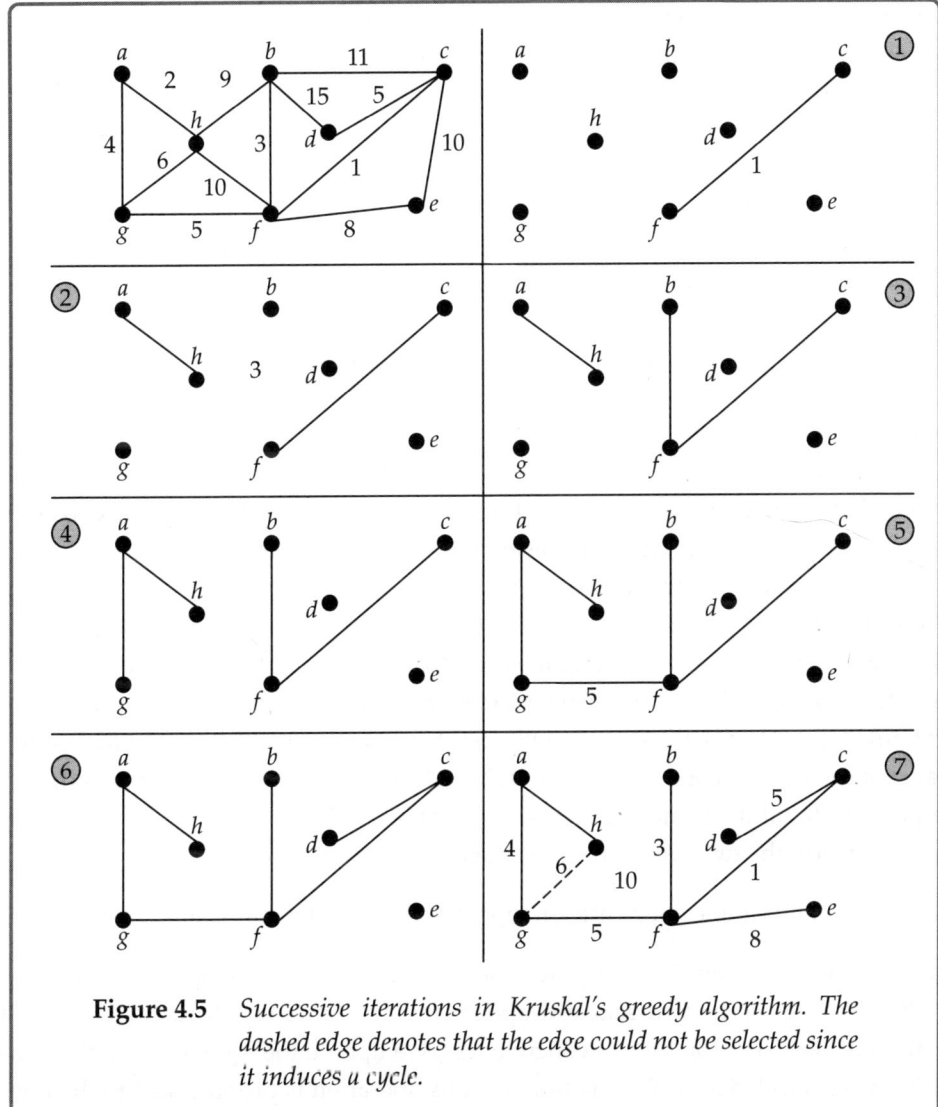

Figure 4.5 *Successive iterations in Kruskal's greedy algorithm. The dashed edge denotes that the edge could not be selected since it induces a cycle.*

Therefore, we can answer the cycle test query, provided we maintain the partition of vertices into trees in T. It should be clear that we need to maintain a data structure that supports the following operations

Find Given a vertex, find out which connected component it belongs to.

Union Combine two connected components into one component.

The *Find* operation corresponds to checking if adding an edge creates a cycle. Indeed, we just need to check if the endpoints of a vertex belong to the same connected component.

The *Union* operation is needed to update the set of connected components. When we add an edge, two components in T merge into one tree.

For obvious reasons, such a data structure is called a *Union–Find* data structure. In fact, we can view the data structure in a more general context. Given a set S, we shall maintain a family of disjoint subsets that are connected components. The Union-Find data structure supports two operations – given an element of S, it finds the subset in this family which contains this element, and replaces two subsets in this family by their union. Next, we explore how to implement this data structure.

4.3.1   A simple data structure for Union–Find

We will use a more general context where we are given a set with n elements, labeled $1, 2, \ldots, n$. We are also given a family of subsets of $\{1, 2, \ldots, n\}$ – the subsets in this family form a disjoint partition of $\{1, 2, \ldots, n\}$. Initially, we assume that the family consists of n singleton sets, one for each element (this corresponds to the case when the set T is empty in the minimum spanning tree algorithm).

We use an array A of size n to represent the sets. For each element i, $A(i)$ contains the label of the set containing i – we assign each set in the family a unique label. Initially, we set $A(i)$ to i. Thus, all labels are distinct to begin with. As we merge sets (during union operation), we create a new set – we will need to assign a new label. We will ensure that the label of each set remains in the range $1, 2, \ldots, n$ at all times. For each set (label), we also have pointers to all its elements, that is, the indices of the array that belongs to the set. Now we perform the two *Find* and *Union* operations as follows:

Find   This is really simple – for vertex i, report $A(i)$. This takes $O(1)$ time.

Union   To perform union(S_j, S_k), where S_j and S_k are labels of two sets, we first consider the elements in the two sets and update the $A[i]$ value for all such elements i to a unique label. For the sake of concreteness, we update the $A[i]$ values of all elements in the set labeled S_j to S_k. The time for this operation is proportional to the number of elements in set S_j. Note that we had a choice here – we could have changed the labels of all elements in S_k to S_j. For obvious reasons, we would change labels of the smaller subset (this is called *union-by-rank* heuristic).

Note that the time for a single union operation can be quite high. Instead, as in Kruskal's algorithm, we shall analyze the total time taken by a sequence of union operations. Consider a fixed element x. The key to the analysis lies in the answer to the following question.

How many times can the label of x change?

Every time there is a label change, the size of the set containing x increases by a factor of two because of the union-by-rank heuristic. Since the size of any set in our set system is at most n, this implies that the maximum number of label changes is $\log_2 n$ for any element x and a total of $O(n \log n)$ for all the n elements.. Kruskal's algorithm involves $|E|$ finds and at most $|V| - 1$ unions; it follows that this can be done in $O(m + n \log n)$ steps using the array data structure described earlier.

4.3.2 A faster scheme

The previous data structure gives optimal performance for $m \in \Omega(n \log n)$ – indeed, any algorithm for the minimum spanning tree (MST) must look at every edge. Hence, theoretically, we want to design better schemes for graphs with fewer edges. For this, we will explore faster schemes implementing the Union–Find data structure.

Instead of arrays, it is easier to visualize the data structure using trees.[4] We represent every subset using a rooted tree. We will maintain the invariant that every such tree will have as many nodes as the number of elements in the corresponding connected component – each node of the tree will be labeled by a unique element of the subset. An example is given in Fig. 4.6. We can label the three sets by the labels of their roots, that is, 6, 12, and 5 respectively.

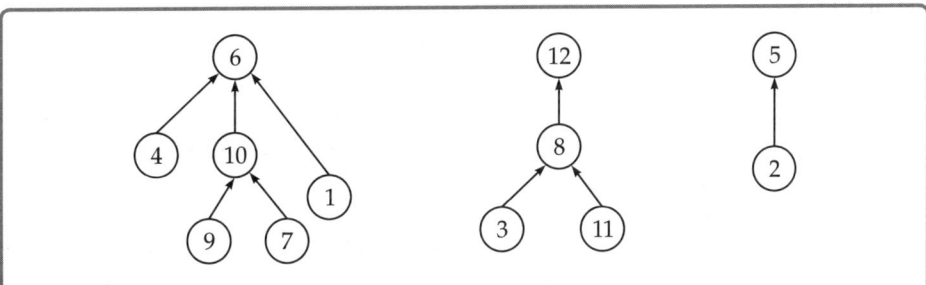

Figure 4.6 *An example of a Union–Find data structure storing elements $\{1, 2, \ldots, 12\}$. The three sets are $\{6, 4, 10, 1, 9, 7\}$, $\{12, 8, 3, 11\}$, $\{5, 2\}$.*

Initially, all trees are singleton nodes (which represent singleton sets). The root of each tree is associated with a label (of the corresponding subset) and a rank which denotes the maximum depth of any leaf node in this tree. To perform the operation Find(x), we traverse the tree starting from the node x till we reach the root and report its label. Hence, the cost of a Find operation is the maximum depth of a node.

---

[4] This tree should not be confused with the MST that we are trying to construct.

To perform Union (T_1, T_2), where T_1 and T_2 are the roots of two trees, we make the root of one tree the child of the root of the other tree. To minimize the depth of a tree, we attach the root of the smaller rank tree to the root of the larger rank tree. This strategy is known as the *union by rank* heuristic. The rank of the resulting tree is determined as follows: if both T_1 and T_2 have the same rank, then the rank of the new tree is one more than the rank of T_1 (or T_2); otherwise, it is equal to the maximum of the ranks of the two trees. Note that once a root node becomes a child of another node, its rank does not change anymore. Clearly, the union operation takes $O(1)$ steps. We leave it as an exercise to show that a tree of rank r has at least 2^r nodes, and so, the depth of any tree in this data structure will be bounded above by $\log n$. It follows that a Find operation will take $O(\log n)$ time. We have already argued that a Union operation takes $O(1)$ time. Therefore, Kruskal's algorithm can be implemented in $O(m \log n + n)$ time. This seems to be worse than the array implementation mentioned earlier. Seemingly, we have not quite gained anything; so let us use the following additional heuristic.

Path compression heuristic In this heuristic, we try to reduce the height of a tree even below $\log n$ by *compressing* any sequence of nodes which lie on a path starting from the root. When we do a Find(x) operation, let $x, x_1, x_2, \ldots, x_r = $ root of x be the sequence of nodes visited (in the reverse order starting from x). In this heuristic, we update x_r as the parent of $x_1, x_2, \ldots, x$ (i.e., x_r has all the other nodes in this path as its children). An example is given in Fig. 4.7.

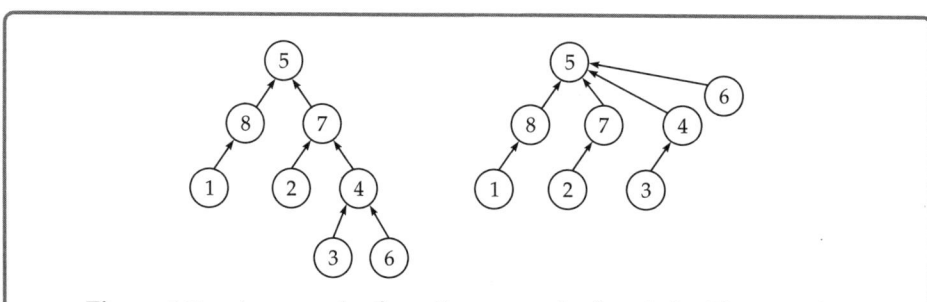

Figure 4.7   *An example of a path compression heuristic. The operation Find(6) causes 6, 4, and 7 to become children of the root node.*

Clearly, the motivation is to bring more nodes closer to the root node so that the time for subsequent Find operations involving these nodes decrease. Note that the time spent during the path compression heuristic is not much – it only doubles the cost of the current Find operation.

While it is intuitively clear that this method should give us an advantage, we have to rigorously analyze if it indeed leads to any asymptotic improvement. Before we get into the analysis, we first introduce a very slowly growing function which will be used for expressing the heights of such trees.

4.3.3 The slowest growing function?

Let us look at a very rapidly growing function, namely the *tower of two* which looks like

$$2^{2^{2^{\cdot^{\cdot^{2}}}}} \Big\}i$$

This can defined more formally as the following function

$$B(i) = \begin{cases} 2^1 & i = 0 \\ 2^2 & i = 1 \\ 2^{B(i-1)} & \text{otherwise for } i \geq 2 \end{cases}$$

We now define the operation which applies log iteratively i times. More formally, we define

$$\log^{(i)} n = \begin{cases} n & i = 0 \\ \log(\log^{(i-1)} n) & \text{for } i \geq 1 \end{cases}$$

The inverse of $B(i)$ is defined as

$$\log^* n = \min\{i \geq 0 | \log^{(i)} n \leq 1\}$$

In other words,

$$\log^* 2^{2^{2^{\cdot^{\cdot^{2}}}}} \Big\}n = n + 1$$

We will use the function $B()$ and $\log^*()$ to analyze the effect of path compression. We will say that two integers x and y are in the same *block* if $\log^* x = \log^* y$.

Although $\log^*$ appears to be slower than anything we can imagine (for example $\log^* 2^{65536} \leq 5$), there is a generalized family of functions called the inverse Ackerman function that is even slower!

Ackerman's function is defined as

$$\begin{aligned} A(1, j) &= 2^j & \text{for } j \geq 1 \\ A(i, 1) &= A(i-1, 2) & \text{for } i \geq 2 \\ A(i, j) &= A(i-1, A(i, j-1)) & \text{for } i, j \geq 2 \end{aligned}$$

Note that $A(2, j)$ is similar to $B(j)$ defined earlier. The inverse Ackerman function is given by

$$\alpha(m, n) = \min\{i \geq 1 | A(i, \lfloor \frac{m}{n} \rfloor) > \log n\}$$

To get a feel for how slowly it grows, verify that

$$\alpha(n, n) = 4 \text{ for } n = 2^{2^{2^{\cdot^{\cdot^{2}}}}} \Big\}16$$

4.3.4   Putting things together

Clearly the cost of *Find* holds a key to the analysis of this *Union–Find* data structure. Since the rank of any root node is at most $\log n$, we already have an upper bound of $\log n$ for any individual Find operation. We now show that the path compression heuristic further reduces the cost of Find operations to $O(\log^* n)$. Before we do this, we need to recall the definition of rank function. The earlier rank of a node v was defined as the maximum distance from the node v to a leaf in the sub-tree rooted below v. We cannot use this definition now, but will continue to use the definition which was used to *define* these values. In other words, every node maintains a rank. Whenever we need to make a root u a child of a root v, the rank of v is updated as before – it is rank(v) if rank(u) < rank(v), otherwise it is rank$(v) + 1$. We state below some simple properties of the rank function. The proof is left as an exercise problem.

Lemma 4.1 *The rank function has the following properties:*

- *Property 1: The rank of a root node is strictly larger than that of any children nodes.*

- *Property 2: There are at most $n/2^r$ nodes of rank r.*

- *Property 3: For a node v, the rank of its parent node never decreases (note that the parent node could change because of union operations followed by path compression).*

- *Property 4: If the root node of a tree changes from w to w', then the rank of w' is strictly larger than that of w.*

We shall use $\log^*(rank(v))$ for a node v to refer to the *block number* of v. We will adopt the following strategy for counting the cost of Find operations. Let us refer to every visit that a Find operation makes to a node, as *charging* the node. Clearly, the total cost of all Find operations is bounded by the total number of charges. We distinguish between three kinds of *charges*.

Base charge If the parent of v is the root node (of the tree containing v), then v receives a base charge. Clearly, each Find operation incurs at most one base charge, resulting in a total of m charges.

Block charge If the block number of the parent node $p(v)$ is strictly greater than that of the node v, that is, $\log^*(rank(p(v))) > \log^*(rank(v))$, then we assign v a block charge. Clearly, the maximum number of block charges for a single Find operation is $O(\log^* n)$

Path charge Any charge incurred by a Find operation that is not a block charge or a base charge is the path charge.

From our previous observation, we will focus on counting the path charges. Consider a node v. Whenever it gets a path charge, its parent is the root node. For it to incur a path charge again, the root node of this tree needs to change. But then the rank of the root node will go up (Property 4). Consequently, v, whose rank lies in block j (say), will continue to incur path charge for at most $B(j) - B(j-1) \le B(j)$ Find operations.

Since the number of elements with rank r is at most $\frac{n}{2^r}$ (Property 2), the number of elements having ranks in block j is

$$\frac{n}{2^{B(j-1)+1}} + \frac{n}{2^{B(j-1)+2}} + \ldots + \frac{n}{2^{B(j)}} = n \left(\frac{1}{2^{B(j-1)+1}} + \frac{1}{2^{B(j-1)+2}} + \ldots \right)$$

$$\le 2n \cdot \frac{1}{2^{B(j-1)+1}} = \frac{n}{2^{B(j-1)}}$$

Therefore, the total number of path charges for elements in block j is at most $\frac{n}{2^{B(j-1)}} \cdot B(j)$ which is $O(n)$. For all the $\log^* n$ blocks the cumulative path charges is $O(n \log^* n)$. Further, the total number of block charges will be $O(m \log^* n)$. Therefore, the total time taken by the Find and Union operations is $O((m+n) \log^* n)$.

4.3.5 Path compression only*

To gain a better understanding of role of path compression, let us analyze the use of path compression without the union-by-rank heuristic. We can define the rank of a node similar to the previous version with a subtle change. If a tree T_1 with rank r_1 links to a tree T_2 having a smaller rank r_2, then the rank of the root of T_2 becomes $r_1 + 1$. If T_2 links to T_1 then the ranks remain unchanged. Without the union by rank heuristic, both options are permissible and so we cannot bound the rank of a node by $\log n$ and it can be $n - 1$ in the worst case.

Let us denote the parent of a node x as $p(x)$. The *level* of a node x, denoted by $\ell(x)$, is an integer i such that $2^{i-1} \le \text{rank}(p(x)) - \text{rank}(x) \le 2^i$. Therefore, $\ell(x) \le \log n$. Note that $\ell(x)$ is defined for non-root vertices only.

We account for the cost of a Find(x) operation by charging one unit of cost to all the nodes in the path from x to the root (except the root). The only exception is that for any level i, $1 \le i \le \log n$, the last node in the path to the root in level i is not charged. Instead, the cost is charged to the Find operation. Clearly, the number of charges to the Find operation is $O(\log n)$.

Claim 4.1 *For any other node y, we claim that whenever it gets charged by the Find operation, $\ell(y)$ increases by at least one.*

Since $\ell(y)$ is bounded by $\log n$, this will imply that any node y is charged at most $\log n$ times.

Let us now see why the claim is correct. Since y is not the last node in its level, there is another node v above y in the path to the root such that $\ell(v) = \ell(y) = i$ (say). By definition of level,

$$\text{rank}(p(v)) - \text{rank}(y) = \text{rank}(p(v)) - \text{rank}(v) + \text{rank}(v) - \text{rank}(y)$$
$$\geq \text{rank}(p(v)) - \text{rank}(v) + \text{rank}(p(y)) - \text{rank}(y) \geq 2 \cdot 2^{i-1} = 2^i.$$

The second last inequality here follows from the fact that v lies above $p(y)$ in the path from y to the root, and so, rank of v will be at least that of $p(y)$. Let w be the parent of v (before this Find operation) and r be the root of this tree. Again, by rank monotonicity, $\text{rank}(r) \geq$ $\text{rank}(w)$. We have shown earlier that $\text{rank}(w) - \text{rank}(y) \geq 2^i$, and so, $\text{rank}(r) - \text{rank}(y) \geq 2^i$ as well. Since r will now be the parent of y, it follows that $\ell(y) \geq i + 1$. This proves our claim.

Therefore, over the course of all the Union–Find operations, a node can get charged at most $\log n$ times resulting in a total cost of $O(m \log n)$ for all the Find operations.

4.4 Greedy in Different Ways

The matroid structure is closely related to the form of the greedy algorithm described in Figure 4.2. But there may be other variations that attempt to choose the next best element without necessarily picking them in reverse sorted order of their weights. One such classic algorithm is Prim's MST algorithm described in Figure 4.8. Recall that Kruskal's algorithm maintains several connected components – at each step, it picks an edge and merges two of these components into one. In Prim's algorithm, we maintain only one connected component which is initially just the cheapest edge. At each step, the algorithm finds the least weight edge which can *extend* this connected component by one more edge.

Procedure Prim(G, w)

1 **Input** Graph $G = V, E$, a weight function $w : E \Rightarrow \mathbb{R}$;
2 $T = e_1$ where $e_i \in E$ is the smallest weighted edge;
3 **for** $|T| \leq n - 1$ **do**
4 $\quad\quad$ Let (u, v) be the least weight edge in $V_T \times (V - V_T)$;
5 $\quad\quad$ $T \leftarrow T \cup \{(u, v)\}$;
6 Output T as MST of G.

Figure 4.8 *Prim's minimum spanning tree algorithm*

Although Prim's algorithm intuitively seems to do the right thing, note that the sequence of edges that it picks could be different from Kruskal's and therefore it requires a separate proof of correctness. It is clear that it outputs a tree. Indeed, it always picks an edge, one of whose endpoints is not in the current component T. Therefore, the added edge cannot induce a cycle.

Let us first address the running time. The algorithm needs to pick the least weighted edge going out of the current tree T. For this, we can maintain a label for every vertex denoting its distance from the tree T. If $v \in T$, then its label is 0. Let $N(v)$ denote the neighbors of v in G. Then, label of v is $\ell(v) = \min_{u \in N(v)} w(u,v) + \ell(u)$. (If no neighbor of v belongs to T, then its label is ∞.) The labels maintain the shortest distance of any vertex to the nearest vertex in T. The details of maintaining this data structure and update time are left as exercises (see also Dijkstra's algorithm in Chapter 10). We can use a heap data structure to store these labels so that finding the minimum weight edge will take $O(\log n)$ time. Consequently, the algorithm can be implemented in $O(m \log n)$ time.

To prove correctness, we will invoke a useful result whose proof is left as an exercise. The setting is an undirected weighted graph $G = (V, E)$, whose edges are colored red, blue (or left uncolored) according to the following rules:

(i) **Red rule**: An edge is colored red if it is the heaviest (i.e., highest weight) edge in a cycle[5].

(ii) **Blue rule**: An edge is colored blue if it is the lightest edge across any *cut* of the graph. A cut is a partition of the vertices V; an edge across the cut has one endpoint in each partition.

(iii) The two rules can be applied in any order.

Theorem 4.2 (Red–blue rule) *There exists an MST of G that includes all the blue edges and none of the red edges.*

The proof is left as an exercise. This theorem has profound connections to all the known MST algorithms. Prim's algorithm can be seen as coloring edges blue where each blue edge is the lightest cut defined by the tree vertices and the remaining graph.

Kruskal's algorithm can be viewed as coloring an edge red if the two endpoints are within the same component and the order of adding the edges ensures that it is the heaviest edge (in the cycle formed by this edge and the edges in the component). On the other hand, if an edge connects two components, then it must be a cut-edge if the two components are in different partitions (other components can be arbitrarily assigned to either partitions). Moreover, it is the lightest edge among the edges not added, and so it must be colored blue by definition.

A lesser known algorithm called Borůvka's algorithm is described in Figure 4.9. The algorithm maintains several connected components at any point of time as in Kruskal's algorithm. The set $\mathcal{F}$ denotes the set of these components. At any point of time, it picks a component C in $\mathcal{F}$ and chooses the least weight edge which has exactly one endpoint in C – such an edge would have its other endpoint in some other component C' in $\mathcal{F}$. The

---

[5] Assume all edges have unique weight.

algorithm picks this edge e and replaces C and C' with $C \cup C' \cup \{e\}$. Note that the choice of C is arbitrary. The algorithm terminates when there is one connected component in $\mathcal{F}$.

The correctness of the algorithm follows from the use of the blue rule and an additional assumption that the edge weights are unique.[6] Indeed, whenever we add an edge joining C and C', it is the cheapest edge in the cut formed by C and the rest of the vertices. There are several advantages of this algorithm. It is inherently parallel in nature as all the components can simultaneously choose the nearest neighboring vertex. Moreover, the fastest known linear MST algorithm is based on an adaptation of this algorithm with clever use of randomization.

Procedure Boruvka(G, w)

1 **Input** Graph $G = V, E$, a weight function $w : E \Rightarrow \mathbb{R}$;
2 $\mathcal{F} = \{\{v_1\}, \{v_2\}, \ldots\}$, where $v_i \in V$ are initial components without any edges;
3 $T = \phi$;
4 **while** $|\mathcal{F}| > 1$ **do**
5   Pick a component C in $\mathcal{F}$;
6   Let $(v, w) \in E$ be the least weight edge out of component C ;
7   Suppose w lies in component C' in $\mathcal{F}$;
8   Replace C and C' by $C \cup C' \cup \{(v, w)\}$ in $\mathcal{F}$.
9 Output the single component in $\mathcal{F}$ as MST of G.

Figure 4.9   *Boruvka's minimum spanning tree algorithm*

4.5 Compromising with Greedy

So far, we have shown that the greedy strategy yields an optimal solution for a large class of problems. However, in many cases, the greedy strategy does not always yield an optimal solution. It is still attractive because of its simplicity and efficiency. What if we compromise our objective of finding an optimal with a *near* optimal solution? We touch on this aspect of algorithm design in a later chapter more formally – here we illustrate this with an example.

Recall the **maximum matching** problem discussed in Example 4.4. Although the example discussed the special case of bipartite graphs, the same definition extends to general graphs. More formally, we are given an undirected graph $G = (V, E)$. We want to find a subset $E' \subset E$ such that no two edges in E' share any endpoints (the degree of the

---

[6] One can add lower significant bits based on edge labels to break ties.

induced subgraph is exactly 1) and we want to maximize the number of edges in E'. For a weighted graph, we want to maximize $\sum_{e \in E'} w(e)$, where $w(e)$ is the weight of e. We had shown in Example 4.4 that the subset system corresponding to matchings is not a matroid.

Nevertheless, let us persist with the greedy strategy for finding a matching and analyze the outcome. Consider the following algorithm: sort the edges in decreasing order of the weights. Maintain a solution G initialized to empty. We look at edges in this order, and add an edge to G if it does not have a common endpoint with any of the edges chosen in G so far. It is easy to show that this algorithm may not give an optimal solution. Still it turns out that the total weight of edges in G is always at least half of that of an optimal solution. We now prove this claim.

Let O denote an optimal solution. Let $w(O)$ and $w(G)$ denote the total weight of edges in O and G respectively. Clearly, $w(O) \geq w(G)$. Consider an edge $e = (x, y) \in O \setminus G$. When the greedy algorithm considered e, there must have been an edge $e' \in G$ which shared a common endpoint with e. Further, $w(e') \geq w(e)$. Thus, we can define a mapping $B : O \setminus G \to G \setminus O$ (mapping edges $e \in O \setminus G$ to $e' \in G \setminus O$). How many edges can map to an edge e' in G using this mapping B? We claim that there can be at most 2 such edges, and both these edges have weight at most $w(e')$. Indeed, e' has two endpoints, and if $B(e) = e'$, then e must have a common endpoint with e'. The claim now follows from the fact that no two edges in O (and so, in $O \setminus G$) share a common endpoint. Therefore, the total weight of edges in $O \setminus G$ is at most twice that of edges in $G \setminus O$. Therefore, $w(O) = w(O \setminus G) + w(O \cap G) \leq 2w(G \setminus O) + w(G \cap O) \leq 2w(G)$ or equivalently $w(G) \geq \frac{w(O)}{2}$.

Thus, the greedy strategy can have some provable guarantees even though it does not yield an optimal solution.

4.6 Gradient Descent*

So far we have used the greedy strategy to solve 'discrete' optimization problems, that is, problems where a decision variable can take a finite set of values. For example, in the minimum spanning tree problem, we have one variable with each edge – should we include this edge in the tree solution? This variable is a binary variable, because it can take only two values – true or false. Similarly, in the more general setting of finding the maximum weight independent set in a matroid, we have to decide whether to add an element in the independent set or not. We now consider a different application of the greedy strategy where the variables can have values from a continuous interval.

We are given a continuous (and differentiable) function $f : \Re^n \to \Re$, where the domain of the function, denoted by $\mathrm{dom}(f)$, is a convex and compact set. We would like to find a minimizer of f, that is, a point $x^* \in \mathrm{dom}(f)$ such that $f(x^*) \leq f(x)$ for all $x \in \mathrm{dom}(f)$. Because the domain is a compact set, we know that there will be at least one such point x^*, though it may not be unique. The problem is not very well defined because it may not be possible to express the point x^* using finite precision. Consider, for example, the function $f(x) = x^2 - 2$. Clearly, $x^* = \sqrt{2}$, and it cannot be expressed using a finite number of bits. We shall assume that the input instance also provides an error parameter ε; we are required to find a point x such that $f(x) - f(x^*) \leq \varepsilon$.

For a general function f, we do not know how to solve this problem in polynomial time (where the polynomial may include terms depending on the diameter of the domain of f and bounds on the slope of f). We focus on a special class of functions called convex functions. It turns out that any local minimum for such functions is also a global minima, and so, it suffices to run a greedy algorithm to find a local minimum of such a function. Let f be a function of one variable, that is, $f : \Re \to \Re$. We say that f is convex if for every $x, y \in \text{dom}(f)$, and parameter λ, where $0 \leq \lambda \leq 1$,

$$f(\lambda x + (1 - \lambda)y) \leq \lambda f(x) + (1 - \lambda)f(y). \tag{4.6.1}$$

Graphically, it means that if we look at the plot of f, then the line joining the points $(x, f(x))$ and $(y, f(y))$ should lie above the curve f in the interval $[x_1, x_2]$ (see Figure 4.10). We say that f is strictly convex if the aforementioned inequality is strict. It is not clear how we can use this definition to easily check if a function is convex. Fortunately, if we make some mild assumptions on f, then there are other equivalent definitions which turn out to be easier to work with. It turns out that if f is differentiable, then it is enough to check that for every pair of points $x, y \in \text{dom}(f)$,

$$f(y) \leq f(x) + (y - x) \cdot f'(x) \tag{4.6.2}$$

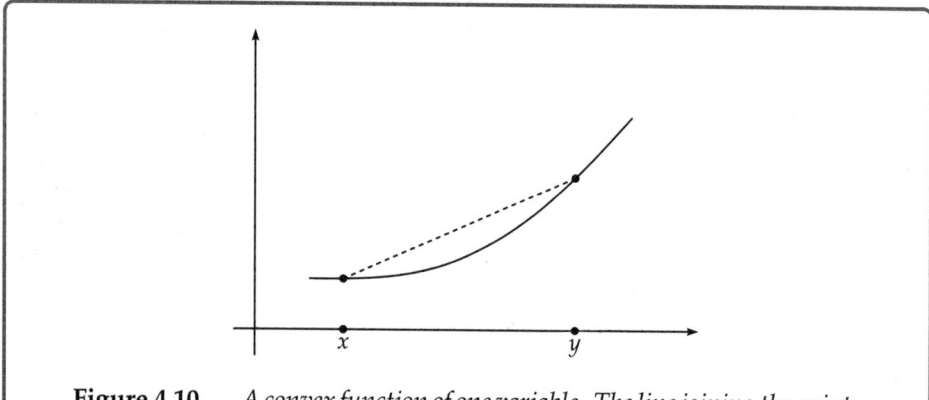

Figure 4.10   *A convex function of one variable. The line joining the points $(x, f(x))$ and $(y, f(y))$ stays above the plot of f.*

where $f'(x)$ denotes the derivative of f w.r.t. x. This result gives another way of thinking about convex functions: if we draw the tangent at any point on the curve corresponding to

f, then the entire curve lies above the tangent line. If f happens to be twice differentiable, there happens to be another intuitive definition of convex functions: the second derivative of f is always non-negative (see Exercise 4.23).

We now extend these definitions to a function of multiple variables. We say that a function $f : \Re^n \to \Re$ is convex if the *restriction* of f on any line is a convex function. Recall that a line can be specified by two vectors: a point x_0 on the line, and a direction d. Any point on this line can be described by a single parameter t: $x_0 + t \cdot d$. Thus, we can define a function $h(t) = f(x_0 + t \cdot d)$, and think of h as the restriction of f on this line. According to our definition, f is convex iff every such function h is convex.

As in the case of functions of one variable, we now want to define convexity in terms of first and second derivatives of f. Let ∇f denote the gradient of f. Exercise 4.25 shows the analogous statements. It turns out that a local minimum of a convex function is also a global minimum. More precisely,

Lemma 4.2 *Suppose $x \in \mathrm{dom}(f)$ and x is a local minimum of f, that is, there is a radius $r > 0$ such that $f(x) \leq f(y)$ for all y satisfying $\|y - x\| \leq r$. Then, x is also a global minimum of f, that is, $f(x) \leq f(y)$ for all $y \in \mathrm{dom}(f)$.*

The intuitive reason for this statement is as follows. Suppose a convex function has a local minimum at x. Let x' be any other point in the domain of f. Consider the one-dimensional projection of f along the line joining x and x'. Suppose, for the sake of contradiction, that $f(x') < f(x)$. Then, by convexity of f, the curve corresponding to f lies below the line joining x and x'. Therefore, we can find a point x'' in the vicinity of x such that $f(x'') < f(x)$. This contradicts the fact that x is a local minimum.

Thus, if we want to find the minimum of a convex function, it is enough to find a local minimum of f – note that in general, a convex function may not have a unique local minimum; but a strictly convex function has a unique local minimum. The gradient descent algorithm is a popular greedy algorithm for minimizing a convex function. Intuitively, it starts from an arbitrary point in the domain of f, and tries to move along the 'steepest direction' at the current point.

The algorithm starts with an initial guess $x^{(0)}$, and iteratively moves to points x which have smaller $f(x)$ values. The intuition can be described as follows. Suppose we are currently at a point x and want to make a small step of size η along a direction d, that is, we want to move to a point $x + \eta d$, where d is a unit vector. What should be the best choice for d? If η is a small quantity, then we can approximate f by a linear approximation using Taylor expansion, where we assume that d and $\nabla f(x)$ are column vectors and d^T denotes the transpose of d.

$$f(x + \eta d) \approx f(x) + \eta d^T \nabla f(x).$$

Now, we know that $|d^T \nabla f(x)| \leq ||d|| ||\nabla f(x)||$,[7] with equality if and only if d is along $\nabla f(x)$ (using Cauchy Schwarz inequality). Therefore, we should pick d along the negative gradient direction. This motivates the gradient descent algorithm described in Figure 4.11. The parameter η, which is often called the 'learning rate', should be chosen carefully: if it is too small, then the progress toward a local minimum will be slow, whereas if we pick η to be large, we may not converge to the desired point. Similarly, the time τ till which we need to run the algorithm depends on how close we want to come to the optimal solution.

Procedure Gradient descent(f, η, x_0)

1 **Input** Convex function f, step size η and initial point $x^{(0)}$;
2 **for** $t = 1, \ldots, \tau$ **do**
3 $\quad \lfloor \; x_t \leftarrow x_{t-1} - \eta \nabla f(x)$
4 Output x_T.

Figure 4.11   *Gradient descent algorithm*

Let us see an example. Consider $f(x) = x^2 - 1$. Clearly, $x^\star = 0$ is the global minimum. Now, if we start at $x = 1$, and set $\eta = 10$, it is easy to see that the successive points will diverge from $x^\star$. Therefore, it is important to keep η small, preferably much smaller than the distance between the current point and the desired minimum. However, as Exercise 4.27 shows, even a very small value of η can lead to an oscillatory behavior if we do not assume the smoothness properties of f. The reason for the oscillatory behavior of the gradient descent in this exercise is because the derivative of the function changes suddenly at $x = 0$. We now assume that the derivative of the function cannot change very fast, that is, there is a parameter L such that for all $x, y \in \text{dom}(f)$,

$$||\nabla f(x) - \nabla f(y)|| \leq L \cdot ||x - y||.$$

Such a convex function is said to be L-smooth. One consequence of L-smoothness is that a convex function cannot deviate from the tangent line at a point too fast. Let x and y be two points in the domain of a real valued L-smooth function f. Then,

$$0 \leq f(y) - f(x) - f'(x) \cdot (y - x) \leq \frac{L}{2} \cdot (y - x)^2. \tag{4.6.3}$$

The first inequality follows by the definition of convexity. For the other inequality, note that

---

[7]We have used the notation $|x|$ to denotet the absolute value and $||x||$ to denote the length of the vector.

$$f(y) - f(x) - f'(x) \cdot (y - x) = \int_0^1 (f'(x + t(y-x)) \cdot (y-x) - f'(x) \cdot (y-x)) dt$$

$$= \int_0^1 (f'(x + t(y-x)) - f'(x)) \cdot (y-x) dt$$

$$\leq L \cdot \int_0^1 t(y-x)^2 dt = \frac{L}{2} \cdot (y-x)^2.$$

The first inequality follows from the fact that if $g(t)$ denotes $f(x + t(y-x))$ for a real t, then $\int_0^1 g'(t) = g(1) - g(0)$. The last inequality uses the fact that f is smooth. The following theorem states that the gradient descent converges in a small number of steps provided we pick η suitably. We prove this for a function f of one variable only – the ideas for the more general proof when f is a function of several variables are similar, though the details require a bit more work. We first observe a simple consequence of L-smoothness.

Theorem 4.3 *Let f be an L-smooth convex function. Let $x^\star$ denote a global minimum of f. If we run the gradient descent with $\eta = 1/L$, then $f(x_t) - f(x^\star) \leq \varepsilon$ for $t \geq \frac{LR^2}{\varepsilon}$, where R denotes $|x_0 - x^\star|$.*

To prove this result, we show that the gradient descent algorithm makes some progress at each step – the progress is higher if we are farther from $x^\star$ and it slows down as we start approaching $x^\star$. From the description of the gradient descent algorithm, we know that $x_s - x_{s+1} = \eta f'(x_s) = \frac{f'(x_s)}{L}$. Using this fact and substituting $x = x_s$ and $y = x_{s+1}$ in the inequality in Eq. (4.6.3), we see that

$$f(x_s) - f(x_{s+1}) \geq (x_s - x_{s+1}) f'(x_s) - \frac{L}{2}(x_s - x_{s+1})^2 = \frac{1}{2L} \cdot f'(x_s)^2.$$

Thus, if $f'(x_s)$ is large, we make more progress; as we approach $x^\star$, $f'(x_s)$ gets closer to 0, and so, our progress also slows down. Now, we show how to make this argument more formal. Assume without loss of generality that $x_0 > x^\star$. We will show that x_s will also be at least $x^\star$ for all values of $s \geq 1$. Assuming this is the case, it follows that $x_s - x^\star \leq x_0 - x^\star \leq R$. Now, if δ_s denotes $f(x_s) - f(x^\star)$, then the aforementioned observation can be restated as

$$\delta_s - \delta_{s+1} \geq \frac{f'(x_s)^2}{2L}. \tag{4.6.4}$$

Convexity of f implies that

$$\delta_s = f(x_s) - f(x^\star) \leq f'(x_s)(x_s - x^\star) \leq R f'(x_s).$$

Substituting this in the inequality in Eq. (4.6.4), we get

$$\delta_s - \delta_{s+1} \geq \frac{\delta_s^2}{2LR^2}.$$

It only remains for us to solve this recurrence relation. Observe that

$$\frac{1}{\delta_{s+1}} - \frac{1}{\delta_s} = \frac{\delta_s - \delta_{s+1}}{\delta_s \delta_{s+1}} \geq \frac{\delta_s - \delta_{s+1}}{\delta_s^2} \geq \frac{1}{2LR^2}.$$

Adding this for $s = 1, \ldots, \tau$, we see that

$$\frac{1}{\delta_\tau} - \frac{1}{\delta_1} \geq \frac{\tau - 1}{2LR^2}.$$

Finally, notice that the inequality in Eq. (4.6.3) implies that

$$\delta_1 = f(x_1) - f(x^\star) \leq f'(x^\star)(x_1 - x^\star) + \frac{LR^2}{2} = \frac{LR^2}{2}.$$

Substituting this in the aforementioned inequality, we see that δ_τ is $O(LR^2/\tau)$. This proves the theorem.

It remains to show that x_s always stays at least $x^\star$, that is, the iterates never cross from the right of $x^\star$ to the left of $x^\star$. This happens because our step sizes are small enough – in fact this illustrates the concept that the step size should be large enough to make enough progress but small enough to avoid 'overshooting' the desired point. From the definition of L-smoothness, we know that $f'(x_s) - f'(x^\star) \leq L(x_s - x^\star)$, and so, $f'(x_s) \leq L(x_s - x^\star)$. Therefore, $x_{s+1} = x_s - f'(x_s)/L \geq x_s - (x_s - x^\star) \geq x^\star$. This completes the analysis of the greedy algorithm and shows that under mild conditions, it converges to the optimal solution.

Remarks:

1. In practice, the parameter η is often chosen in an ad-hoc manner by figuring out the right trade-off between convergence rate and accuracy.

2. The decision to stop the iterations of the gradient descent algorithm can also be based on several criteria: (i) there could be an upper bound on the number of iterations, (ii) the value $\|x_t - x_{t-1}\|$ becomes smaller than a given threshold, (iii) the values $f(x_t) - f(x_{t+1})$ become smaller than a given threshold.

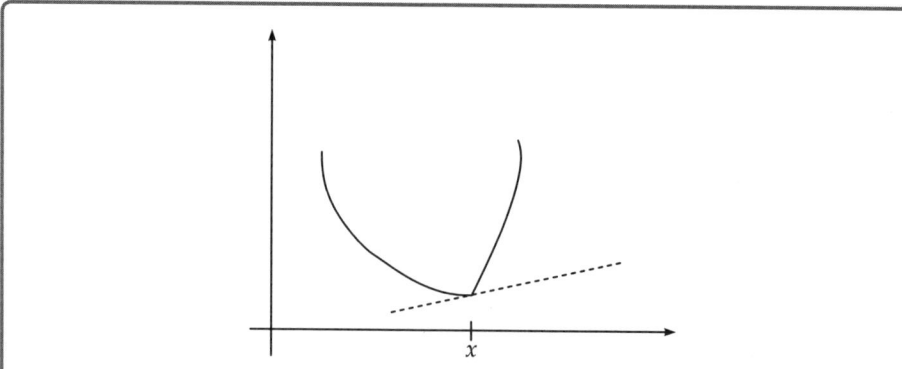

Figure 4.12   *The convex function is non-differentiable at x. We can instead use the slope of the dotted line as sub-gradient at x.*

3. Sometimes, the function f may not be differentiable at the current point x_t. Consider, for example, the function in Figure 4.12 – this function is convex, but not differentiable at the point x. It turns out that one can still use the gradient descent algorithm at x provided one uses a vector v instead of the gradient $\nabla f(x)$ and the following condition holds for all points y in the domain of f:

$$f(y) \geq f(x) + (y-x)^T v.$$

Such a vector v is called a *sub-gradient* at x – note that there is no unique choice for v here.

4.6.1 Applications

Gradient descent is a very popular general purpose algorithm for optimizing a function. In practice, it is used even if the function is not convex – the hope is that one would instead converge to a local optimum. We now give some sample applications.

Locating a point by multiple measurements

Suppose we want to find the location of an object P in the two-dimensional plane. There are three observation points O_1, O_2, O_3. For each of the observation points O_i, you can measure the distance r_i between P and O_i. As shown in Figure 4.13, you can find the location of P by finding the common intersection point of the circles of radii r_1, r_2, r_3 centered at O_1, O_2, O_3 respectively. But the measurements incur some error, and so we only know approximations to r_1, r_2, r_3 – call these $\tilde{r}_1, \tilde{r}_2, \tilde{r}_3$. Given these three approximate values, we would like to find the best possible estimate for the location of P (see Figure 4.13).

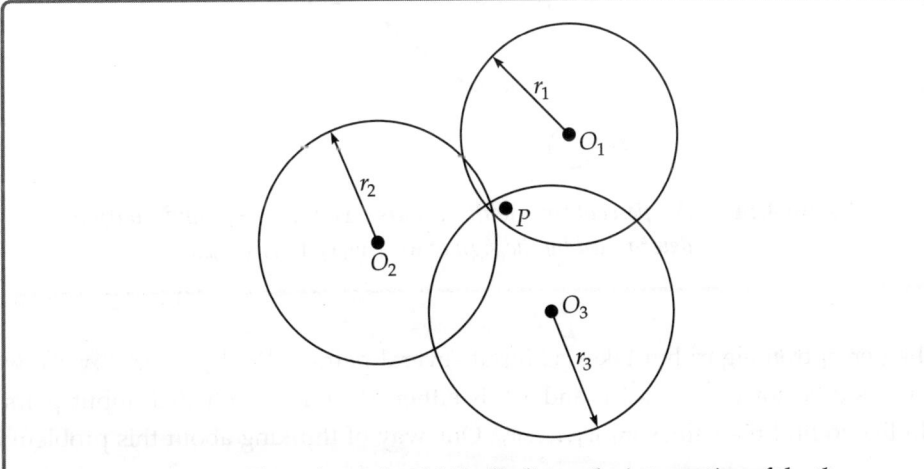

Figure 4.13   *The point P should ideally lie on the intersection of the three circles, but there are some measurement errors.*

Such problems are often solved by solving an appropriate optimization problem. Suppose the coordinates of O_i are (a_i, b_i), $i = 1, 2, 3$. Let (x, y) be the coordinates of P. Note that (a_i, b_i) are known quantities, whereas we would like to find (x, y). Assuming the errors in the measurements are small, one way of framing this problem would be to find the values of (x, y) such that the overall error is as small as possible. In other words, let $f_i(x, y)$ denote $\left(\tilde{r}_i - \sqrt{(a_i - x)^2 + (b_i - y)^2}\right)^2$. Observe that $f_i(x, y)$ denotes (square of) the error in the measurement of r_i. And so we would like to find the value of (x, y) such that $f(x, y) = \sum_{i=1}^{3} f_i(x, y)$ is minimized. We can solve this by the gradient descent algorithm. It is easy to write down the gradient of $f(x, y)$ and so, one can run the gradient descent algorithm till the values converge.

Perceptron algorithm

A neuron is often modeled as a unit with a threshold w_0. When the input to the neuron exceeds w_0, it outputs 1. Otherwise, it outputs -1.[8] Consider the situation shown in Figure 4.14. There are n input variables $x_1, x_2, \ldots, x_n$, and weights $w_1, w_2, \ldots, w_n$ (shown on the 'input' edges in the figure). Therefore, the input to the neuron is $w_1 x_1 + \ldots + w_n x_n$ – if this exceeds w_0, output is 1; otherwise, output is –1. In other words (replacing w_0 by $-w_0$), the output is determined by the sign of $w_0 + w_1 x_1 + \ldots + w_n x_n$.

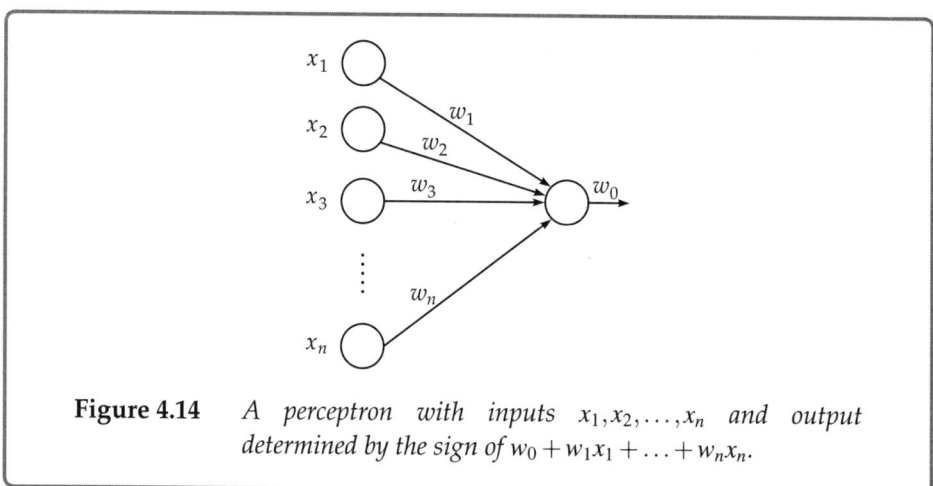

Figure 4.14   *A perceptron with inputs $x_1, x_2, \ldots, x_n$ and output determined by the sign of $w_0 + w_1 x_1 + \ldots + w_n x_n$.*

The perceptron algorithm takes as input several pairs $(x^{(1)}, y^{(1)}), \ldots, (x^{(m)}, y^{(m)})$, where each $x^{(j)}$ is a vector $(x_1^{(j)}, \ldots, x_n^{(j)})$, and $y^{(j)}$ is either –1 or 1. Given such input pairs, we would like to find the values $w_0, w_1, \ldots, w_n$. One way of thinking about this problem is as

---

[8] Inputs and outputs are electrical signals in nature. Also note that this is an ideal model of a neuron. In reality, there will be a 'gray' area where it outputs something between -1 and 1. The output -1 refers to the fact that there is no output signal, and so, it corresponds to the zero signal.

follows: consider the hyperplane $w_0 + w_1 x_1 + \ldots + w_n x_n = 0$ in n dimensions (where the coordinates are given by $(x_1, x_2, \ldots, x_n)$). The points $x^{(j)}$ for which $y^{(j)}$ is 1 lie on one side of this hyperplane, whereas the rest lie on the other side. Thus, we can frame this problem as follows: given a set of points where each point is labeled either '+' or '−' (depending on whether the y-coordinate is 1 or 0), find a hyperplane separating the points.

We can express this as an optimization problem as follows. Let $\mathbf{w}$ denote the vector $(w_0, w_1, \ldots, w_n)$. Given such a solution $\mathbf{w}$, we can count the number of *mis-classified* inputs – an input $(x^{(j)}, y^{(j)})$ is mis-classified if the sign of $\mathbf{w}^T x^{(j)}$ is positive if and only if $y^{(j)}$ is −1. More formally, define a function sgn as $\mathrm{sgn}(z)$ to be 1 if $z < 0$, 0 otherwise. Notice that the quantity $\mathrm{sgn}(y^{(j)} \cdot \mathbf{w}^T x^{(j)})$ is 1 if and only if we make a mistake for the input $(x^{(j)}, y^{(j)})$. Therefore, we can state this problem as minimizing $f(\mathbf{w}) := \sum_j \mathrm{sgn}(y^{(j)} \cdot \mathbf{w}^T x^{(j)})$. However, the function f is not convex – it is easy to see that the function sgn (as a function of one variable) is not convex. Instead, we replace f by another function which is convex and approximates the sgn function. There are various ways to do this. We will use a simple approach which replaces sgn by the following function g: $g(x) = x$ is $|x|$ if $x < 0$, otherwise it is 0. This function is shown in Figure 4.15. Note that the function g is convex, but it is not differentiable at $x = 0$. Recall from the discussion on the gradient descent algorithm that we can run it on a non-differentiable convex function as long as we can define the notion of a sub-gradient. It is easy to see that we can define any number between 0 and −1 as the sub-gradient at $x = 0$ – we shall define it to be 0 here. If $x < 0$, the derivative of g is −1. If $x \geq 0$, the derivative is 0. Now, we can define the function f as $f(\mathbf{w}) := \sum_j g(y^{(j)} \cdot \mathbf{w}^T x^{(j)})$. Observe that if $\mathbf{w}$ indeed represents a separating hyperplane, then f would be 0. In this sense, we are justified in replacing sgn by the function g. It is now easy to write down the derivative of f at a point $\mathbf{w}$. Let $N(\mathbf{w})$ be the set of input points which are incorrectly classified by $\mathbf{w}$, that is, $y^{(j)} \cdot \mathbf{w}^T x^{(j)} < 0$. Then, the derivative is given by $-\sum_{j \in N(\mathbf{w})} y^{(j)} x^{(j)}$. Thus, we get the following simple rule for finding the separating hyperplane: if $\mathbf{w}^t$ is the estimate for $\mathbf{w}$ at iteration t, then:

$$\mathbf{w}^{t+1} = \mathbf{w}^t + \sum_{j \in N(\mathbf{w}^t)} y^{(j)} x^{(j)}.$$

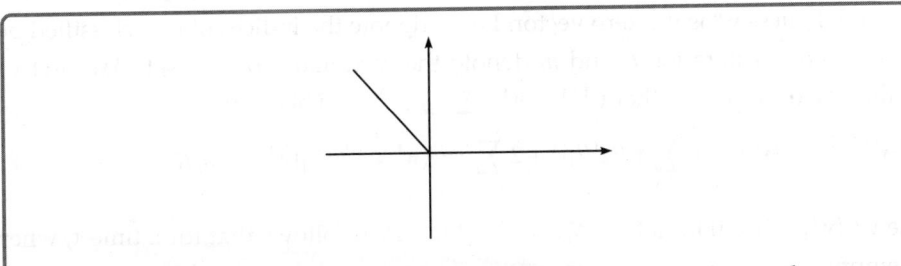

Figure 4.15   *A plot of the function g as an approximation to the* sgn *function.*

Geometrically, this rule says that we *tilt* the vector w^t based on the points which are getting mis-classified. Observe that the learning rate η is 1 here.

We now analyze this algorithm. We cannot apply Theorem 4.3 directly because the slope of the function g changes instantaneously from -1 to 0 at $x = 0$. Such functions can still be analyzed, but we will consider a simpler case and show that the algorithm stops after a small number of iterations. Let P denote the points for which $y^{(j)}$ is 1 and N denote the points for which this quantity is -1. We will assume that there is a hyperplane which separates P and N, and there is a margin around this hyperplane which does not contain any point (see Figure 4.16). We can state this condition formally as follows: there is a unit vector $\mathbf{w}^\star$ such that for all points $x^{(j)}$,

$$y^{(j)}\langle x^{(j)}, \mathbf{w}^\star\rangle \geq \gamma,$$

where γ is a positive constant. Note that γ denotes the margin between the hyperplane and the nearest point. We now show that the gradient descent algorithm, with step size $\eta = 1$, terminates in $O(R^2/\gamma^2)$ iterations, where R is an upper bound on $||x^{(j)}||$ for all points $x^{(j)}$.

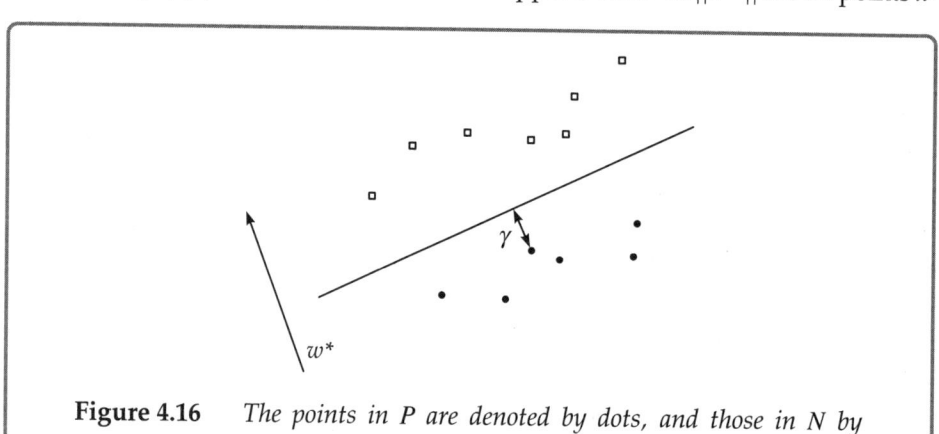

Figure 4.16   *The points in P are denoted by dots, and those in N by squares.*

The idea behind the proof is to bound the size of $\mathbf{w}^t$ in two different ways. We assume that the initial guess $\mathbf{w}^0$ is the zero vector. Let N_t denote the indices of mis-classified points at the beginning of iteration t, and n_t denote the cardinality of this set. We first upper bound the size of $\mathbf{w}^t$. Recall that $\mathbf{w}^{t+1} = \mathbf{w}^t + \sum_{j \in N_t} y^{(j)} x^{(j)}$. Therefore,

$$||\mathbf{w}^{t+1}||^2 = ||\mathbf{w}^t||^2 + ||\sum_{j \in N_t} y^{(j)} x^{(j)}||^2 + 2\sum_{j \in N_t} y^{(j)}\langle \mathbf{w}^t, x^{(j)}\rangle \leq ||\mathbf{w}^t||^2 + n_t^2 R^2,$$

because $y^{(j)}\langle \mathbf{w}^t, x^{(j)}\rangle < 0$ for all $j \in N_t$, and $||x^{(j)}|| \leq R$. It follows that for a time τ, where N_τ is non-empty,

$$||\mathbf{w}^{\tau+1}||^2 \leq R^2 \sum_{t=1}^{\tau} n_t^2.$$

Now we lower bound the size of $\mathbf{w^t}$ by considering the quantity $\langle \mathbf{w^t}, \mathbf{w^\star} \rangle$. Note that

$$\langle \mathbf{w^{t+1}}, \mathbf{w^\star} \rangle = \langle \mathbf{w^t}, \mathbf{w^\star} \rangle + \sum_{j \in N_t} \langle y^{(j)} x^{(j)}, \mathbf{w^\star} \rangle \geq \langle \mathbf{w^t}, \mathbf{w^\star} \rangle + \gamma n_t.$$

Therefore, it follows that

$$||\mathbf{w^{\tau+1}}|| \geq \langle \mathbf{w^{\tau+1}}, \mathbf{w^\star} \rangle \geq \gamma \sum_{t=1}^{\tau} n_t \geq \gamma \sqrt{\tau} \cdot \left(\sum_{t=1}^{\tau} n_t^2 \right)^{1/2},$$

where the last inequality follows from the Cauchy–Schwarz inequality. Comparing the upper and lower bounds on $||\mathbf{w^{\tau+1}}||$, it follows that $\tau \leq R^2/\gamma^2$. Thus, the algorithm will not find any mis-classified points after R^2/γ^2 iterations.

Further Reading

Solving an optimization problem using a brute force search is a basic instinct that should be avoided for obvious reasons such as prohibitive running time. Use of heuristics like $\alpha\beta$ pruning or heuristic counterparts like $A*$ algorithm are widely used but without guarantees. The randomized AND–OR tree evaluation algorithm was given by [135]. The theory of matroids was developed by Whitney [153] and many extensions are known – Lawler [88] provides a comprehensive account of the theory with many applications. The minimum spanning tree has been a celebrated problem for more than a hundred years now, where the algorithms of Kruskal and Prim are among the best known. Boruvka's algorithm [110] turned out to be the basis of the linear time algorithm of Karger, Klein, and Tarjan [75] which is randomized. The best deterministic algorithm runs in $O(n\alpha(n))$ time and is due to Chazelle [29]. The characterization of the MST algorithms using red–green rule is from Tarjan [140]. The Union–Find data structure has a long history starting with Hopcroft and Ullman [64] including the path compression heuristics; it culminates with Tarjan [139] who gives a matching lower bound in the pointer model. Many variations of the basic heuristics of Union–Find are discussed by [138]. The Ackerman function is well known in computability theory as a primitive recursive function that is not μ-recursive.

   Gradient descent algorithms form an important class of first order algorithms for convex optimization and many variations have been studied, both in theory and practice. Convex optimization is a rich topic of research for many applications [23]. In this book, we have covered the case of unconstrained optimization only. There are many settings where there are additional constraints for a feasible solution. This is called constrained optimization. For example, in linear programming, the objective function is a linear function (and hence convex), but any feasible point must also satisfy a set of linear inequality or equality constraints [31]. Another way of visualizing this problem is that we are given a polytope (i.e., a convex figure obtained by intersection of several half-spaces) and we want to minimize a linear function over all points in this polytope. Starting from a feasible point, gradient descent would take us in the direction of negative gradient (note that the gradient of the objective function is a constant vector). But we cannot move indefinitely along this direction as we may move out of the polytope. One of the most popular algorithms, called the simplex algorithm, maintains a feasible point on the *boundary* of this polytope, and always moves in a direction which improves the objective function.

Exercise Problems

4.1 Construct an instance of a knapsack problem that visits every leaf node, even if we use branch and bound. You can choose any well defined way of pruning the search space.

4.2 Recall the knapsack problem where we have n objects $x_1, \ldots, x_n$, with object x_i having volume w_i and profit p_i. The capacity of the knapsack is C. Show that if we use a greedy strategy based on profit/volume, that is, if we choose the objects in decreasing order of this ratio, then the resultant final profit is at least half of the optimal solution. For this claim, we need to make one change, namely, if x_k is the last object chosen, such that $x_1, x_2, \ldots, x_k$ is in decreasing order of their ratios that can fit in the knapsack, then eventually we choose $\max\{\sum_{i=1}^{i=k} p_i, p_{k+1}\}$. Note that x_{k+1} is such that $\sum_{i=1}^{i=k} w_i \leq C < \sum_{i=1}^{i=k+1} w_i$.

4.3 Consider the special case of $k = 1$ in the analysis of the AND–OR tree. Show that the expected number of evaluations is 3. (We must consider all cases of output and take the worst, since we are not assuming any distribution on input or output).

4.4 Complete the analysis of the AND–OR tree when the root is an AND node.

4.5 Consider the following special case of *Union–Find*. There are three phases where in each phase, all the *Unions* precede all the *Find* operations. Can you design a more efficient implementation for this scenario?

4.6 We are given a sequence of integers in the range $[1, n]$, where each value occurs at most once. An operation called EXTRACT-MIN occurs at arbitrary places in the sequence which detects the minimum element up to that point in the sequence and discards it. Design an efficient algorithm for this any given sequence of EXTRACT-MIN operations.

For example, in $4, 3, 1, E, 5, 8, E, \ldots$ the output is $1, 3, \ldots$.

4.7 Prove that Borůvka's algorithm outputs an MST correctly.

4.8 Given an undirected graph $G = (V, E)$, consider the subset system (E, M), where M consists of those subsets of edges which induce a subgraph of G with at most one cycle. Prove that this subset system is a matroid.

4.9 Without using the rank property, show that the exchange property holds for the MST problem. In other words do not invoke the matroid theorem to prove the exchange property.

4.10 For implementing Prim's algorithm, design a suitable data structure to choose the minimum label as well as update the labels.

4.11 Suppose you are given an MST for a graph. Now suppose we increase the weight of one of the edges e in this MST from w_e to w'_e. Give a linear time algorithm to find the new MST.

4.12 Discuss appropriate data structures to implement Borüvka's algorithm efficiently.

4.13 The second minimal spanning tree is one that is distinct from the minimal spanning tree (has to differ by at least one edge) and is an MST if the original MST is ignored although they may even have the same weight. Design an efficient algorithm to determine the second MST.

Hint: Show that the second MST differs from the original MST by exactly one edge.

4.14 A *bottleneck* spanning tree (BST) minimizes the maximum weight edge among all spanning trees of a weighted undirected graph $G = (V, E)$. The *value* of BST $= \min_{T \in \mathcal{T}} (\max_{e \in T} \{\text{weight}(e)\})$, where $\mathcal{T}$ is the set of all spanning trees of G.

(a) Design a linear time algorithm to determine if the BST has value $\leq b$ for a given b.

(b) Design an efficient, preferably linear time algorithm for finding a BST.

4.15 Given a set J of unit duration jobs with deadlines, how would you determine if all the jobs can be scheduled within the deadlines. Describe an algorithm that either determines a feasible schedule or concludes that it is not possible.

4.16 Consider a set of jobs J_i, $1 \leq i \leq n$ such that every job J_i consists of two subtasks (a_i, b_i), where a_i units of time is required on a single common resource and b_i units can be done independently. Both a_i, b_i are non-negative integers and the second subtask can be started strictly after the completion of the first subtask for each job. For example, if $J_1 = (4, 7)$ and $J_2 = (5, 5)$, then one possible schedule is to start J_1 with subtask 1 requiring 4 units. Following which, subtask 2 of J_1 can be done and subtask 1 of J_2 can be started. So J_1 is completed after 11 units while J_2 finishes both subtasks after 14 units. Hence, both jobs are completed after 14 units if we start with J_1. For the case where we schedule J_2 before J_1, these jobs complete after 10 and 16 units respectively. Therefore, the first schedule completes faster.

Given n jobs, how would you schedule the jobs so as to minimize the completion time of the longest job? Let s_i denote the starting time for job J_i. Then we want to minimize $\max_i \{s_i + a_i + b_i\}$.

4.17 Consider a job scheduling problem where each job J_i has a start and a finish time (s_i, f_i). Two jobs cannot run simultaneously and once started, a job must run to its completion (i.e., we cannot split a job into parts). Given a set of jobs,

(i) If we schedule greedily in increasing order of finish times, can we maximize the number of jobs completed? Justify.

(ii) If job J_i is associated with a profit p_i (≥ 0), can you apply a greedy algorithm to maximize the profit (of all completed jobs)? Justify.

4.18 We are given a set of events (with starting and finishing times) that have to be scheduled in a number of halls without conflicts. Design an algorithm to find the minimum number of halls needed for this purpose. Note that the timings are fixed and no two events can happen at the same time in the same hall.

You can think of the events as intervals on the real line such that we have to assign a color to each interval in a way that no two overlapping intervals are assigned the same color. What is the minimum number of colors required?

4.19 Prove Lemma 4.1.

4.20 Prove Theorem 4.2.

4.21 Consider a long straight road from left to right with houses scattered along the road (you can think of houses as points on the road). You would like to place cell phone towers at some points on the road so that each house is within 4 km of at least one of these towers. Describe an efficient algorithm which achieves this goal and uses as few cell phone towers as possible.

Hint: Consider a solution where each tower is located as much to its right as possible (without changing the number of towers). How would you construct such a solution?

4.22 Suppose $f : \Re \to \Re$ is differentiable. Prove that f is convex if and only if for every pair of points $x, y \in \text{dom}(f)$,

$$f(y) \leq f(x) + (y - x) \cdot f'(x), \qquad (4.6.5)$$

where $f'(x)$ denotes the derivative of f.

4.23 Suppose $f : \Re \to \Re$ is twice differentiable. Prove that f is convex if and only if $f''(x) \geq 0$ for all $x \in \text{dom}(f)$. Use this to prove that the functions x^2, e^x and e^{x^2} are convex.

4.24 Consider the following functions on n variables $x_1, \ldots, x_n$: (i) $a_1 x_1 + \ldots + a_n x_n$, where $a_1, \ldots, a_n$ are constants, (ii) $\log(e^{x_1} + \ldots + e^{x_n})$, (iii) $x_1^2 + x_2^2 + \ldots + x_n^2$. Prove that these functions are convex.

4.25 Let f and h be defined as in Section 4.6. Prove that $h'(t) = d^T \nabla f(x_0 + td)$. Conclude that a differentiable function f is convex if and only if for all points $x, y \in \text{dom}(f)$,

$$f(y) \geq f(x) + (y - x)^T \nabla f(x) \qquad (4.6.6)$$

Suppose the second derivative, that is, the Hessian of f, denoted by $H(x)$ exists. Show that f is convex if and only if the matrix $H(x)$ is positive semi-definite[9] at all points x in the domain of f.

4.26 Show that a strictly convex function has a unique local minimum in its domain.

4.27 Consider the convex function $f(x) = x^2 - 1$. Starting from $x_0 = 1$, run the gradient descent algorithm with $\eta = 0.1$ for 10 steps (you may need to use a calculator). How close is the final estimate to the minimizer of f?

4.28 Consider the function $f(x) = |x|$ and $\eta = 0.1$ with starting point 3.05. Show that the gradient descent algorithm will never converge to the minimum.

---

[9] an $m \times m$ matrix H is said to be positive semi-definite if $x^T H x \geq 0$ for all vectors x.

5

Optimization II:
Dynamic Programming

The idea behind dynamic programming is very similar to the concept of divide and conquer. In fact, one often specifies such an algorithm by writing down the recursive sub-structure of the problem being solved. If we directly use a divide and conquer strategy to solve such a problem, it can lead to an inefficient implementation. Consider the following example: the Fibonacci series is given by the sequence 1,1,2,3,5,8, ... If F_n denotes the nth number in this sequence, then $F_0 = F_1 = 1$, and subsequently, $F_n = F_{n-1} + F_{n-2}$. This immediately gives a divide and conquer algorithm (see Figure 5.1) for the problem of computing F_n for an input number n. However, this algorithm is very inefficient – it takes exponential time (see Section 1.1 regarding this aspect), even though there is a simple linear time algorithm for this problem. The reason why the divide and conquer algorithm performs so poorly is because the same recursive call is made multiple times. Figure 5.2 shows the recursive calls made while computing F_6. This is quite wasteful and one way of handling this would be to store the results of a recursive call in a table so that multiple recursive calls for the same input can be avoided. Indeed a simple way of fixing this algorithm would be to have an array $F[]$ of length n, and starting from $i = 0$ onward, we fill the entries $F[i]$ in this array.

Thus, dynamic programming is a divide and conquer strategy done in a careful manner. Typically, one specifies the table which should store all possible recursive calls that the algorithm will make. In fact, the final algorithm does not make any recursive calls. The entries in the table are computed such that whenever we need to solve a sub-problem, all the sub-problems appearing in the recursive call needed for this have already been solved and stored in the table. For example, in the Fibonacci series example, the table corresponds to the array F, and when we need to compute $F[i]$, the values $F[i-1]$ and $F[i-2]$ have already been computed.

Procedure Computing(F_n)

---

1 **Input** Positive integer n ;
2 $F(n)\{$
3 　If $n = 1$ or $n = 0$ Output 1
4 　Else Output $F(n-1) + F(n-2)\}$;

Figure 5.1　*Recursive Fibonacci sequence algorithm*

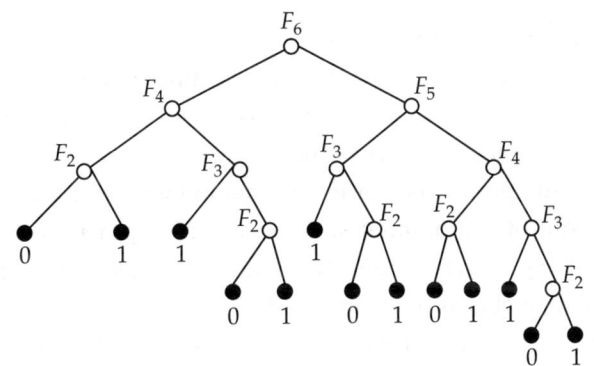

Figure 5.2　*The recursive unfolding of computing F_6. Each internal node represents an addition of the children nodes. The leaf nodes are indicated in solid circles and correspond to terminating conditions. The reader may note the multiplicity of F_2, F_3, etc; this gets worse as we compute bigger Fibonacci numbers.*

A generic dynamic programming approach can be summarized as follows. We begin with a recurrence (or an inductive) relation. In a typical recurrence, you may find repeated subproblems as we unfold the recurrence relation. There is an interesting property that the dynamic programming problems need to satisfy. The overall optimal solution can be described in terms of the optimal solution of the subproblems, sometimes known as the *optimal substructure* property. This is what enables us to write an appropriate recurrence for the optimal solution.

Following this, we describe a table that contains the solutions to the various subproblems. Each entry of the table $\mathcal{T}$ must be computable using only the previously computed entries. This sequencing is very critical to carry the computation forward. The running time is proportional to

$$\sum_{s \in \mathcal{T}} t(s), \text{ where } t(s) \text{ is the time taken to compute an entry } s$$

In the Fibonacci series example, $t(s) = O(1)$. The space bound is proportional to part of the table that must be retained to compute the remaining entries. This is where we can make substantial savings by sequencing the computation cleverly. Dynamic programming is often seen as a trade-off between space and running time, where we reduce the running time at the expense of extra space. By storing the solutions of the repeated subproblems, we save the time for re-computation.

5.1 Knapsack Problem

In the knapsack problem, we are given a set of n objects, and a knapsack of size B. Object x_i has profit p_i and weight (or size) w_i. We want to find a subset of objects whose total weight is at most B and the total profit is maximized. As described earlier, we begin with a recurrence for this problem, which reduces the optimum value for this problem to finding the optima of sub problems. For integer parameters i, y, $1 \leq i \leq n$ and $0 \leq y \leq B$, let $F(i,y)$ denote the profit of the optimal solution for a knapsack of capacity y and using only the objects in $\{x_1, x_2, \ldots, x_i\}$. Under this notation, $F(n,B)$ is the optimal value to the knapsack problem with n objects and capacity M (we are assuming that all weights and B are integers). As a base case, consider $F(1,y)$ – this is easy to define. If $y \geq w_1$, then $F(1,y)$ is p_1, otherwise it is 0. Assuming $i \geq 1$, we can write the following equation

$$F(i,y) = \max\{F(i-1,y), F(i-1,y-w_i) + p_i\}$$

where the two terms correspond to inclusion or exclusion of the object i in the optimal solution (if $y < w_i$, we do not have the second term). Also note that once we decide about the choice of x_i, the remaining choices must be optimal with respect to the remaining objects and the residual capacity of the knapsack.

This algorithm can be easily implemented where we can store F using a two-dimensional table. The rows of F can be indexed by i and columns by y. Note that the computation of row i requires us to know the entries in row $i-1$. Since we can compute row 1 as described here, we can implement this algorithm by computing entries row-wise from row 1 to row n. Since computation of each entry $F(i,y)$ requires constant time, this algorithm can be implemented in $O(nB)$ time. As stated earlier, the algorithm requires $O(nB)$ space. But as outlined here, the computation of entries in row i only requires entries in row $i-1$. Therefore, we can reduce the space to $O(B)$. Note that this may not be polynomial time. The parameter B requires only $O(\log B)$ bits for its representation.

Hence, if B happens to be 2^n, the running time of this algorithm would be 2^n, even though the input size is $O(n)$. Table 5.1 illustrates this for the example given earlier in Chapter 4.

So far, we have shown how to compute the *value* of the optimal solution. But we may want to find the actual subset of objects which are selected by an optimal solution. This can be easily gleaned from the table once we have computed all its entries. Indeed, each entry $F(i, y)$ is a choice between two options. Besides storing the optimal value of the corresponding problem, we will also store which choice was selected while computing this entry. From this, it is easy to compute an optimal solution. We start with the table entry $F(n, B)$. If its value was $F(n - 1, B)$ (first choice), then we proceed from the table entry $F(n - 1, B)$; otherwise (second choice), we select object n in our solution, and proceed to the table entry $F(n - 1, B - w_n)$. We repeat this process till we exhaust looking at all the rows. Note that now we need to store the entire table, we cannot just use $O(B)$ storage. Although there are tricks which even allow us to compute the optimal solution using $O(B)$ storage (and $O(nB)$ time), we will not cover them in this chapter.

Table 5.1   *The dynamic programming table for Knapsack*

$$p_1 = 10, w_1 = 2 \ \ p_2 = 10, w_2 = 4 \ \ p_3 = 12, w_3 = 6 \ \ p_4 = 18, w_4 = 9 \ \ B = 15$$

i	$y = 1$	2	3	4	5	6	7	8	9	10	11	12	13	14	15
1	0	10	10	10	10	10	10	10	10	10	10	10	10	10	10
2	0	10	10	10	10	20	20	20	20	20	20	20	20	20	20
3	0	10	10	10	10	20	20	22	22	22	22	32	32	32	32
4	0	10	10	10	10	20	20	22	22	22	28	32	32	32	38 **

5.2 Context Free Parsing

Given a context free grammar G in a Chomsky normal form (CNF) and a string $X = x_1 x_2 \dots x_n$ over some alphabet Σ, we want to determine if X can be derived from the grammar G.

Recall that a grammar in CNF has the following production rules

$$A \rightarrow BC \ \ A \rightarrow a$$

where A, B, C are non-terminals and a is a terminal (symbol of the alphabet). All derivations must start from a special non-terminal S which is the start symbol. We will use the notation $S \overset{*}{\Rightarrow} \alpha$ to denote that S can derive the sentence α in a finite number of steps by applying the production rules of the grammar.

The basis of our algorithm is the following observation

Observation 5.1 $A \overset{*}{\Rightarrow} x_i x_{i+1} \dots x_k$ *iff* $A \rightarrow BC$ *and there exists an* $i < j < k$ *such that* $B \overset{*}{\Rightarrow} x_i x_{i+1} \dots x_j$ *and* $C \overset{*}{\Rightarrow} x_{j+1} \dots x_k$.

There are $k-1$ possible partitions of the string and we must check all partitions to see if the aforementioned condition is satisfied. More generally, for the given string $x_1 x_2 \ldots x_n$, we consider all substrings $X_{i,k} = x_i x_{i+1} \ldots x_k$, where $1 \leq i < k \leq n$ – there are $O(n^2)$ such substrings. For each substring, we try to determine the set of non-terminals A that can derive this substring. To determine this, we use the observation above. Note that both B and C derive substrings that are strictly smaller than $X_{i,k}$. For substrings of length one, it is easy to check which non-terminals derive them, so these serve as base cases.

We define a two-dimensional table T such that the entry $T(s,t)$ corresponds to all non-terminals that derive the substring starting at x_s of length t. For a fixed t, the possible values of s are $1, 2, \ldots, n-t+1$ which makes the table triangular. Let N denote the number of non-terminals in the grammar. Then $T(s,t)$ consists of all non-terminals A such that one of the following conditions are satisfied: (i) If $t=1$ then there should be a rule $A \rightarrow x_s$ in the CNF grammar, or (ii) there is an index $k, s < k < t$ and a rule $A \rightarrow BC$ in the grammar such that $T(s,k)$ contains B and $T(k,t)$ contains C. Note that such an entry can be computed in $O(tN)$ time.

Each entry in the table can be filled up in $O(t)$ time for column t. This yields a total running time of $\sum_{t=1}^{n} O((n-t) \cdot t)$, which is $O(n^3)$. The space required is the size of the table which is $O(n^2)$. This algorithm is known as the CYK (Cocke–Young–Kassimi) algorithm after the discoverers.

Example 5.1 *Given the following grammar*

$$S \rightarrow AB \qquad S \rightarrow BA$$
$$A \rightarrow BA \qquad A \rightarrow a$$
$$B \rightarrow CC \qquad B \rightarrow b$$
$$C \rightarrow AB \qquad C \rightarrow a$$

determine if strings $s_1 = aba$ and $s_2 = baaba$ are generated by this grammar. The tables in Figure 5.3 corresponds to the two input strings.

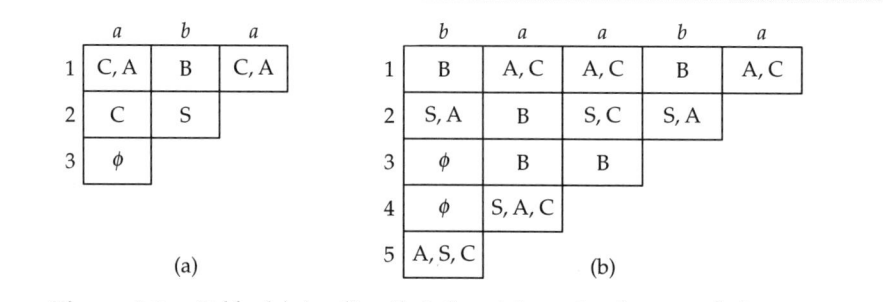

Figure 5.3 *Table (a) implies that the string aba does not belong to the grammar whereas Table (b) shows that baaba can be generated from S*

5.3 Longest Monotonic Subsequence

We are given a sequence S of numbers $x_1, x_2, \ldots, x_n$, a subsequence $x_{i_1}, x_{i_2}, \ldots, x_{i_k}$, where $i_{j+1} > i_j$ is *monotonic* if $x_{i_{j+1}} \geq x_{i_j}$. We want to find the longest (there may be more than one) monotonic subsequence.

Claim 5.1 *For any sequence of length n, either the longest increasing monotonic subsequence or the longest decreasing subsequence has length at least* $\lceil \sqrt{n} \rceil$.

This is known as the Erdos–Szekeres theorem. The proof is based on a clever application of the pigeon-hole principle and is left as an exercise. The previous result is only an existential result but here we would like to find the actual sequence. Let L_i denote the length of the largest monotonic subsequence in $x_1, \ldots, x_i$ which ends at x_i. Clearly, L_1 is just 1. We can write an easy recurrence for $L_i, i > 1$. Consider such a longest subsequence ending at x_i. Let x_k be the element in this subsequence before x_i (clearly x_k must be less than or equal to x_i). Then, L_i must be $L_k + 1$. Thus, we get

$$L_i = \max_{k:1 \leq k < i, x_k \leq x_i} L_k + 1$$

One can see that computing L_i for all i takes $O(n^2)$ time. The length of the longest monotonic subsequence is just $\max_i L_i$. It is also easy to see that once we have computed this table, we can also recover the actual longest monotonic subsequence in $O(n)$ time. Note that the space requirement is only $O(n)$.

Can we improve the running time? For this, we will actually address a more general problem[1], namely for each j, we will compute a monotonic subsequence of length j (if it exists). For each $j \leq i \leq n$, let $M_{i,j}$ denote the set of monotonic subsequences of length j in $x_1 x_2 \ldots x_i$. Clearly, if $M_{i,j}$ exists, then $M_{i,j-1}$ exists as well and the maximum length of the subsequence is given by the largest value of j for which $M_{n,j}$ is non-empty.

Further, among all subsequences of length j, we will like to focus on the subsequence $m_{i,j} \in M_{i,j}$ which has the minimum terminating value. For example, among the subsequences 2,4,5,9 and 1,4,5,8 (both length 4), we will choose the second one, since $8 < 9$.

Let $\ell_{i,j}$ be the last element of $m_{i,j}$. Here is a simple property of the $\ell_{i,j}$ values.

Observation 5.2 *For any fixed index i, the* $\ell_{i,j}$*s form a non-decreasing sequence in j.*

To prove this, we argue by contradiction. Fix an index i and let $j_1 < j_2 \leq i$ be such that $\ell_{i,j_1} > \ell_{i,j_2}$. But now, look at the monotonic subsequence m_{i,j_2}. This will contain a monotonic subsequence of length j_1 in which the last element is at most ℓ_{i,j_2} and hence, less than ℓ_{i,j_1}.

---

[1]See Exercise 5.3 for an alternate approach.

We now write a recurrence for $\ell_{i,j}$. As a convention, we will set $\ell_{i,j}$ to infinity if the set $M_{i,j}$ is empty. As base cases, $\ell_{i,1}$ is just the smallest number in $x_1, \ldots, x_i$, for all i, and $\ell(i,j)$ is infinity for all $j > 1$. For $i \geq 1, j > 1$,

$$
\ell_{i+1,j} = \begin{cases} x_{i+1} & \text{if } \ell_{i,j-1} \leq x_{i+1} < \ell_{i,j} \\ \ell_{i,j} & \text{otherwise} \end{cases}
$$

This follows, since, $m_{i+1,j}$ is either $m_{i,j}$ or x_{i+1} must be the last element of $m_{i+1,j}$; in the latter case, it must satisfy the previous observation. As an example, consider the sequence: 13, 5, 8, 12, 9, 14, 15, 2, 20. Starting from index 1, $m_{1,1} = 13$. Then, successively we have $m_{2,1} = 5$, and $m_{3,1} = 5$, $m_{3,2} = 5,8$. Subsequently, we get $m_{4,1} = 5$ $m_{4,2} = 5,8$ $m_{5,3} = 5,8,12$, etc. Therefore, $\ell_{4,2}$ would be 8, $\ell_{5,3}$ would be 12, and so on. If we compute $\ell_{i,j}$ using the recurrence given here, this will take $O(n^2)$ time because there are n^2 table entries. Instead, we use the aforementioned observation to update these table entries in a fast manner.

For a fixed i, the observation shows that the sequence $\ell_{i,j}$ is a non-decreasing sequence – call this sequence D_i. We now show how to update D_i to D_{i+1} quickly. Indeed, the previous recurrence shows that if $\ell_{i,k-1} \leq x_{i+1} < \ell_{i,k}$, then D_{i+1} is obtained from D_i by just inserting x_{i+1} after $\ell_{i,k-1}$ and removing $\ell_{i,k}$. We can implement these two operations, namely, search for x_{i+1} in a sorted sequence, and insert x_{i+1} in this sequence efficiently using a dynamic dictionary data structure, like an AVL tree, in $O(\log n)$ time. Consider the following example sequence. The sequences $D_1, \ldots, D_9$ are as given here:

D_1	13	∞	∞	∞	∞	∞	∞	∞	∞
D_2	5	∞	∞	∞	∞	∞	∞	∞	∞
D_3	5	8	∞	∞	∞	∞	∞	∞	∞
D_4	5	8	12	∞	∞	∞	∞	∞	∞
D_5	5	8	9	∞	∞	∞	∞	∞	∞
D_6	5	8	9	14	∞	∞	∞	∞	∞
D_7	5	8	9	14	15	∞	∞	∞	∞
D_8	2	8	9	14	15	∞	∞	∞	∞
D_9	2	8	9	14	15	20	∞	∞	∞

Once we have all these sequences, we can construct the longest monotonic subsequence easily. Consider the given example. We know that there is a subsequence of length 6 ending at 20. Looking at D_8, we can see that there is a subsequence of length 5 ending at 15, and we know from our earlier observation that 20 follows 15 in the input sequence. Hence, there is a subsequence of length 6 ending with 15, 20, and so on. The reader should work out the details of the information maintained in D_i and reconstruction of the longest sequence.

5.4 Function Approximation

Consider an integer valued function $h(i)$ on integers $\{1, 2, \ldots, n\}$. Given a parameter $k \leq n$, we want to define another function $g(i)$ with a maximum of k steps such that the difference between g and h, $\Delta(g, h)$ is minimized according to some measure Δ. One of the most common measures is the sum of the squares of the differences of the two functions that we will denote by L_2^2, that is, $L_2^2(g, h) = \sum_{i=1}^{n}(g(i) - h(i))^2$. We will also denote $L_2^2(g, h)$ by the *sum of squares error* between g and h.

Given indices $i \leq k$ and $j \leq n$, let $g_{i,j}^*$ denote the optimal i-step function for this problem restricted to the points $1, \ldots, j$ – we are interested in computing $g_{k,n}^*$. Note that $g_{i,j}^*$ for $i \geq j$ is identical to h restricted to points $1 \ldots j$.

Claim 5.2 $g_{1,j}^* = \frac{1}{j}\sum_{i=1}^{j} h(i)$, *that is, it is a constant function equal to the mean.*

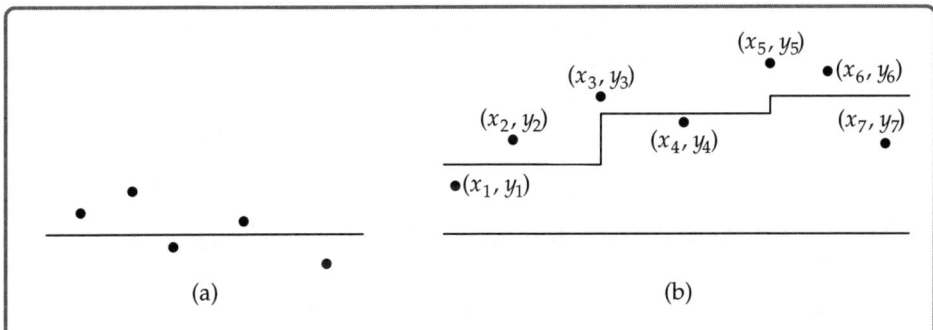

Figure 5.4 *In (a), the constant function is an average of the y values which minimizes the sum of squares error. In (b), a 3 step function approximates the 7 point function.*

The proof of this claim is left as an exercise.

For indices $i, j, 1 \leq i \leq j \leq n$, let $D(i, j)$ denote $\sum_{l=i}^{j}(h(l) - A_{i,j})^2$, where $A_{i,j}$ is the mean of $h(i), \ldots, h(j)$. In other words, it is the smallest value of the sum of squares error between h restricted to the range $\{i, i+1, \ldots, j\}$ and a constant function. We can now write a recurrence for the $g_{i,\ell}^*$ as follows. Let $t(i, j)$ denote the smallest $s \leq j$ such that $g_{i,j}^*$ is constant for values $\geq s$, viz., $t(i, j)$ is the last step of $g_{i,j}^*$. Then,

$$t(i, j) = \min_{s < j}\{L_2^2(h, g_{i-1,s}^*) + D_{s,j}\}$$

We can now write

$$g_{i,\ell}^*(s) = \begin{cases} g_{i-1,t(i,\ell)}^*(s) & s < t(i,\ell) \\ A_{t(i,\ell),\ell} & \text{otherwise} \end{cases}$$

The recurrence captures the property that an optimal k step approximation can be expressed as an optimal $k-1$ step approximation till an intermediate point followed by the best 1 step approximation of the remaining interval (which is the mean value in this interval from our previous observation). Assuming that $D_{j,\ell}$ are pre-computed for all $1 \leq j < \ell \leq n$, we can compute the $g_{i,j}^*$ for all $1 \leq i \leq k$ and $1 \leq j \leq n$ in a table of size kn. The entries can be computed in increasing order of i and thereafter in increasing order of js. The base case of $i = 1$ can be computed directly from the result of the previous exercise. We simultaneously compute $t(i,j)$ and the quantity $L_2^2(h, g_{i,j}^*)$. Each entry can be computed from $j-1$ previously computed entries yielding a total time of

$$\sum_{i=1}^{i=k} \sum_{j=1}^{n} O(j) = O(k \cdot n^2)$$

The space required is proportional to the previous row (i.e., we need to keep track of the previous value of i), *given that $D_{j,\ell}$ can be stored/computed quickly*. Note that an i-step function can be stored as an i-tuple; so the space in each row is $O(k \cdot n)$, since $i \leq k$.

To complete the analysis of this algorithm, the computation of $D_{i,j}$ is left as an exercise. The reader is also encouraged to explore alternate dynamic programming recurrence to compute the optimal function.

5.5   Viterbi's Algorithm for Maximum Likelihood Estimation

In this problem, we have a *labeled* directed graph $G = (V, E)$, where the edges are labeled with symbols from an alphabet Σ. Note that more than one edge can share the same label. Further, each edge (u, v) has a weight $W_{u,v}$, where the weights are related to probabilities and the sum of the probabilities on outgoing edges with the same label from any given vertex is 1. Given a string $\sigma = \sigma_1 \sigma_2 \ldots \sigma_n$ over Σ, find the most probable path in the graph starting at v_o with label equal to σ. The label of a path is the concatenation of labels associated with the edges. To find the most probable path, we can actually find the path that achieves the maximum probability with label σ. By assuming independence between successive edges, we want to choose a path that maximizes the product of the probabilities. Taking the log of this objective function, we can instead maximize the sum of the probabilities. So, if the weights are negative logarithms of the probability – the objective is to *minimize* the sum of the weights of edges along a path (note that the log of probabilities are negative numbers).

We can write a recurrence based on the following observation.

The optimal least-weight path $x_1, x_2, \ldots, x_n$ starting at vertex x_1 with label $\sigma_1, \sigma_2, \ldots, \sigma_n$ is such that the path $x_2, x_3, \ldots, x_n$ is optimal with respect to the label $\sigma_2, \sigma_3, \ldots, \sigma_n$. For paths of lengths one, it is easy to find the optimal labeled path. Let $P_{i,j}(v)$ denote the optimal labeled path for the labels $\sigma_i, \sigma_{i+1}, \ldots, \sigma_j$ starting at vertex v. We are interested in $P_{1,n}(v_o)$.

$$P_{i,j}(v) = \min_{w|(v,w)\in E} \{P_{i+1,j}(w) + W_{v,w}|\text{label of } (v,w) = \sigma_i\}$$

Starting from the base case of length 1 path, we build length 2 paths from each vertex and so on. Note that the length $i+1$ paths from a vertex v can be built from length i paths from w (computed for all vertices $w \in V$). The paths that we compute are of the form $P_{i,n}$ for all $1 \leq i \leq n$. Therefore, we can compute the entries of the table starting from $i = n - 1$. From the previous recurrence, we can now compute the entries of the $P_{n-2,n}$, etc. by comparing at most $|V|$ entries (more specifically the outdegree) for each starting vertex v.

More precisely, we can argue that each of this is proportional to d_v which is the outdegree of any vertex v. Therefore, the total time for each iteration is $\sum_v d_v = O(|E|)$ steps, where $|E|$ is the number of edges. Hence, the total time to fill up the table is $O(n \cdot |E|)$. Although the size of the table is $n \cdot |V|$, the space requirement can be reduced to $O(|V|)$ from the observation that only the $(i - 1)$ length paths are required to compute the optimal i length paths.

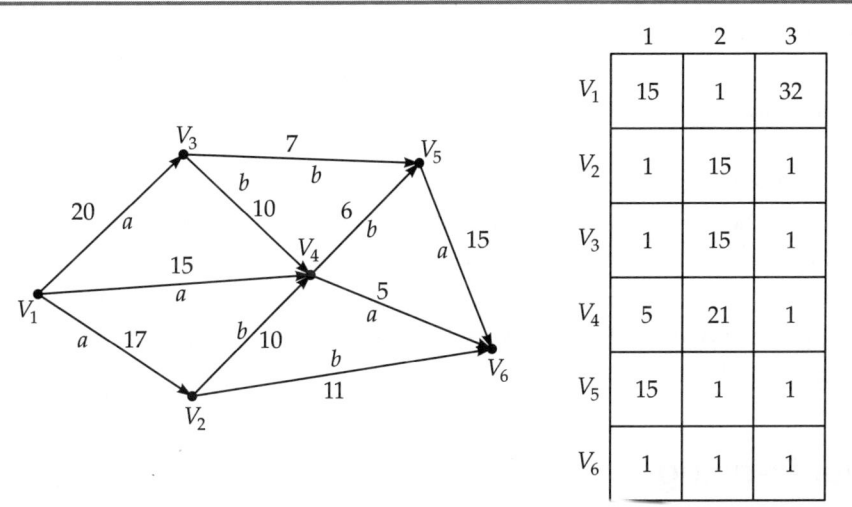

Figure 5.5   *For the label aba and starting vertex v_1, there are several possible labeled paths like $[v_1, v_3, v_4, v_6], [v_1, v_4, v_5, v_6]$, etc. The weights are normalized instead of taking logarithm of probability.*
The entry (v_i, j) corresponds to the optimum path corresponding to the labels $\sigma_j \sigma_{j+1} \ldots \sigma_n$ starting from vertex v_i. In this case, the entry corresponding to $(v_1, 3)$ is the answer. The entry $(v_1, 2)$ corresponds to the labeled path ba from v_1 that does not exist and therefore it is ∞.

5.6   Maximum Weighted Independent Set in a Tree

We are given a rooted tree T and each vertex v in T has a weight w_v. An independent set in T is a subset of vertices such that no two of them have an edge between them (i.e., no two of them have a parent–child relation). We would like to find an independent set whose weight is as large as possible. For a node v in the tree, let T_v denote the sub-tree rooted below v. The dynamic programming algorithm is based on the following idea: suppose we want to find the maximum weighted independent set in the tree T_v. There are two options at v – (i) If we do not pick v, we can recursively solve each of the subproblems defined by the children of v; (ii) If we pick v, then we again need to solve these subproblems, but now we cannot pick the children of v. Motivated by this, we define a table $I(v, b)$, where v is a node in the tree, and the parameter b is either 0 or 1. $I(v, 1)$ denotes the optimal value of the subproblem defined by T_v under the restriction that we are not allowed to pick v, and $I(v, 0)$ is the same quantity with the restriction that we must pick v.

As a base case, if v is a leaf, then $I(v, 1)$ is w_v (assuming weights are non-negative) and $I(v, 0)$ is 0. If v is not a leaf, let $w_1, \ldots, w_k$ denote the children of v. Then, we have the following recurrences:

$$I(v, 1) = w_v + \sum_{i=1}^{k} I(w_i, 0), \qquad I(v, 0) = \sum_{i=1}^{k} \max(I(w_i, 1), I(w_i, 0))$$

If r is the root of the tree, then we output $\max(I(r, 0), I(r, 1))$. In order to compute $I(v, b)$, we need to know the values $I(w, b')$ for all children w of v. Therefore, we can compute them using post-order traversal of the tree. Computing each entry takes time proportional to the number of children, and so, the optimal value can be computed in $O(n)$ time, where n is the number of nodes in the tree. Note that $\sum_{v \in V} d(v) = n - 1$ for any tree, where $d(v)$ is the number of children of a node v.

Further Reading

Dynamic programming is one of the fundamental techniques of algorithm design, and is covered in many classical textbooks [7, 37]. The knapsack problem has been well-studied by the algorithms community both for its simplicity and wide range of applications. The fact that we could not get a polynomial time algorithm is not surprising because it happens to be NP-hard (see Chapter 12). But one can get polynomial time algorithms which come very close to the optimal solution (see Chapter 12). We could save space in the dynamic programming table for the knapsack problem by keeping only row $i - 1$ while computing row i. This is typical of many dynamic programming algorithms. However, if we also want to reconstruct the actual solution (and not just the value of the optimal solution) and still save space, then it requires more clever tricks (see for example,

Hirschberg's algorithm [62]). The CYK algorithm is named after its discoverers Cocke, Younger, and Kasami [7], and remains one of the most efficient algorithms for CNF parsing. With careful choice of data structures, it can be implemented in $O(n^3 t)$ time, where t denotes the size of the CNF grammar. Viterbi's algorithm is named after its discoverer Viterbi [149] and finds many applications in digital communication, machine learning, and other areas. Many graph theoretic optimization problems like vertex cover, independent set, clustering become easier on trees. Often dynamic programming is the principal technique for solving these problems on trees. Use of additional properties like the quadrangle inequality by Yao [158] would lead to non-trivial improvements over the straightforward dynamic programming formulations like matrix-chain product and constructing optimal binary search trees – see exercise problems.

Exercise Problems

5.1 For any sequence of length n prove that either the longest increasing monotonic subsequence or the longest decreasing subsequence has length at least $\lceil \sqrt{n} \rceil$.

5.2 Given a sequence of n real numbers $[x_1, x_2 \ldots x_n]$, we want to find integers $1 \leq k \leq \ell \leq n$, such that $\sum_{i=k}^{\ell} x_i$ is maximum. Note that x_is may be negative, otherwise the problem is trivial. Design a linear time algorithm for this problem.

5.3 If you could design a data structure that would return the maximum value of L_j for all $x_j \leq x_i$ in $O(\log n)$ time then we may be able to obtain a better bound with the simpler recurrence for the longest monotonic sequence. Note that this data structure must support insertion of new points as we scan from left to right. Since the points are known in advance, we can pre-construct the BST skeleton (see Figure 5.6) and fill in the actual points as we scan from left to right, thereby avoiding dynamic restructuring. As we insert points, we update the heap data along the insertion path. Work out the details of this structure and the analysis to obtain an $O(n \log n)$ algorithm.

You may also want to refer to Section 7.3 for a similar data structure.

5.4 A *bitonic* sequence of numbers $x_1, x_2, x_3 \ldots x_k$ is such that there exists an i, $1 \leq i \leq k$ such that $x_1, x_2, \ldots, x_i$ is an increasing sequence and $x_i, x_{i+1}, \ldots, x_n$ is a decreasing sequence. It is possible that one of the sequences is empty, i.e., strictly increasing (decreasing) sequences are also considered bitonic. For example, 3, 6, 7 , 5, 1 is a bitonic sequence where 7 is the discriminating number.

Given a sequence of n numbers, design an efficient algorithm to find the longest *bitonic subsequence*. For the sequence, 2, 4, 3 , 1, −10, 20, 8, the reader can verify that 2, 3, 20, 8 is such a sequence of length 4.

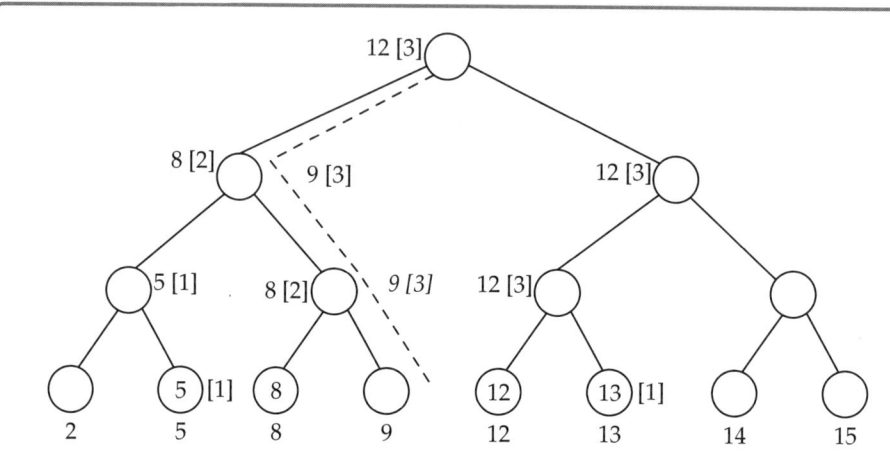

Figure 5.6   *In the sequence 13, 5, 8, 12, 9, 14, 15, 2 we have predefined the tree structure but only the first four numbers have been scanned, i.e., 13, 5, 8, 12. The internal node contains the tuple a[b] to denote that the longest increasing subsequence corresponding to all the scanned numbers in the subtree has length b ending at a. When 9 is inserted along the dotted path, we only need to consider the subtrees to the left of the path, as these are smaller than 9 and choose the largest b value among them.*

5.5 Recall the definition of $g(i,j)$ in the function approximation problem. Show that $g^*_{1,j} = \frac{1}{n}\sum_{i=1}^{j} h(i)$, i.e., it is a constant function equal to the mean.

Further show that $L^2_2(h, g^*_1 - \delta) = L^2_2(h, g^*_1) + \delta^2 \cdot n$, i.e., for $\delta = 0$, the sum of squares of deviation is minimized.

5.6 In the algorithm for function approximation, design an efficient algorithm for the prior computation of all the $D_{i,j}$s.

5.7 Instead of partitioning $g^*_{i,j}$ in terms of an optimal $i-1$ step approximation and a 1 step (constant) approximation you can also partition as i' and $i - i'$ step functions for any $i - 1 \geq i' \geq 1$.

Can you analyze the algorithm for an arbitrary i'?

5.8 **Matrix chain product** Given a chain $(A_1, A_2 \ldots A_n)$ of matrices where matrix A_i has dimensions $p_{i-1} \times p_i$, we want to compute the product of the chain using minimum number of multiplications.

  (i) In how many ways can the matrices be multiplied?

  (ii) Design an efficient algorithm that does not exhaustively use (i).

5.9 Given two character strings $S_1 = x[1..n]$ and $S_2 = y[1..m]$, over some alphabet Σ, the *edit distance* is the cost of transforming the string x to y using a minimum number of operations from the set $\{copy, replace, insert, delete\}$. Design an efficient algorithm to find the minimum edit distance between two given strings. For example, the string cat can be transformed to kite by the following sequence of operations.

(1) replace c with k; (2) replace a with i; (3) copy t; (4) insert e.

There are specific costs associated with each of the operation and we need to minimize the total cost. This has direct application to DNA sequencing problems, i.e., how close two strings are to each other.

5.10 *Typesetting problem* The input is a sequence of n words of lengths $l_1, l_2, \ldots, l_n$ measured in characters. We want to print it nicely on a number of lines that can hold a maximum of M characters each. The criterion for 'niceness' is as follows. No word can be split across lines with a blank separating words and each line should be as full as possible. The penalty for a trailing space of s is s^3. If s_i is the trailing space left in line i, we want to minimize $\sum_i s_i^3$.

If the penalty function is $\sum_i s_i$, would a greedy approach work?

5.11 An ordered subset of a word is a *subsequence*, for example, xle is a subsequence of the string example. For the strings length and breadth, the longest common subsequence is eth, that occurs in both strings. Given two strings s_1 and s_2 of lengths m and n respectively, design an efficient algorithm to find their longest common subsequence.

5.12 *Optimal BST* We are given a sequence $K = \{k_1, k_2, \ldots, k_n\}$ of n distinct keys in sorted order with associated probability p_i that the key k_i will be accessed. Moreover, let q_i represent the probability that the search will be for a value (strictly) between k_i and k_{i+1}. So $\sum_i p_i + \sum_j q_j = 1$. How would you build a binary search tree so as to optimize the expected search cost?

Note that we are not trying to build a balanced tree but to optimize the weighted length of paths – the more probable value should be closer to the root.

5.13 A non-deterministic finite automaton (NFA) is known to be equivalent to a deterministic finite automaton (DFA) in terms of the languages that can be accepted. Given an NFA, how do you find out an equivalent regular expression? What is the running time of your algorithm if a regular expression of length ℓ can be output in (i) constant time (ii) time proportional to ℓ?

Recall that a regular expression represents a set (possibly infinite) of strings over an alphabet which is called a regular set and NFA/DFAs accept precisely this class of languages. An NFA may be thought of as a directed, labeled transition graph where states correspond to vertices. We want to characterize all the strings using a regular expression that take the automaton from an initial state to one of the final accepting states.

Remark This problem can be attempted only if you have some familiarity with regular expressions and finite automata.

5.14 A taxi-driver has to decide about a schedule to maximize his profit based on an estimated profit for each day. Due to some constraint, he cannot go out on consecutive days. For example, over a period of 5 days, if his estimated profits are 30, 50, 40, 20, 60, then by going out on 1st, 3rd and 5th days, he can make a profit of 30+40+60 =130. Alternately, by going out on 2nd and 5th days, he can earn 110. First, convince yourself that by choosing alternate days (there are two such schedules), he won't maximize his profit. Design an efficient algorithm to pick a schedule to maximize the profit for an n days estimated profit sequence.

5.15 A *vertex cover* of a graph $G = (V, E)$ is a subset $W \subseteq V$ such that for all $(x, y) \in E$ at least one of the endpoints $x, y \in W$.

(i) For a given tree $\mathcal{T}$, design an efficient algorithm to find the minimum cardinality *vertex cover* of $\mathcal{T}$. The tree is not necessarily balanced, nor is it binary.

(ii) If every vertex has a non-negative real number weight, find the minimum weight vertex cover of a given tree.

5.16 You are given a stick of (integral) length n and you are required to break into pieces of (integral) lengths $\ell_1, \ell_2, \ldots, \ell_k$ such that a piece having length ℓ_i fetches profit $p_i > 0$ – for all other lengths, the profit is zero. How would you break the stick into pieces so as to maximize the cumulative profit of all the pieces.

5.17 There are n destinations D_i, $1 \leq i \leq n$ with demands d_i. There are two warehouses W_1, W_2 that have inventory r_1 and r_2 respectively such that $r_1 + r_2 = \sum_i d_i$. The cost of transporting $x_{i,j}$ units from W_i to D_j is $c_{i,j}(x_{i,j})$. We must ensure that $x_{i,j} + x_{2,j} = d_j$ in a way so as to minimize $\sum_{i,j} c_{i,j}(x_{i,j})$.

Hint: Let $g_i(x)$ be the cost incurred when W_1 has an inventory of x and supplies are sent to D_j in an optimal manner – the inventory at W_2 is $\sum_{1 \leq j \leq i} d_j - x$.

5.18 An $n \times n$ grid has integer (possibly negative) labels on each square. A player starts from any square at the left end and travels to the right end by moving to one of the 3 adjacent squares in the next column in one step. From the top and bottom row there are only 2 adjacent squares in the next column. The *reward* collected by the player is the sum of the integers in all the squares traversed by the player. Design an efficient (polynomial time) algorithm that maximizes the reward collected by the player.

5.19 Given a convex n-gon (number of vertices is n), we want to triangulate it by adding non-crossing diagonals. Recall that $n - 3$ diagonals are required to triangulate. The *cost* of triangulation is the sum of the lengths of the diagonals added. For example, in a parallelogram, we will choose the shorter diagonal for minimizing cost. Design an efficient algorithm to find the minimum cost diagonalization of a given n-gon.

5.20 Suppose you want to replicate a file over a collection of n servers, labeled $S_1, \ldots, S_n$. Placing a copy of the file at server S_i results in a placement cost of c_i for an integer $c_i > 0$.

Now if a user requests the file from server S_i, and no copy of the file is present at S_i, then the servers $S_{i+1}, S_{i+2}, \ldots$ are searched in order until a copy of the file is finally found, say at server S_j, $j > i$. This results in an access cost of $j - i$. Thus, the access cost of S_i is 0 if S_i holds a copy of the file, otherwise it is $j - i$, where $j > i$ is the smallest integer greater than i such that S_j has a copy of the file. We will require that a copy of the file be always placed at S_n, the last server, so that all such searches terminate.

Now you are given the placement cost c_i for each server S_i. We would like to decide which servers should contain a copy of the file so that the sum of the placement cost and the sum of access costs for all the servers is minimized. Give an efficient algorithm which solves this problem.

5.21 The classical traveling salesman problem (TSP) involves finding a shortest *tour* in a directed weighted graph $G = (V, E)$ that visits every vertex exactly once. A *brute force* method would try all permutations of $[1, 2, \ldots, n]$, where $V = \{1, 2, \ldots, n\}$ that results in an $\Omega(n!)$ running time with $O(n \log n)$ space to count all the permutations. Design a faster dynamic programming algorithm based on the following idea.

Let $T(i, W)$ denote the shortest path in the graph which starts from vertex i, visits only the vertices in W, and ends in vertex 1. Then, the shortest tour of G can be expressed as

$$\min_k \{w(1, k) + T(k, V - \{1\})\}$$

Show how to compute $T(i, W)$ using dynamic programming and also analyze the time and space complexity.

5.22 You are given a set of points (i.e., real numbers) $x_1, x_2, \ldots, x_n$, and a parameter k. In the k-means clustering problem, we wish to partition the set of points into k disjoint intervals $I_1, \ldots, I_k$ such that the objective function

$$\sum_{i=1}^{k} \sum_{x_j \in I_i} |x_j - \mu_i|^2$$

is minimized, where μ_i denotes the average of all the points in I_i. Give an efficient algorithm for solving this problem.

5.23 Knapsack cover problem You are given a knapsack of size B. You are also given a set of n items, where the ith item has size s_i and cost c_i. We want to select a minimum cost subset of items whose total size is **at least** B. Give an efficient algorithm for this problem. You can assume that all quantities are positive integers.

5.24 You are given n (closed) intervals $I_1, \ldots, I_n$ on a line. Each interval I_i has an associated weight w_i. Give an efficient algorithm to select a maximum weight subset of intervals such that the selected intervals are pair-wise disjoint.

5.25 Consider the same setting as in the previous exercise. But now, we would like to select a maximum weight subset of intervals such that for any point p on the line, there are at most two selected intervals containing p. Give an efficient algorithm for this problem.

6
CHAPTER

Searching

The problem of searching is basic in the computer science field and vast amount of literature is devoted to many fascinating aspects of this problem. Starting with searching for a given key in a pre-processed set to the more recent techniques developed for searching documents, the modern civilization forges ahead using *Google Search*. Discussing the latter techniques is outside the scope of this chapter, so we focus on the more traditional framework. Knuth [83] is one of the most comprehensive sources of the earlier techniques; all textbooks on data structures address common techniques like binary search and balanced tree-based dictionaries like AVL (Adelson-Velsky and Landis) trees, red–black trees, B-trees, etc. We expect the reader to be familiar with such basic methods. Instead, we focus on some of the simpler and lesser known alternatives to the traditional data structures. Many of these rely on innovative use of randomized techniques, and are easier to generalize for a variety of applications. They are driven by a somewhat different perspective of the problem of searching that enables us to get a better understanding including practical scenarios where the universe is much smaller. The underlying assumption in comparison-based searching is that the universe may be infinite, that is, we can be searching real numbers. While this is a powerful framework, we miss out on many opportunities to develop faster alternatives based on hashing in a bounded universe. We will address both these frameworks so that the reader can make an informed choice for a specific application.

6.1 Skip-Lists – A Simple Dictionary

Skip-list is a data structure introduced by Pugh [119] as an alternative to balanced binary search trees for handling dictionary operations on ordered lists. The reader may recall that linked lists are very amenable to modifications in $O(1)$ time although they do not support fast searches like binary search trees. We substitute complex book-keeping information used for maintaining balance conditions for binary trees by random sampling techniques. It has been shown that given access to random bits, the *expected* search time in a skip-list of n elements is $O(\log n)$, which compares very favorably with balanced binary trees. The basic idea is to add shortcut pointers to the original sorted list so that we can quickly narrow the search to a much smaller interval and develop this idea recursively. Moreover, it retains the simplicity of insertion and deletion procedures of linked lists, which makes this data structure a very attractive alternative to balanced binary trees.

6.1.1 Construction of skip-lists

A skip-list is maintained as a hierarchy of sorted linked-lists. The bottom-most level is the entire set of keys S. We denote the linked list at level i from the bottom as L_i and let $|L_i| = n_i$. By definition, $L_0 = S$ and $|L_0| = n$. For every $i > 0$, we shall maintain the invariant that $L_i \subset L_{i-1}$ and the topmost level, say level k, has a constant number of elements. Moreover, correspondences are maintained between common elements of lists L_i and L_{i-1} using (vertical) links. We define a pair $T_i = (l_i, r_i)$, $l_i \leq E \leq r_i, l_i, r_i \in L_i$ corresponding to E in the list L_i. We call this tuple the *straddling pair* (of E) in level i that is well-defined for any element E in the universe. The reader may note that this can be easily ensured by notionally adding the elements $-\infty$ and $+\infty$ to the list.

We first describe the procedure for searching for an element E in the set S. The search begins from the topmost level L_k, where T_k can be determined in constant time (see Figure 6.1). If $l_k = E$ or $r_k = E$, then the search is successful else we recursively search among the elements $[l_k, r_k] \cap L_0$. Here, $[l_k, r_k]$ denotes the closed interval bound by l_k and r_k. This is done by searching the elements of L_{k-1} which are bounded by l_k and r_k. Since both $l_k, r_k \in L_{k-1}$, the *descendence* from level k to $k-1$ is easily achieved in $O(1)$ time. In general, at any level i, we determine the tuple T_i by walking through a portion of the list L_i. If l_i or r_i equals E, then we are done; else, we repeat this procedure by *descending* to level $i-1$.

In other words, we refine the search progressively until we find an element in S equal to E or we terminate when we have determined (l_0, r_0). This procedure can also be viewed as searching in a tree that has variable degree (not necessarily two as in a binary tree).

Of course, to be able to analyze this algorithm, one has to specify how the lists L_i are constructed and how they are dynamically maintained under deletions and additions. Roughly, the idea is to have elements in the ith level point to approximately 2^i nodes ahead (in S) so that the number of levels is approximately $O(\log n)$. The time spent in each level i depends on $[l_{i+1}, r_{i+1}] \bigcap L_i$ and hence, the objective is to keep this small. To achieve these goals on-line, we use the following intuitive strategy. The nodes from the bottom-most layer (level 0) are chosen with probability p (for the purpose of our discussion, we shall assume $p = 0.5$) to be in the first level. Subsequently, in any level i, the nodes of level i are chosen to be in level $i+1$ independently with probability p and at any level, we maintain a simple linked-list, where the elements are in sorted order. For $p = 0.5$, then it is not difficult to verify that for a list of size n, the *expected* number of elements in level i is approximately $n/2^i$ and are spaced about 2^i elements apart. The expected number of levels is $O(\log n)$, and the expected space requirement is $O(n)$ as the expected number of levels that each element moves up is 2 when $p = 1/2$. It follows from the linearity of expectation that the expected total number of nodes is $2n$.

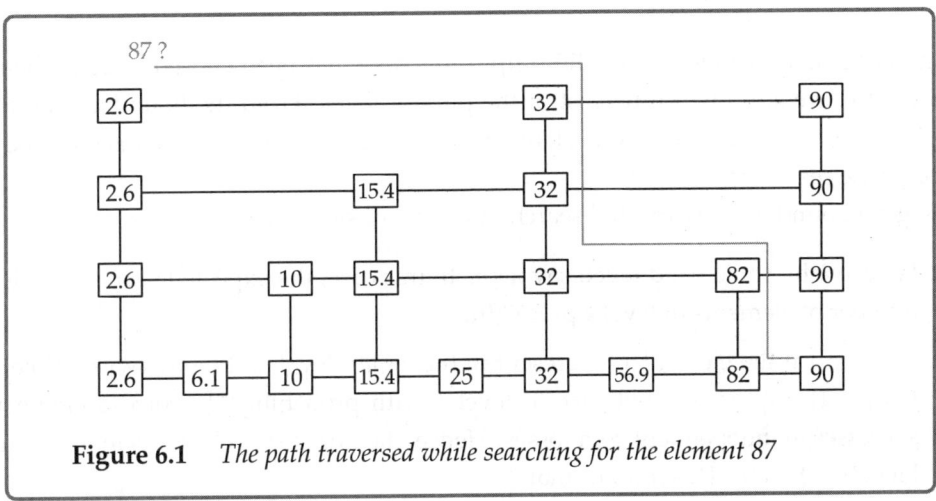

Figure 6.1   *The path traversed while searching for the element 87*

To insert an element, we first locate its position using the search strategy described previously. Note that a byproduct of the search algorithm is the knowledge of all the T_is. At level 0, we choose it to be in level L_1 with probability p. If it is selected, we insert it in the proper position (which can be trivially done from the knowledge of T_1), update the pointers and repeat this process from the present level. Deletion is the exact reversal of this process and it can be readily verified that deletion and insertion have the same asymptotic run time as the search operation. So we shall focus on the search operation.

6.1.2   Analysis

To analyze the run-time of the search procedure, we look at it backward, that is, we retrace the path from level 0. The search time is clearly the length of the path (number of links) traversed over all the levels. So one can count the number of links one traverses before climbing up a level. In other words, the expected search time can be expressed by the following recurrence

$$C(i) = (1-p)(1+C(i)) + p(1+C(i+1))$$

where $C(i)$ is the expected number of steps starting from level i to reach the topmost level k. This recurrence can be justified as follows. Suppose we start from a node v in level 0. With probability $(1-p)$, we stay at this level and move to the next node in level 0. Now, the expected number of steps needed to reach level k is still $C(i)$. With probability p, we move a copy of node v to level $i+1$. In this case, the expected number of steps is $C(i+1)$. The recurrence implies that

$$C(i) = C(i+1) + \frac{1}{p}$$

At any node of a given level, we climb up if this node has been chosen to be in the next level or else, we add one to the cost of the present level. From the boundary condition $C(k) = 0$, where k is the topmost level, one readily obtains $C(0) = k/p$, that is the expected length of the search path.

To get a bound on the expected search time, we consider two variations.

(i) We cap k by some fixed level, say $\log n$. In this case, the expected search time is the number of elements in level k plus $C(0)$.

Let $U_i = 1$ if the element x_i is present in the $\log n$th level and 0 otherwise. Note that $\Pr[U_i = 1] = \frac{1}{n}$ since it will rise to level k with probability $\frac{1}{2^k}$, corresponding to k successes in independent coin tosses. Hence, the number of elements in the topmost level $U = \sum_{i=1}^{n} U_i$. This implies that

$$\mathbb{E}[U] = \mathbb{E}[\sum_i U_{i=1}^n U_i] = \sum_{i=1}^{n} \mathbb{E}[U_i] = \sum_{i=1}^{n} \frac{1}{n} = 1$$

So the expected search time is $O(\log n)$.

(ii) We construct the skip-list until the number of elements in the topmost level is bounded by 4 (it is possible to choose some other suitable constant also).

Let L be the number of levels; we would like to compute $\mathbb{E}[L]$. From the definition of expectation, it follows that

$$\mathbb{E}[L] = \sum_{i \geq 0} i \Pr[L = i] = \sum_{i \geq 0} \Pr[L \geq i]$$

Therefore,

$$\mathbb{E}[L] \;=\; \sum_{i=0}^{\log n} \Pr[L \geq i] + \sum_{i > \log n} \Pr[L \geq i] \tag{6.1.1}$$

$$\leq \; \log n + \frac{1}{2} + \frac{1}{4} + \frac{1}{2^3} \cdots \tag{6.1.2}$$

$$\leq \; \log n + 1 \tag{6.1.3}$$

since $\Pr[L \geq \log n + j] \leq \frac{1}{2^j}$. This can be verified as follows. Consider a single element x. The probability that it gets promoted up to level $\log n + j$ is at most $\frac{1}{n \cdot 2^j}$. Using union bound, the probability that any element gets promoted up to level $\log n + j$ is at most $\frac{1}{2^j}$. But this is exactly the probability of the event $[L \geq \log n + j]$.

The expected number of steps for searching is bounded by the expected number of traversals in each level multiplied by the expected number of levels which is $2(\log n + 1)$.[1]

6.1.3 Stronger tail estimates

If one is not satisfied with expected bounds (after all, there could be constant probability that the search time will exceed the expected bounds), one can get tighter estimates of deviation from expected performance. The search procedure (again looking backward), either moves left or moves up. Whenever it visits a node, it moves up with probability p – we consider such an event a success event. Note that this event will happen $O(\log n)$ times because the height of the data structure is $O(\log n)$.

Thus, the entire search procedure can be viewed in the following alternate manner. We are tossing a coin which turns up heads with probability p – how many times should we toss to come up with $O(\log n)$ heads? Each head corresponds to the event of climbing up one level in the data structure and the total number of tosses is the cost of the search algorithm. We are done when we have climbed up $O(\log n)$ levels (there is some technicality about the number of levels being $O(\log n)$ but that will be addressed later). The number of heads obtained by tossing a coin N times is given by a binomial random variable X with parameters N and p. Using Chernoff bounds (see Eq. 2.2.6), for $N = 15 \log n$ and $p = 0.5$, $\Pr[X \leq 1.5 \log n] \leq 1/n^2$ using $\delta = 9/10$ in this equation. Using appropriate constants, we can get rapidly decreasing probabilities of the form $\Pr[X \leq c \log n] \leq 1/n^\alpha$ for $c, \alpha > 0$ with α increasing with c. These constants can be fine-tuned although we shall not bother with such an exercise here.

We thus state the following lemma.

---

[1] This is based on a very common stochastic phenomenon called the *random sum*. Consider a random variable $X = X_1 + X_2 + \ldots + X_T$, where X_is are identically distributed and T itself is an integral random variable that has finite expectation. Then, it can be shown easily that $\mathbb{E}[X] = \mathbb{E}[X_i] \cdot \mathbb{E}[T]$.

Lemma 6.1 *The probability that access time for a fixed element in a skip-list data structure of length n exceeds $c \log n$ steps is less than $O(1/n^2)$ for an appropriate constant $c > 1$.*

Proof: We compute the probability of obtaining fewer than k (the number of levels in the data structure) heads when we toss a fair coin ($p = 1/2$) $c \log n$ times for some fixed constant $c > 1$. That is, we compute the probability that our search procedure exceeds $c \log n$ steps. Recall that each head is equivalent to climbing up one level and we are done when we have climbed k levels. To bound the number of levels, it is easy to see that the probability that any element of S appears in level i is at most $1/2^i$, that is, it has turned up i consecutive heads. So the probability that any fixed element appears in level $3 \log n$ is at most $1/n^3$. The probability that $k > 3 \log n$ is the probability that at least one element of S appears in $L_{3 \log n}$. This is clearly at most n times the probability that any fixed element survives and hence, probability of k exceeding $3 \log n$ is less than $1/n^2$.

Given that $k \leq 3 \log n$, we choose a value of c, say c_0 (to be plugged into Eq. (2.2.8) of Chernoff bounds) such that the probability of obtaining fewer than $3 \log n$ heads in $c_0 \log n$ tosses is less than $1/n^2$. The search algorithm for a fixed key exceeds $c_0 \log n$ steps if one of the aforementioned events fail – either the number of levels exceeds $3 \log n$ or we get fewer than $3 \log n$ heads from $c_0 \log n$ tosses. This is clearly the summation of the failure probabilities of the individual events which is $O(1/n^2)$.   □

Theorem 6.1 *The probability that the access time for any arbitrary element in a skip-list exceeds $O(\log n)$ is less than $1/n^\alpha$ for any fixed $\alpha > 0$.*

Proof: A list of n elements induces $n + 1$ intervals. From the previous lemma, we know that the probability P that the search time for a fixed element exceeding $c \log n$ is less than $1/n^2$. Note that all elements in a fixed interval $[l_0, r_0]$ follow the same path in the data structure. It follows that for any interval, the probability of the access time exceeding $O(\log n)$ is n times P. As mentioned before, the constants can be chosen appropriately to achieve this.   □

It is possible to obtain even tighter bounds on the space requirement for a skip-list of n elements. We can show that the expected space is $O(n)$ since the expected number of times a node moves to upper levels is 2.

6.2 Treaps: Randomized Search Trees

The class of binary (dynamic) search trees is perhaps the first introduction to non-trivial data structures in computer science. However, update operations, although asymptotically very fast, are not the easiest to remember. The rules for *rotations* and *double-rotations* of the AVL trees, the splitting/joining in B-trees, and the color changes of red–black trees are often complex; so are their correctness proofs. *Randomized search trees* (also known as randomized treaps) provide a practical alternative to the balanced BST (binary search tree). We still rely on rotations, but no explicit balancing rules are used. Instead, we rely on the magical properties of random numbers.

The randomized search tree (RST) is a binary tree that has the keys in an in-order ordering. In other words, for every node, the key stored in it is larger than or equal to the keys stored in the nodes in the left subtree rooted to it, and less than or equal to that in the right subtree rooted to it. In addition, each element is assigned a priority when it arrives; the nodes of the tree are heap-ordered based on the priorities with the highest priority as the parent. In other words, the priority of any element is larger than the priorities of elements stored as its children. Because the key values follow an in-order numbering, insertion follows the normal procedure (as in a binary search tree). After insertion, the heap-ordering may not hold and subsequently, it needs to be restored. This is done by rotations, since rotation preserves in-order numbering.

Claim 6.1 *For a given assignment of (distinct) priorities to elements, show that there is a* unique *treap.*

The proof is left as an exercise problem. The priorities are assigned randomly (and uniquely) from a sufficiently large range[2]. The priorities induce a random ordering of the N nodes. By averaging over the random ordering, the expected height of the tree is *small*. This is the crux of the following analysis of the performance of RSTs.

Let us first look at the search time (for an element Q) based on a technique known as *backward analysis*. A more elaborate exposition of this technique can be found in Section 7.7. For that, we (hypothetically) insert the N elements in decreasing order of their *priorities*. Note that whenever we insert a node (as in a binary search tree), it gets inserted at a leaf node. Therefore, the heap priority is satisfied (because we are inserting them in decreasing order of priorities). At any time during this process, the set of inserted nodes will form a sub-tree of the overall search tree. For an element N_i, we say that Q can *see* this element if no previously inserted element has a key which lies between the keys corresponding to Q and N_i. We count the number of elements that Q can see during this process.

Claim 6.2 *The tree constructed by inserting nodes in order of their priorities (highest priority is the root) is the same as the tree constructed on-line.*

This follows from the uniqueness of the treap.

Claim 6.3 *The number of nodes Q sees during the insertion sequence is exactly the number of comparisons performed for searching for Q. In fact, the order in which it sees these nodes corresponds to the search path of Q.*

---

[2] For N nodes, $O(\log N)$ bit random numbers suffice – see Section 2.3.3.

Let $N_1, N_2, \ldots$ denote the nodes arranged in decreasing order of priorities. Let us prove the aforementioned claim by induction on i, that is, among nodes $N_1, N_2, \ldots, N_i$, the total number of comparisons performed equals the number of nodes Q sees. For the base case, N_1 corresponds to the root and every element can see N_1 and must be compared to the root. Assuming this is true for i, what happens when we consider N_{i+1}? Node Q will see N_{i+1} iff they are in the same interval induced by the nodes $N_1, N_2, \ldots, N_i$. From the previous observation, we know that we can construct the tree by inserting the nodes in decreasing order of priorities. If node Q and N_{i+1} are in the same interval, then Q is in the subtree rooted at N_{i+1} (node N_{i+1} has one of the previous nodes as its parent) and so it will be compared to N_{i+1} in the search tree. Conversely, if Q and N_{i+1} are not in the same interval, then there must be some node $Y \in \{N_1, \ldots, N_i\}$ such that Q and N_{i+1} are in different subtrees of Y (the least common ancestor). So, N_{i+1} cannot be a node that Q will be compared with (recall that the tree is unique).

Theorem 6.2 *The expected length of a search path in RST is $O(H_N)$, where H_N is the Nth harmonic number.*

To prove this property, let us arrange the elements in increasing order of keys (note that this is deterministic and different from the ordering $N_1, N_2, \ldots$). Let this ordering be $M_1, M_2, \ldots, M_n$. Suppose Q happens to be M_j in this ordering. Consider an element $M_{j+k}, k \geq 0$. What is the probability that M_j sees M_{j+k}? For this to happen, the priorities of all the elements between M_j and M_{j+k} must be smaller than the priority of M_{j+k}. Since priorities are assigned randomly, the probability of this event is $1/k$. Using linearity of expectation, and summing over all k, both positive and negative, we see that the expected number of elements that M_j sees is at most 2 times H_N. When $Q \neq M_j$ then the same argument holds for the closest element M_j and therefore true for Q.

Insertions and deletions require changes in the tree to maintain the heap property and rotations are used to push up or push down some elements as per this need. Recall that rotations do not violate the binary search tree property. Figure 6.2 shows the sequence of changes including rotations required for the insertion sequence 25, 29, 14, 18, 20, 10, 35 having priorities 11, 4, 45, 30, 58, 20, 51 respectively that are generated on the fly by choosing a random number between 1 and 100.

A similar technique can be used for counting the number of rotations required for RST during insertion and deletions. In the cases of both skip-lists and RSTs, implicitly we are relying on a simple property of random numbers referred to as the *principle of deferred decision*. The random priorities in the case of RSTs (or the number of copies of an element in skip-lists) are not known to the algorithm in advance but revealed at the time it is needed. This does not affect the performance or the analysis of the algorithm and allows us to deal with the dynamic updates seamlessly without any additional complications compared to the static case – as if the present set of elements were given to us all at once. The high probability bounds for RSTs is left as an exercise problem.

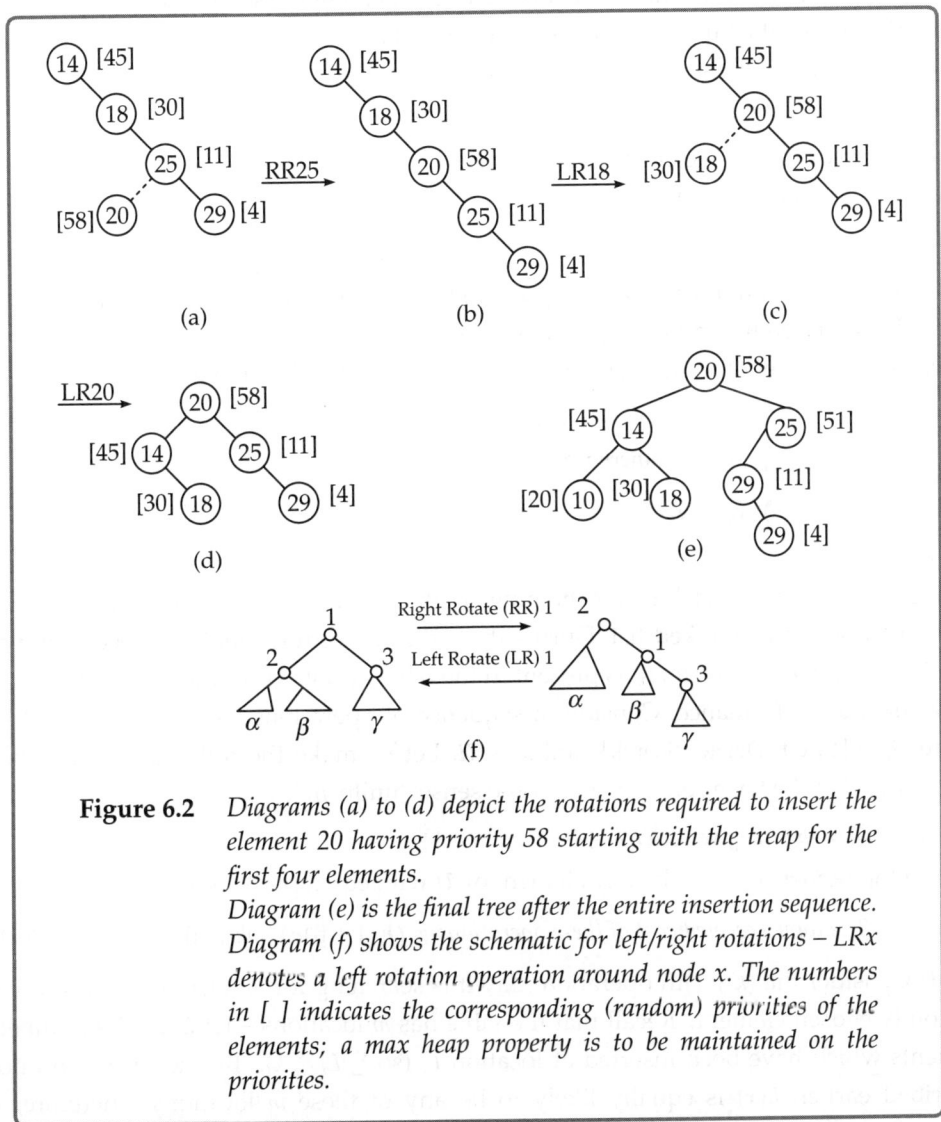

Figure 6.2   *Diagrams (a) to (d) depict the rotations required to insert the element 20 having priority 58 starting with the treap for the first four elements.*
Diagram (e) is the final tree after the entire insertion sequence. Diagram (f) shows the schematic for left/right rotations – LRx denotes a left rotation operation around node x. The numbers in [ ] indicates the corresponding (random) priorities of the elements; a max heap property is to be maintained on the priorities.

6.3   Universal Hashing

Consider that we are given a set of n keys which take values from a large universe $\mathcal{U}$. Recall that hashing maps these n keys to values in a small range (often called a hash table) such that each of these keys is (hopefully) mapped to a distinct value. If we can achieve this, we can perform operations like search in $O(1)$ time.

For the simple reason that the number of possible key values (i.e., size of $\mathcal{U}$) is much larger than the table size, it is inevitable that more than one key is mapped to the same location in the hash table. The number of conflicts increase the search time. If the keys are randomly chosen, then it is known that the expected number of conflicts is $O(1)$. However, this may be an unrealistic assumption; so we must design a scheme to handle any arbitrary subset of keys. We begin with some useful notations.

- Universe : $\mathcal{U}$, Let the elements be $0, 1, 2, \ldots, N-1$
- Set of elements : S, where $|S| = n$
- Hash locations : $\{0, 1, \ldots, m-1\}$ usually, $m \geq n$

Collision: In collision, $x, y \in \mathcal{U}$ are mapped to the same location by a hash function h. We define $\delta_h(x, y)$ to be the indicator variable when x and y are hashed to the same location. Similarly, $\delta_h(x, S)$ denotes the number of elements in S which collide with x.

$$\delta_h(x, y) = \begin{cases} 1 & : \quad h(x) = h(y), x \neq y \\ 0 & : \quad \text{otherwise} \end{cases}$$

$$\delta_h(x, S) = \sum_{y \in S} \delta_h(x, y)$$

Hash by chaining: In hashing by chaining, all the elements that get mapped to the same location are stored in a linked-list. During the search procedure, one may need to traverse such lists to check if an element is present in the table or not. Thus, the more the collision, the worse the performance. Consider a sequence of operations $O_1(x_2), O_2(x_2), \ldots, O_n(x_n)$, where $O_i \in \{\text{Insert, Delete, Search}\}$ and $x_i \in \mathcal{U}$. Let us make the following assumptions, which says that the hash function is in some sense "uniform".

1. $|h^{-1}(i)| = |h^{-1}(i')|$, where $i, i' \in \{0, 1, \ldots, m-1\}$.
2. In the sequence, x_i can be any element of $\mathcal{U}$ with equal probability.

Claim 6.4 *The total expected cost of these operations is $O((1+\beta)n)$, where $\beta = \frac{n}{m}$ (load factor).*

Proof: Consider the $(k+1)$th operation. Say it is searching for an element x, where x is a randomly chosen element. Recall that the table has m locations – let L_i be the number of elements which have been inserted at location L_i (so $\sum_i L_i \leq k$). By the second property described earlier, $h(x)$ is equally likely to be any of these m locations. Therefore, the expected time to search would be

$$\sum_{i=1}^{m} \frac{1}{m} \cdot (1 + L_i) \leq 1 + k/m$$

So total expected cost over all the operations would be at most $\sum_{k=1}^{n}(1 + \frac{k}{m})$ $= n + \frac{n(n+1)}{2m} = (1 + \frac{\beta}{2})n$. Note that this is the *worst* case over *operations* but not over *elements* – the elements are assumed to be randomly chosen from the universe. $\qquad\square$

Note that the commonly used hash function $h(x) = x \bmod m$ satisfies the first property and therefore, it behaves well for randomly chosen keys. However, it is easy to construct a bad example by choosing keys that map to same modulo class.

Universal hash functions: We now define a family of hash functions which have the property that a randomly chosen hash function from this family will have small number of collisions in expectation.

Definition 6.1 *A collection $H \subset \{h | h : [0...N-1] \to [0...m-1]\}$ is c-universal if for all $x, y \in [0...N-1]\ x \neq y$,*

$$|\{h | h \in H \text{ and } h(x) = h(y)\}| \leq c\frac{|H|}{m}$$

for some (small) constant c. Equivalently, $\sum_h \delta_h(x, y) \leq c\frac{|H|}{m}$.

This definition states that if we pick a hash function from H uniformly at random, then the probability that x and y collide is at most c/m. Thus, assuming m is large enough compared to n, the expected number of collisions would be small. Note the subtle difference from the previous analysis – now x and y are *any* two elements; they may not be randomly chosen ones. The following claim says that given any set S of size n, the probability that a randomly chosen hash function causes a collision for an element x is small.

Claim 6.5

$$\frac{1}{|H|} \sum_{h \in H} \delta_h(x, S) \leq c\frac{n}{m}$$

where $|S| = n$.

Proof: Working from the LHS, we obtain

$$= \frac{1}{|H|} \sum_{h \in H} \frac{1}{|H|} \sum_h \sum_{y \in S} \delta_h(x, y)$$

$$= \frac{1}{|H|} \sum_y \sum_h \delta_h(x, y)$$

$$\leq \frac{1}{|H|} \sum_y c\frac{|H|}{m}$$

$$= \frac{c}{m}n$$

$\square$

The expected length of a chain (containing an element x) for *any* choice of S can be denoted by $\mathbb{E}[\delta_h(x,S)]$ for any $x \in S$. This can be calculated as $\frac{1}{|H|} \sum_h \delta_h(x,S)$ for a random choice of $h \in H$. From the previous claim, this is bounded by $\frac{cn}{m}$ and so the expected cost of any operation is bounded by $1 + \frac{cn}{m}$, where the additional 1 is the cost of the actual list operation of insert or delete. So, the expected cost of t operations is bounded by $(1 + c\beta)t$ for *any* set S of n elements, that is, it holds for worst case input.

6.3.1 Existence of universal hash functions

The idea behind universal hash functions is to map a given element $x \in \mathcal{U}$ uniformly into any of the m locations with equal likelihood by choosing a hash function randomly from a given set H. If this is true for all elements in S, then the behavior is similar to the previous case of a random subset S implying that the expected length of a chain is $O(\frac{n}{m})$. So let us try a very simple family of the form

$$h_a(x) = (x + a) \bmod N \bmod m, \text{ where } a \in \{0, 1, 2, \ldots, (N-1)\}$$

The reader should verify that for a random choice of a, an element x will be mapped to any of the m locations with equal probability.

However, this is not enough as we can verify that this family of functions is not universal. Suppose for elements $x, y \in \mathcal{U}$, $x \neq y$, the following holds: $x - y = m \bmod N$. Then,

$$(x + a) \bmod N = (y + m + a) \bmod N = (y + a) \bmod N + m \bmod N$$

Taking mod m, we obtain $h_a(x) = h_a(y)$, for *any* a and therefore, it violates the key property of the universal hash family. This reasoning can be extended to $|x - y| = k \cdot m$ for $k \in \{1, 2, \ldots, \lfloor \frac{N}{m} \rfloor\}$.

Next, we try another variant H' in our quest for universal hash functions. Let $H' : h_{a,b}; h_{ab}(x) \rightarrow ((ax + b) \bmod N) \bmod m$, where $a, b \in 0 \ldots N - 1$ (N is prime[3]).

If $h_{ab}(x) = h_{ab}(y)$, then for some $q \in [0 \ldots m-1]$ and $r, s \in [0 \ldots \frac{N-1}{m}]$.

$$ax + b = (q + rm) \bmod N$$
$$ay + b = (q + sm) \bmod N$$

This is a unique solution for a, b once q, r, s are fixed (note that we are using the fact that N is prime, in which case, the numbers $\{0, 1, \ldots, N - 1\}$ form a field and therefore each element has an inverse modulo N). So there are a total of $m(\frac{N^2}{m})$ solutions $= \frac{N^2}{m}$. Also, since $|H'| = N^2$, H' is '1' universal.

---

[3] This is not as restrictive as it may seem, since from Bertrand's postulate there is at least one prime between an integer i and $2i$.

6.4 Perfect Hash Function

Universal hashing is a very useful property but may not be acceptable in a situation where we do not want any collisions. *Open addressing* is a method that achieves this at the expense of an increased search time. In case of conflicts, we define a sequence of probes that is guaranteed to find an empty location (if there exists one).

We will extend the scheme of universal hashing to one where there is no collision without increasing the expected search time. Recall that the probability that an element x collides with another element y is less than $\frac{c}{m}$ for some constant c (when we choose a random function from a universal family of hash functions). Therefore, the expected number of collisions in a subset of size n by considering all pairs is $f = \binom{n}{2} \cdot \frac{c}{m}$. By Markov's inequality, the probability that the number of collisions exceeds $2f$ is less than $1/2$. For $c = 2$ and $m \geq 4n^2$, the value of $2f$ is less than $\frac{1}{2}$, that is, there are no collisions. Thus, if we use a table of size $\Omega(n^2)$, it is unlikely that there will be any collisions. However, we end up wasting a lot of space. We now show that it is possible to achieve low collision probability and $O(n)$ space complexity.

We use a two-level hashing scheme. In the first level, we hash it to locations $1, 2, \ldots, m$ in a hash table. For each of these locations, we create another hash table which will store the elements which get hashed to this location. In other words, if there are n_i keys that get mapped to location i, we subsequently map them to a hash table of size $4n_i^2$. From our previous discussion, we know that we can avoid collisions with probability at least $1/2$. If a collision still happens (in this second-level hash table), we create another copy of it and use a new hash function chosen randomly from the family of universal hash functions. So we may have to repeat the second-level hashing a number of times (expected value is 2) before we achieve zero collision for the n_i keys. Clearly, the expected search time is $O(1)$ for both the levels.

The expected space bound is a constant times $\sum_i n_i^2$. We can write

$$
\begin{aligned}
n_i^2 &= n_i + 2 \sum_{\substack{x,y \mid h(x) \\ =h(y)=i}} 1 \\
\sum_i n_i^2 &= \sum_i n_i + 2 \cdot \sum_{\substack{x,y \mid h(x) \\ =h(y)=i}} 1 \\
&= 2 \sum_i n_i + 2 \sum_i \sum_{\substack{x,y \mid h(x) \\ =h(y)=i}} 1 \\
&= 2n + \sum_{x,y} \delta(x,y)
\end{aligned}
$$

Taking expectation on both sides (with respect to any choice of a random hash function), the RHS is $2E[\sum_{x,y \in S} \delta(x,y)] + 2n$. The first expression equals $2\binom{n}{2} \cdot \frac{c}{m}$ since $E[\delta(x,y)] = \Pr[h(x) = h(y)] \leq \frac{c}{m}$. Therefore, the total expected space required is only $O(n)$ for $m \in O(n)$.

6.4.1   Converting expected bound to worst case bound

We can convert the *expected* space bound to a worst case space bound in the following manner as can be done for transforming any *Monte Carlo* procedure to a *Las Vegas* algorithm. In the first level, we repeatedly choose a hash function until $\sum_i n_i^2$ is $O(n)$. We need to repeat this twice in the expected sense. Subsequently, at the second stage, for each i, we repeat it till there are no collisions in mapping n_i elements in $O(n_i^2)$ locations. Again, the expected number of trials for each i is two that takes overall $O(n)$ time for n keys. Note that this method makes the space worst case $O(n)$ at the expense of making the time *expected* $O(n)$. But once the hash table is created, for any future query, the time is worst case $O(1)$.

For practical implementation, the n keys will be stored in a single array of size $O(n)$, where the first-level table locations will contain the starting positions of keys with value i and the hash function used in the second-level hash table.

6.5   A log log N Priority Queue*

Searching in a bounded universe is faster when we use hashing. Can we achieve similar improvements for other data structures? Here we consider maintaining a priority queue for elements drawn from universe $\mathcal{U}$. Let $|\mathcal{U}| = N$. The operations supported are *insert*, *minimum*, and *delete*.

Imagine a complete binary tree on N leaf nodes that correspond to the N integers of the universe – this tree has depth $\log N$. Let us color the leaf nodes of the tree black if the corresponding integer is present in the set $S \subset \mathcal{U}$, where $|S| = n$. We would like to design a data structure that supports *predecessor* queries faster than doing a conventional binary search. The elements are from an universe $\mathcal{U} = \{0, 1, 2, \ldots, N-1\}$, where N is a power of 2. Given any subset S of n elements, we want to construct a data structure that returns $\max_{y \in S} y \leq X$ for any query element $X \in \mathcal{U}$.

The elements of S are marked in the corresponding leaf nodes and in addition, we also mark the corresponding leaf to root paths. Each internal node of the binary tree T stores the smallest and the largest element of S in its subtree. If there are none, then these are undefined. This information can be computed at the time of marking out the paths to the node. The smallest and the largest element passing through a node can be maintained easily. See Figure 6.3 for an illustration.

Given a query X, consider the path from the leaf node for X to the root. The predecessor of X can be uniquely identified from the first *marked* node on this path since we can identify the interval of S that contains X. Note that all the ancestors of a marked node are also marked. So from any given leaf node, we need to identify the closest marked ancestor. With a little thought, this can be done using an appropriate modification of binary search on the path of ancestors of X. Since it is a complete binary tree, we can map them to an array and query the appropriate nodes. This takes $O(\log(\log N))$ steps which is superior to $O(\log n)$ for moderately large $N = \Omega(n^{polylog n})$. Clearly, we cannot afford to store the binary tree with N nodes. So, we observe that it suffices if we store only the n paths which takes space $O(n \log N)$. We store these $O(n \log N)$ nodes using a universal hash function so that the expected running time of the binary search procedure remains $O(\log \log N)$. The nodes that are hashed return a successful search: otherwise, it returns a failure.

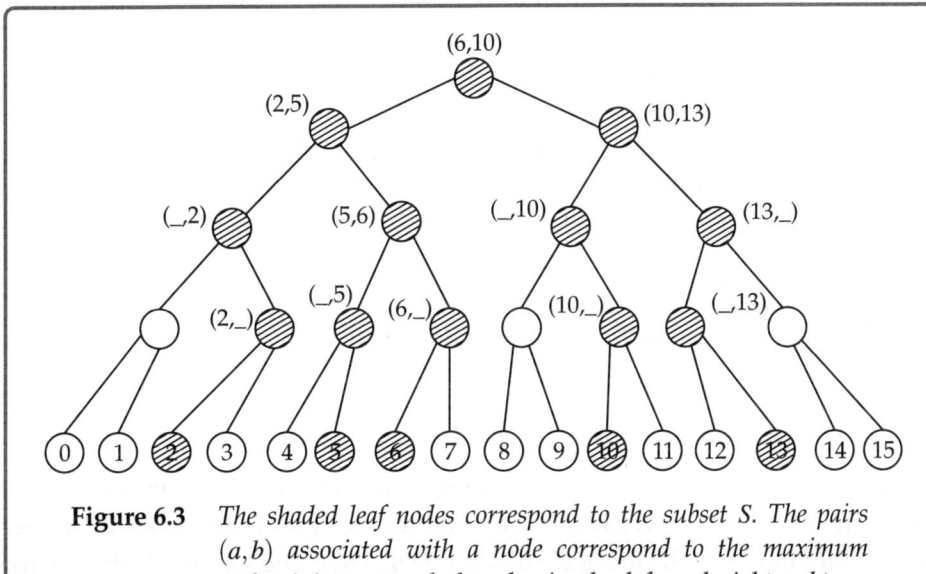

Figure 6.3   *The shaded leaf nodes correspond to the subset S. The pairs (a,b) associated with a node correspond to the maximum and minimum marked nodes in the left and right subtrees respectively and are undefined if there are no marked nodes in the subtree. For example, the path starting at 8 encounters the first shaded node on the path from 10 implying that the successor of 8 is 10 and the predecessor is 6, which can be precomputed*

For a further improvement which does not rely on hash functions, we can do a two-phased search. In the first phase, we build a search tree on a uniform sample of $n/\log N$ keys which are exactly $\log N$ apart, so that the space is $O(\frac{n \log N}{\log N}) = O(n)$. In the second phase, we do a normal binary search on an interval containing at most $\log N$ elements that takes $O(\log \log N)$ steps.

We will now extend this idea to priority queues. We begin by storing the tree with N leaves as done earlier. We again define a notion of coloring of nodes. As before, a leaf is colored if the corresponding element is present in the set S. Let us also imagine that if a leaf node is colored then its *half-ancestor* (halfway from the node to the root) is also colored and is labeled with the smallest and the largest integer in its subtree. The half-ancestor will be present at level $\log N/2$. Denote the set of the minimum elements (in each of the subtrees rooted at level $\log N/2$) by TOP. Note that the set TOP has size $\sqrt{N}$. We will recursively build the data-structure over the elements in TOP. We will denote the immediate predecessor of an element x by $PRED(x)$ and the successor of an element by $SUCC(x)$. The reason we are interested in $PRED$ and $SUCC$ is that when the smallest element is deleted, we must find its immediate successor in set S. Likewise, when we insert an element, we must know its immediate predecessor. Henceforth, we will focus on the operations $PRED$ and $SUCC$ as these will be used to support the priority queue operations.

For a given element x, we will check if its ancestor at depth $\log N/2$ (halfway up the tree) is colored. If so, then we recursively search $PRED(x)$ within the subtree of size $\sqrt{N}$. Otherwise, we recursively search for $PRED(x)$ among the elements of TOP. Note that either we search within the subtree or in the set TOP but not both. Suitable terminating conditions can be defined. The search time can be captured by the following recurrence

$$T(N) = T(\sqrt{N}) + O(1)$$

which yields $T(N) = O(\log\log N)$. The space complexity of the data structure satisfies the recurrence

$$S(N) = (\sqrt{N} + 1)S(\sqrt{N}) + O(\sqrt{N})$$

because we need to recursively build the data structure for the TOP elements and for each of the subtrees rooted at level $\log N/2$ (the additive $O(\sqrt{N})$ is for other book-keeping information; for example, the list of elements in TOP, etc.). The solution to this recurrence is $S(N) = O(N \log\log N)$ that can be verified using induction.

Note that, as long as $\log\log N = o(\log n)$, this data structure is faster than the conventional heap. For example, when $N \leq 2^{2^{\log n/\log\log n}}$, this holds an advantage, but the space is exponential. This is one of the drawbacks of the data structure – the space requirement is proportional to the size of the universe. By using hashing similar to the scheme described before, we can reduce the space to $O(n)$.

Further Reading

The skip-list data structure as an alternative to balanced tree-based dictionary structure was proposed by Pugh [119]. The analysis in the paper was improved to an inverse polynomial tail estimate by Sen [130]. Seidel and Aragon [129] proposed the treap structure as a randomized alternative to maintaining balance in tree-based data structures. Universal hashing was proposed by Carter and Wegman [26] who also proved the optimal worst case bounds for such data structures. An alternate approach called cuckoo hashing was proposed by Pagh and Roddler [113]. The notion of perfect hashing was a subject of intense research, especially the quest for the deterministic techniques [43, 52]. Yao [157] obtained interesting lower bounds for this problem. The reader may note that for random input (or data generated by a known distribution), one can use *interpolation search* that works in expected $O(\log\log n)$ time [116].

The $\log\log n$ priority queue was first proposed by Boas [147]; it demonstrates that one can exploit the size of smaller universe analogous to integer sorting in a smaller range. Mehlhorn's monograph [102] is an excellent source for learning about sophisticated data structures.

Exercise Problems

6.1 Prove the following stronger bound for the skip list space using Chernoff bounds – For any constant $\alpha > 0$, the probability of the space exceeding $2n + \alpha \cdot n$ is less than $\exp^{\Omega(-\alpha^2 n)}$.

6.2 While constructing a skip-list, Professor Thoughtful decided to promote an element to the next level with probability p $(p < 1)$ and calculate the best value of p for which the product $E_q \times E_s$ is minimized, where E_q, E_s are the expected query and expected space respectively. He concluded that $p = 1/2$ is not the best. Justify, giving the necessary calculations.

6.3 To expedite searching in some interval of length ℓ, many data structures provide extra links/fingers that gets us within the proximity for that interval very quickly. In the context of skip-lists, design a strategy for providing such links that can enable us to search a sublist of length ℓ within $O(\log \ell)$ steps.

6.4 For frequently searched items in a skip-list, we want to maintain weights with each element x as $w(x)$ that will help us reduce the search time for such elements. If $W = \sum_x w(x)$, then we want to design a scheme that results in a search time of $O(1 + \log(W/w))$ for an element having weight w.

Hint: A frequently searched element should be closer to the topmost level.

6.5 For a given assignment of priorities, show that there is a *unique* treap.

6.6 *Prove that the probability that the search time exceeds $2 \log n$ comparisons in a randomized treap is less than $O(1/n)$.

Hint: The reader may realize that even if the priorities are assigned independently, the events that M_{j+k} and $M_{j+\ell}$ are visible from M_j for $k \neq \ell$ may require some additional arguments for independence. Without that we may not be able to apply Chernoff bounds.

6.7 A dart game Imagine that an observer is standing at the origin of a real line and throwing n darts at random locations in the positive direction numbered 1 to n. At any point of time, only the closest dart is visible to the observer – so if the next dart lands beyond the closest dart, it will never be visible to the observer.

(i) What is the expected number of darts visible to the observer over the entire exercise of throwing n darts

(ii) Can you obtain a high-probability bound assuming independence between the throws?

(iii) Can you apply this analysis to obtain an $O(n \log n)$ expected bound on quicksort?

6.8 Consider the hash function $h(x) = a \cdot x \bmod N \bmod m$ for $a \in \{0, 1, \dots, (N-1)\}$. Show that it satisfies the properties of a universal family when m is prime.

Hint: For $x \neq y$, $(x - y)$ has a unique inverse modulo m.

6.9 Show that for any collection of hash function H, there exists x, y such that

$$\sum_{h \in H} \delta_h(x, y) \geq |H|(\frac{1}{m} - \frac{1}{n})$$

where n and m are the sizes of the universe and table respectively.

Remark This justifies the definition of universal hash function.

6.10 Assume that the size of the table T is a prime m. Partition a key x into $r + 1$ parts $x = < x_0, x_1, \dots, x_r >$, where $x_i < m$. Let $a = < a_0, a_1, \dots, a_r >$ be a sequence where $a_i \in \{0, 1, \dots, m-1\}$. We define a hash function $h_a(x) = \sum_i a_i x_i \bmod m$. Clearly, there are m^{r+1} distinct hash functions. Prove that $\cup_a h_a$ forms a universal class of hash functions.

A collection of hash functions H is called *strongly* universal if for all keys x, y and any $i, j \in [0, .., m-1]$

$$\Pr_{h \in H} (h(x) = i \wedge h(y) = j) \leq \frac{c}{m^2}$$

How does this differ from the earlier definition in the chapter?

*Can you give an example of a strongly universal family?

6.11 Analyze the preprocessing cost for building the $O(\log \log N)$ search data structure for searching elements in the range $[1, N]$.

6.12 Propose a method to decrease the space bound of the $O(\log \log N)$ search data structure from $O(N \log \log N)$ to $O(N)$ using only worst case deterministic techniques.

Hint: You may want to prune the lower levels of the tree.

6.13 Since the universe N can be much larger than n, describe a method to reduce the space to $O(n)$.

6.14 Show how to implement delete operation in the priority queue to $O(\log \log N)$ steps.

6.15 *Interpolation search** Suppose T is an ordered table of n keys x_i, $1 \le i \le n$ drawn uniformly from $(0,1)$. Instead of doing the conventional binary search, we use the following approach.

Given key y, we make the first probe at the position $s_1 = \lceil y \cdot n \rceil$. If $y = x_{s_1}$, we are through. Else if $y > x_{s_1}$, we recursively search for y among the keys $(x_{s_1}, \ldots, x_n)$. Else recursively search for y among the keys $(x_1, \ldots, x_{s_1})$.

At any stage, when we search for y in a range $(x_l, \ldots, x_r)$, we probe the position $l + \lceil \frac{(y - x_l)(r - l)}{x_r - x_l} \rceil$. We are interested in determining the expected number of probes required by this searching algorithm.

Compare this with the way that we search for a word in the English dictionary.

In order to somewhat simplify the analysis, we modify the aforementioned strategy as follows. In round i, we partition the input into $n^{1/2^i}$ sized blocks and try to locate the block that contains y and recursively search within that block. In the ith round, if the block containing y is $(x_l, \ldots, x_r)$, then we probe the position $s_i = l + \lceil \frac{(y - x_l)(r - l)}{x_r - x_l} \rceil$. We then try to locate the $n^{1/2^i}$-sized block by sequentially probing every $n^{1/2^i}$th element starting from s_i.

Show that the expected number of probes is $O(\log \log n)$.

Hint: Analyze the expected number of probes in each round using Chebychev's inequality.

6.16 **Deferred data structure** When we build a dictionary data structure for fast searching, we expend some initial overheads to build this data structure. For example, we need $O(n \log n)$ time to sort an array so that we can do searching in $O(\log n)$ time. If there were only a few keys to be searched, then the preprocessing time may not be worth it since we can do a brute force search in $O(n)$ time which is asymptotically less that $O(n \log n)$.

If we include the cost of preprocessing into the cost of searching, then the total cost for searching k elements can be written as $\sum_{i=1}^{k} q(i) + P(k)$, where $q(i)$ represents the cost of searching for the ith element and $P(k)$ is the preprocessing time for the first k elements. For each value of k, balancing the two terms would give us the best performance. For example, for $k = 1$, we may not build any data structure but do a brute force search to obtain n. As k grows large, say becomes n, we may want to sort. Note that k may not be known in the beginning, so we may want to build the data structure in an incremental manner. After the first brute force search, it makes sense to find the median and partition the elements.

Describe an algorithm to extend this idea so as to maintain a balance between the number of keys searched and the preprocessing time.

Multidimensional Searching and Geometric Algorithms

Searching in a dictionary is one of the most primitive kind of search problems and it is relatively simple because of the property that the elements can be ordered. One can arrange the elements in an ordered manner (say, sorted array, or balanced search trees) such that one can perform search efficiently. Suppose the points are from the d dimensional space $\mathbb{R}^d$. One way to extend the techniques used for searching in a dictionary is to build a data structure based on lexicographic ordering. That is, given two d dimensional points p and p', there is a total ordering defined by the relation

$$p \prec p' \Leftrightarrow \exists j \leq d : x_1(p) = x_1(p'), x_2(p) = x_2(p') \ldots x_{j-1}(p) = x_{j-1}(p'), x_j(p) < x_j(p')$$

where $x_i(p)$ is the ith coordinate of point p. For example, if $p = (2.1, 5.7, 3.1)$ and $p' = (2.1, 5.7, 4)$, then $p \prec p'$. If we denote a d dimensional point by $(x_0, x_1, \ldots, x_d)$. Then given a set of n d-dimensional points, the immediate predecessor of a query point can be determined in $O(d \cdot \log n)$ comparisons using a straightforward adaption of the binary search algorithm (note that the extra d factor is because comparing two points will take $O(d)$ time). With a little more thought we can try to improve as follows. When two d-tuples are compared, we can keep track of the maximum prefix length (the index j), which is identical in the two tuples. If we create a binary search tree (BST) to support binary search, we also need to keep track of the largest common prefixes between the parent and the children nodes so that we can find the common prefixes between the query tuple and the internal nodes in a way that we do not have to repeatedly scan the same coordinates of the query tuple. For example, if the root node has a common prefix of length 10 with the left child and the query tuple has 7 fields common with the root node (and is smaller), it is clearly smaller than the left child also. The reader is encouraged to solve Exercise 7.1 at the end of this chapter. Queries can of course be far more sophisticated than just point queries but we will focus on the simpler case of searching for points specified by a range (or rectangle).

7.1   Interval Trees and Range Trees

In this section, we describe two commonly used data structures which partition the search space in a suitable manner and store points in each 'cell' of this partition. We first consider the problem of range search in one dimension, and then generalize it to higher dimensions.

Subsequently, we will see how these search algorithms can be used as building blocks for maintaining a convex hull of a set of points.

One-dimensional range searching

Given a set S of n points on a line (without loss of generality, say the x-axis), we have to build a data structure to report the points inside a given query interval $[x_\ell : x_u]$. The *counting* version of range query only reports the number of points in this interval instead of the points themselves.

Let $S = \{p_1, p_2, \ldots, p_n\}$ be the given set of points on the real line. We can solve the one-dimensional range searching problem using a balanced binary search tree T in a straightforward manner. The leaves of T store the points of S and the internal nodes of T store *splitters* to guide the search. Let the splitter-value at node v be denoted by x_v, then the left subtree $T_L(v)$ of a node v contains all the points smaller than or equal to x_v and right subtree $T_R(v)$ contains all the points strictly greater than x_v. Note that x_v can be chosen as any x-coordinate between the rightmost point in $T_L(v)$ and the leftmost point in $T_R(v)$. It is easy to see that we can build a balanced tree by balancing the sizes of $T_R(v)$ and $T_L(v)$.

To report the points in the range query $[x_\ell : x_u]$ we search with x_ℓ and x_u in the tree T. Let ℓ_1 and ℓ_2 be the leaves where the searches end. Then the points in the interval $[x_\ell : x_u]$ are the points stored between the leaves ℓ_1 and ℓ_2.

Another way to view the set of points is the union of the leaves of some subtrees of T[1]. If you examine the search path of x_ℓ and x_u, they share a common path from root to some vertex (may be the root itself), where the paths fork to the left and right – let us call this the *forking* node. The output points correspond to the union of leaves of the right subtrees of the left path and the left subtrees of the right path. This can be formally proved to be the union of at most $2\log n$ subtrees – see Exercise 7.2 and also illustrated in Figure 7.1. If

---

[1] This perspective will be useful for the later extensions.

each node also stores a count of the number of leaf nodes in the subtree then the *counting range query* can be answered as the sum of these $2 \log n$ counters. Since the query interval extends across both sides of the forking node, the left and right paths (also called left and right spines) define a set of subtrees that fall within the span of the query interval. In particular, the right (left) subtrees attached to the left (respectively right) spine correspond to disjoint half-intervals on either side of the forking node. Note that the values x_ℓ or x_u may not be present among the leaf nodes; they can represent some over-hanging portions and be effectively ignored beyond the successor leaf node (of x_ℓ) and predecessor leaf node of x_u.

Complexity The tree uses $O(n)$ space and it can be constructed in $O(n \log n)$ time. Each query takes $O(\log n + k)$ time, where k is the number of points in the interval, that is, the output size. The *counting query* takes $O(\log n)$ time (if we also store the number of nodes stored in the sub-tree rooted at each node). This is clearly the best we can hope for.

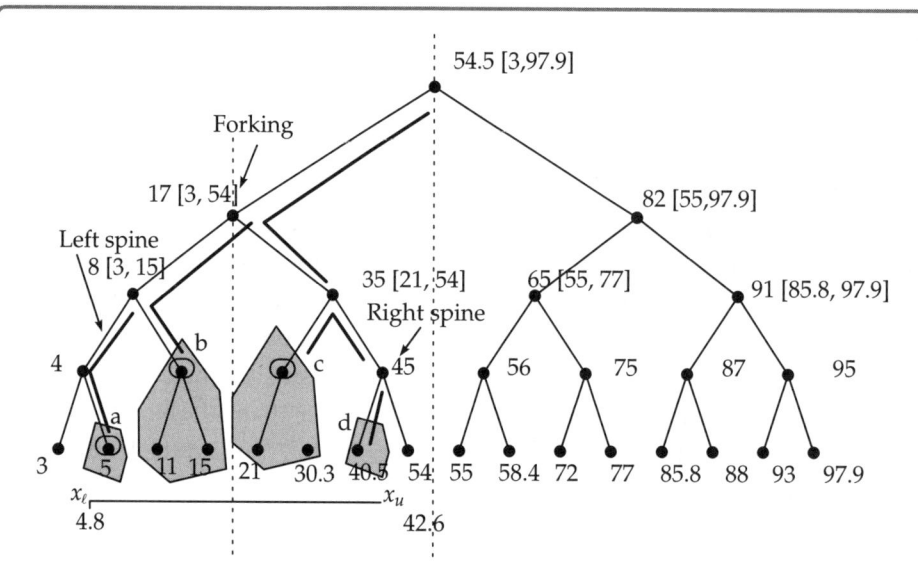

Figure 7.1   *The structure of a one-dimensional range search tree where a query interval is split into at most 2log n disjoint canonical (half)-intervals. Each node is associated with an interval $[\ell, r]$ that corresponds to the leftmost and rightmost coordinates of the points and a splitting coordinate between the left and right subtrees. The query interval $[x_\ell, x_u]$ traces out a search path starting from the root to a forking node, from where it defines a left and a right spine. The union of right subtrees a, b (shaded in the figure) attached to the left path and the left subtrees c, d attached to the right path gives us the disjoint intervals.*

7.1.1 Two-dimensional range queries

We now extend these ideas to the two-dimensional case. Each point has two attributes: its x coordinate and its y coordinate – the two-dimensional range query is a Cartesian product of two one-dimensional intervals. Given a query $[x_\ell : x_u] \times [y_\ell : y_u]$ (a two-dimensional rectangle), we want to build a data structure to report the points inside the rectangular region or alternately, count the number of points in the region.

We extend the previous one-dimensional solution by first considering the vertical slab $[x_\ell : x_u]$. [2] Let us build the one-dimensional range tree identical to the previous scheme (by ignoring the y coordinates of points). Therefore, we can obtain the answer to the *slab query*. As we had observed, every internal node represents an interval in the one-dimensional case and analogously the corresponding vertical slab in the two-dimensional case. The answer to the original query $[x_\ell : x_u] \times [y_\ell : y_u]$ is a subset of $[x_\ell : x_u] \times [-\infty : +\infty]$. Since our objective is to obtain a time-bound proportional to the final output, we cannot afford to list out all the points of the vertical slab. However, if we had the one-dimensional data structure available for this slab, we can quickly find out the final points by doing a range query with $[y_\ell : y_u]$. A naive scheme will build the data structure for all possible vertical slabs that can be $\Omega(n^2)$. We can do much better by observing that we need to worry about only those vertical slabs which correspond to an internal node in the tree – we shall call such vertical slabs as *canonical slabs*.

Each canonical slab corresponds to the vertical slab (the corresponding $[x_\ell : x_u]$) spanned by an internal node. We can therefore build a one-dimensional range tree for all the points spanned by the corresponding vertical slab – this time in the y-direction (and associate it with each of the internal nodes). As in the previous section, we can easily show that each vertical slab is a union of $2\log n$ **canonical** slabs (Figure 7.2). So, the final answer to the two-dimensional range query is the union of at most $2\log n$ one-dimensional range queries, giving a total query time of $\sum_{i=1}^{t} O(\log n + k_i)$, where k_i is the number of output points in slab i among t slabs and $\sum_i k_i = k$. This results in a query time of $O(t \log n + k)$, where t is bounded by $2\log n$. The space is bounded by $O(n \log n)$ since in a given level of the tree T, a point is stored exactly once.

The natural extension of this scheme leads us to d-dimensional range search trees with the following performance parameters.

$$Q(d) \leq \begin{cases} 2\log n \cdot Q(d-1) & \text{for } d \geq 2 \\ O(\log n) & d = 1 \end{cases} \tag{7.1.1}$$

---

[2] We can think about $[y_\ell : y_u]$ as $[-\infty : +\infty]$.

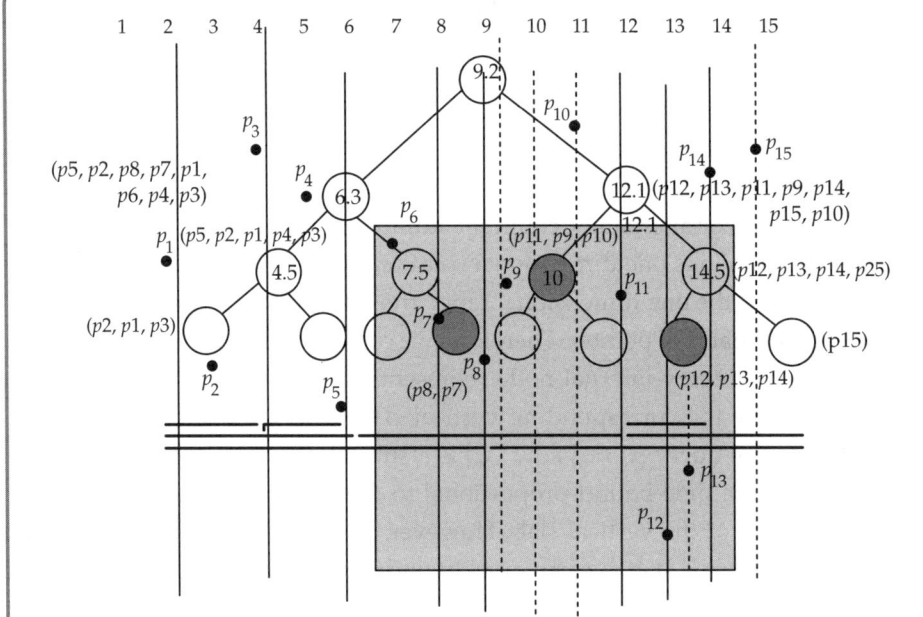

Figure 7.2   *The rectangle is the union of the slabs represented by the darkened nodes plus an overhanging left segment containing p_6. The sorted list of points in the y direction is indicated next to the nodes – not all the lists are shown. The number inside the node of a tree indicates the splitting coordinate of the interval that defines the left and right subintervals corresponding to the left and right children.*

where $Q(d)$ is the query time in d dimensions for n points. This yields $Q(d) = O(2^d \cdot \log^d n)$. A more precise recurrence can be written in terms of n, d.

$$Q(n,d) \leq \begin{cases} 2\sum_i Q(\frac{n}{2^i}, d-1) & \text{for } d \geq 2 \\ O(\log n) & d = 1 \end{cases} \tag{7.1.2}$$

since the number of points in a node at distance i from the root has at most $\frac{n}{2^i}$ points. The reader may want to find a tight solution of this recurrence (Exercise 7.5). The number of output points k can be simply added to $Q(n,d)$ since the subproblems output disjoint subsets.

7.2   *k–d* Trees

A serious drawback of range trees is that both the space and the query time increases exponentially with dimensions. Even for two dimensions, the space is super-linear. For many applications, we cannot afford to have such a large blow-up in space (for a million records, $\log n = 20$).

Let us do a divide-and-conquer on the set of points – we partition the space into regions that contain a subset of the given set of points. The input rectangle is tested against all the regions of the partition. If it does not intersect a region U, then we do not search further. If U is completely contained within the rectangle, then we report all the points associated with U; otherwise we search recursively in U. We may have to search in more than one region – we define a search tree where each region is associated with a node of the tree. The leaf nodes correspond to the original point set. In general, this strategy will work for other (than rectangular) kinds of regions also.

We now give the construction of the k–d tree. Each node v of the tree will have an associated rectangular region $R(v)$ which will consist of at least one input point. The root node of the tree will correspond to a (bounded) rectangle that contains all the n points. Consider a node v. Two cases arise depending on whether the depth of v is odd or even. If depth of v is even, we split the rectangle $R(v)$ by a vertical line into two smaller rectangles, $R_1(v)$ and $R_2(v)$. We add two child nodes v_1 and v_2 to v and assign them the corresponding rectangles – note that if one of these smaller rectangles is empty (i.e., has no input point), we do not partition it further. To create a balanced tree, we choose the vertical line whose x coordinate is the median x coordinate of the points in $R(v)$. Similarly, if the depth of v is odd, we split along a horizontal line. Thus, the levels in the tree alternate between splitting based on x coordinates and y coordinates (Figure 7.3).

Since a point is stored exactly once and the description of a rectangle corresponding to a node takes $O(1)$ space, the total space taken up by the search tree is $O(n)$. Figure 7.4 illustrates k–d tree data structure.

Procedure Search(Q, v)

1   **if** $R(v) \subset Q$ **then**
2   report all points in $R(v)$
3   **else**
4   Let $R(u)$ and $R(w)$ be rectangles associated with the children
      u, w; /* if v is a leaf node then exit */
5   **if** $Q \cap R(u)$ *is non-empty* **then**
6   Search(Q, u)
7   **if** $R \cap R(w)$ *is non-empty* **then**
8   Search (Q, w)

Figure 7.3 *Rectangular range query used in a k–d tree*

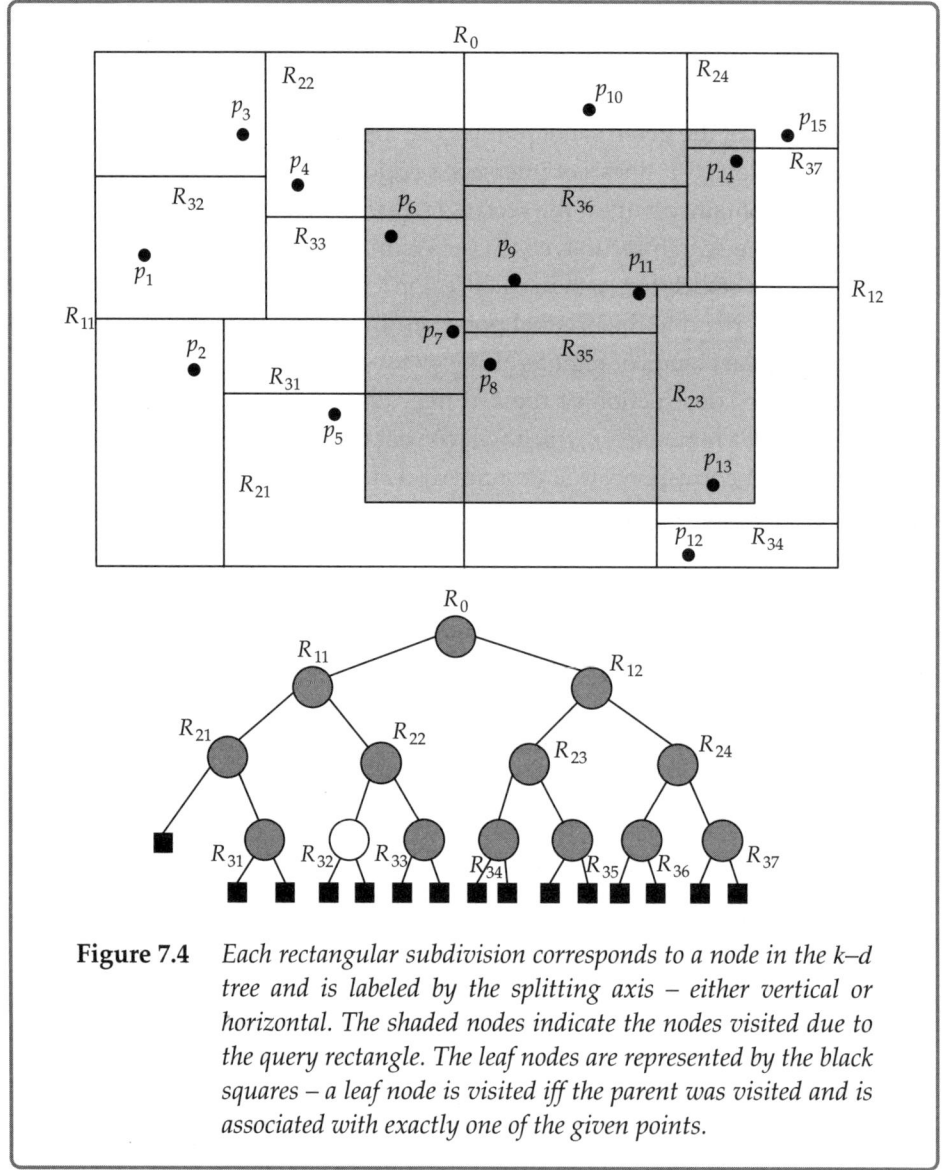

Figure 7.4   *Each rectangular subdivision corresponds to a node in the k–d tree and is labeled by the splitting axis – either vertical or horizontal. The shaded nodes indicate the nodes visited due to the query rectangle. The leaf nodes are represented by the black squares – a leaf node is visited iff the parent was visited and is associated with exactly one of the given points.*

The problem of reporting all points in a query rectangle can be solved as follows (see Figure 7.4). We search a subtree rooted at a node v iff the query rectangle intersects the rectangle $R(v)$ associated with node v. This involves testing if the two rectangles (the query rectangle and $R(v)$) overlap, which can be done in $O(1)$ time. We start at the root and recursively traverse its descendants – if $R(v)$ has no overlap with the query rectangle, we do not proceed to the descendant of v. Similarly, if $R(v)$ is contained in the query rectangle, we can just report all the points contained in $R(v)$. When the traversal reaches a leaf, we have to check whether the point stored at the leaf is contained in the query region and, if so, report it.

Let us now compute the query time of this algorithm. Let Q be the query rectangle. Our algorithm will explore a node v only if $R(v)$ intersects Q, in particular one of the sides of Q. For each of the sides of Q, we will separately count the number of nodes v such that $R(v)$ intersects Q. Consider one of the four sides of Q; assume that is one of the vertical sides, and let L denote the vertical line along this side.

Let $Q(i)$ be the number of nodes v at distance i from the root for which $R(v)$ intersects L. Consider a node v at level i such that $R(v)$ intersects L. Among its four children at level $i+2$, at most two will intersect L, corresponding to the level that partitions using a horizontal cut. Therefore, we get the recurrence:

$$Q(i+2) \leq 2Q(i) + O(1)$$

It is easy to check that $Q(i)$ is $O(2^{\lfloor i/2 \rfloor})$, and so, the total number of nodes which intersect L is $O(\sqrt{n})$. Arguing similarly for each of the sides of Q, we see that the algorithm visits $O(\sqrt{n})$ nodes. For nodes which are completely contained in the query rectangle, we simply report all the points in it. Therefore, the running time is $O(\sqrt{n}+k)$, where k is the number of points which are contained in the query rectangle.

7.3 Priority Search Trees

As we learned in the previous chapter, the combination of BST with heap property resulted in a simple strategy for maintaining balanced search trees called treaps. The heap property was useful to keep a check on the expected height of the tree within $O(\log n)$. What if we want to maintain a heap explicitly on a set of parameters (say the y coordinates) along with a total ordering required for binary search on the x coordinates? Such a data structure would be useful to support a *three sided* range query in linear space.

A *three sided* query is a rectangle $[x_\ell : x_u] \times [y_\ell : \infty]$, that is, a semi-infinite vertical slab.

If we had a data structure that is a BST on x coordinates, we can first locate the two points x_ℓ and x_u to determine (at most) $2\log n$ subtrees whose union contains the points in the interval $[x_\ell : x_u]$. Say, these are $T_1, T_2, \ldots, T_k$. Within each such tree T_i, we want to find the points whose y coordinates are larger than y_l. If T_i forms a *max-heap* on the y coordinates, then we can output the points as described in the procedure Search (v).

The procedure is called with v being the root of the max-heap. In general, the procedure returns the set of nodes in the sub-tree rooted below v whose y-coordinates is at least y_ℓ. Since v is the root of max-heap, if $v_y < y_\ell$, then all the descendents of v also satisfy this property. Therefore, we do not need to search any further. This establishes correctness of the search procedure. Let us mark all the nodes that are visited by the procedure in the second phase. When we visit a node in the second phase, we either output a point or terminate the search. For the nodes that are output, we can charge it to the output size. For the nodes that are not output, let us add a charge to its parent – the maximum charge to a node is two because of its two children. The first phase takes $O(\log n)$ time to determine the canonical sub-intervals and so the total search time is $O(\log n + k)$, where k is the number of output points[3].

Procedure Search(v)

1   Let v_y denote the y coordinate associated with a node v.;
2   **if** $v_y < y_\ell$ *or v is a leaf node* **then**
3 | terminate search
4   **else**
5 | **if** $v_x \geq x_\ell$ **then**
6 | Output the point associated with v ;
7 | Search(u) where u is the left child of v.;
8 | Search (w) where w is the right child of v

Until now, we assumed that such a dual-purpose data structure exists. How do we construct one?

First we can build a leaf-based BST on the x coordinates. Next, we promote the points according to the heap ordering. If a node is empty, we inspect its two children and pull up the larger value. We terminate when no value moves up. Alternately, we can construct the tree as follows.

Procedure Build Priority Search Tree(S)

1   **Input** A set S of n points in plane. ;
2   **Output** A priority search tree ;
3   Let $p \in S$ be the largest y coordinate. Store y in the root r ;
4   **if** $S - p$ *is non-empty* **then**
5 | Let L (respectively R) be the left (respectively right) half of the points in $S - p$ with separating vertical line having x coordinate $m(L, R)$;
6 | Set $X(r) = m(L, R)$ in root r ;
7 | Build Priority Search Tree (L) ;
8 | Build Priority Search Tree (R) ;
9 | **Comment** : The left (right) subtree will be searched iff the query interval extends to the left (right) of $X(r)$

---

[3] This kind of analysis where we are amortizing the cost on the output points is called a *filtering* search.

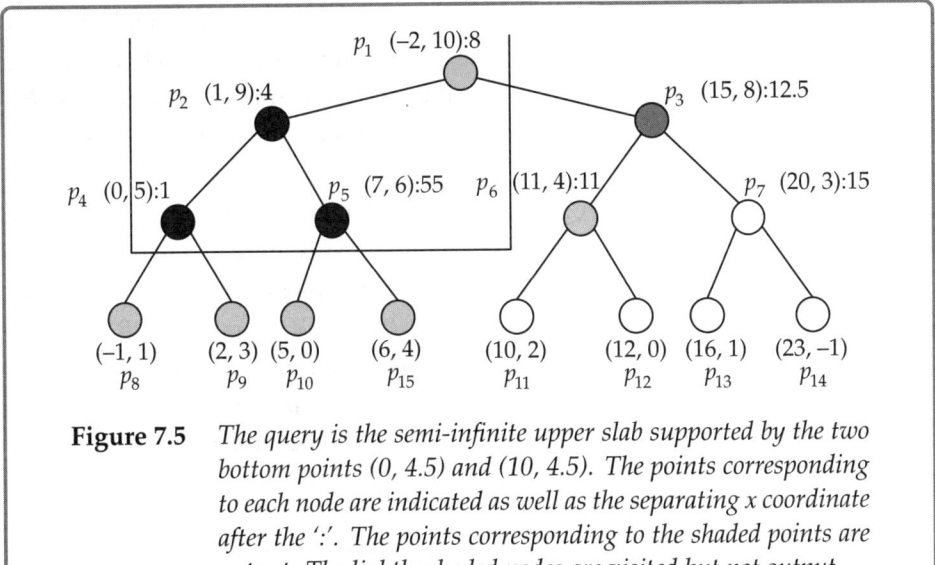

Figure 7.5 *The query is the semi-infinite upper slab supported by the two bottom points (0, 4.5) and (10, 4.5). The points corresponding to each node are indicated as well as the separating x coordinate after the ':'. The points corresponding to the shaded points are output. The lightly shaded nodes are visited but not output.*

Since the number of points reduce by half in every subtree, the height of this tree is clearly $O(\log n)$. This *combo* data structure is known as a *priority search tree*; it takes only $O(n)$ space and supports a $O(\log n + k)$ time *three sided query*. Figure 7.5 illustrates query processing in a priority search tree.

7.4 Planar Convex Hull

In this section, we consider one of the most fundamental problems in computational geometry – computing the convex hull of a set of points. We first define this concept. A subset of points S in the plane (or in any Euclidean space) is said to be convex, if for every pair of points $x, y \in S$, the points lying in the line segment joining x and y also lie in S (see Figure 7.6 for examples of convex and non-convex sets). It is easy to check that if S_1 and

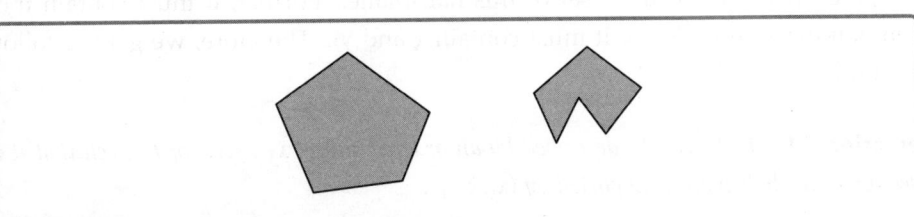

Figure 7.6 *The figure on the left is convex, whereas the one on the right is not convex.*

S_2 are two convex sets, then $S_1 \cap S_2$ is also convex (though $S_1 \cup S_2$ may not be convex). In fact, arbitrary (i.e., even uncountably infinite) intersection of convex sets is also convex. Thus, given a set of points P, we can talk about the *smallest* convex set containing P – this is the intersection of all convex sets containing P. This set is called the convex hull of P (see Figure 7.7 for an example).

This definition is difficult to work with because it involves intersection of an infinite number of convex sets. Here is another definition which is more amenable to a finite set of points in the plane. Let P be a finite set of points in the plane. Then the convex hull of P is a convex polygon whose vertices are a subset of P, and all the points in P are contained inside (or on the boundary of) P. A polygon is said to be convex if it is also a convex set. The two definitions are equivalent for points lying on the plane. Thus, we can compute the convex hull of such a set of points by figuring out the subset of points which lie on the boundary of this convex polygon.

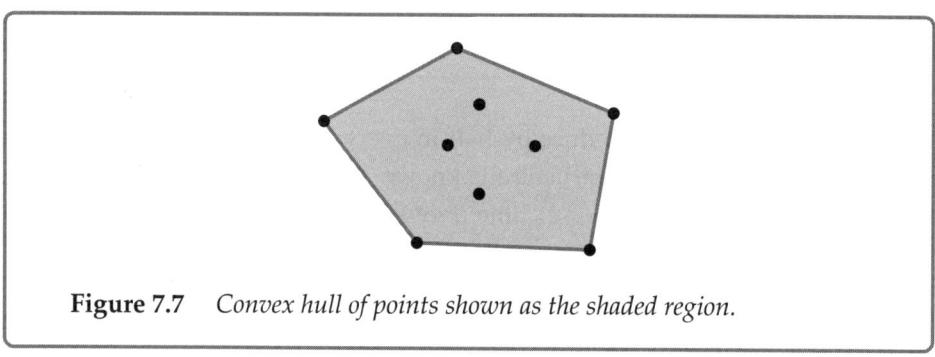

Figure 7.7   *Convex hull of points shown as the shaded region.*

In this section, we restrict our attention to points lying on the plane; in which case, the convex hull is also denoted as a planar convex hull. Given a set of n points P, we use $CH(P)$ to denote the convex hull of P. The equivalence of these two definitions can be seen as follows. Let x, y be two distinct points in P such that all the points in P lie on one side of the line joining x and y. Then, the half-plane defined by this line is a convex set containing P, and so, $CH(P)$ must be a subset of this half-plane. Further, it must contain the line segment joining x and y (since it must contain x and y). Therefore, we get the following observation.

Observation 7.1 *$CH(P)$ can be described by an ordered subset $x_1, x_2, \ldots$ of P, such that it is the intersection of the half-planes supported by (x_i, x_{i+1}).*

We can also assume that any three consecutive points, x_i, x_{i+1}, x_{i+2}, are not collinear; otherwise, we can remove x_i from the list. Under this assumption, each of the points x_i form an extreme point of the convex hull – a point $x \in CH(P)$ is said to be extreme if it does not lie on the (interior of) a line segment joining any two points $CH(P)$. Therefore, $CH(P)$ is a sequence of extreme points and the edges joining those points.

Let x_L and x_R be the points with the smallest and the largest x coordinate (we assume by a simple tie-breaking rule that no two points have the same x coordinate) – these are also called the left-most and the right-most points. For building the convex hull, we divide the points by a diagonal joining x_L and x_R. The points above the diagonal form the *upper hull* and the points below form the *lower hull*. We also rotate the hull so that the diagonal is parallel to the x-axis. We will describe algorithms to compute the upper hull – computing the lower hull is analogous.

The planar convex hull is a two-dimensional problem and it cannot be done using a simple comparison model. While building the hull, we will need to test whether three points $(x_0, y_0), (x_1, y_1)$, and (x_2, y_2) are clockwise (counter-clockwise) oriented. Since the x coordinates of all the points are ordered, all we need to do is test whether the middle point is above or below the line segment formed by the other two. A triplet of points (p_0, p_1, p_2) is said to form a *right turn* iff the determinant

$$\begin{vmatrix} x_0 & y_0 & 1 \\ x_1 & y_1 & 1 \\ x_2 & y_2 & 1 \end{vmatrix} < 0$$

where (x_i, y_i) are the coordinates of p_i. If the determinant is positive, then the triplet points form a left turn. If the determinant is 0, the points are collinear.

We now describe some simple algorithms before delving into the more efficient quickhull algorithm.

7.4.1 Jarvis march

A very intuitive algorithm for computing convex hulls simply simulates wrapping a rope around the boundary points (*or gift wrapping*). It starts with any extreme point, say x_L, and repeatedly finds the successive points in clockwise direction by choosing the point with the least polar angle with respect to the positive horizontal ray from the first vertex. The algorithm is described in Figure 7.8. It maintains the invariant that for all i, $p_0, \ldots, p_i$ form contiguous vertices on the upper convex hull.

The algorithm runs in $O(nh)$ time, where h is the number of extreme points in $CH(P)$. Note that we actually never compute angles; instead, we rely on the determinant method to compare the angle between two points, to see which is smaller. To the extent possible, we only rely on algebraic functions when we are solving problems in $\mathbb{R}^d$. Computing angles require inverse trigonometric functions that we avoid.

When h is $o(\log n)$, Jarvis march is asymptotically faster than Graham's scan, which is described next.

1 **Input** A set of points in the plane such that x_L and x_R form a horizontal line.

2 Initialize $p_0 = x_L, p_{-1} = x_R$.

3 Initialize $i = 0$.

4 **Repeat**

5 find the input point p such that the angle $p_{i-1} p_i p_j$ is largest.

6 Set $p_{i+1} \leftarrow p, i \leftarrow i+1$.

7 Until $p_i = x_R$.

Figure 7.8 *Jarvis March algorithm for convex hull*

7.4.2 Graham's scan

We assume that we need to produce the upper convex hull only, and x_L, x_R form a horizontal line. In this algorithm, points in the input set P are first sorted in order of their x coordinate. Let this order be $p_1, p_2, \ldots, p_n$. It considers the points in this order, and after considering p_i, it stores a subsequence of $p_1, \ldots, p_i$ which form the convex hull of the points $\{p_1, \ldots, p_i\}$. It stores this subsequence in a stack – the bottom of the stack will be p_0 and the top will be p_i. When it considers p_{i+1}, the invariant is maintained as follows. It starts popping points from the stack as long as p_{i+1} and the top two points in the stack form a left turn. When the top two points in the stack and p_{i+1} form a right turn, it pushes p_{i+1} on the stack. It is left as an exercise to show that this algorithm outputs the upper convex hull.

The points in the set P are first sorted using their x coordinate in $O(n \log n)$ time and then inductively a convex chain of extreme points is constructed. For the upper hull, it can be easily seen that a convex chain is formed by successive right turns as we proceed in the clockwise direction from the left-most point. When we consider the next point (in increasing x coordinates), we test if the last three points form a convex sub-chain, that is, they make a right turn. If so, we push it into the stack. Otherwise, the middle point among the triplet is discarded (Why?) and the last three points (as stored in the stack) are tested for right-turn. It stops when the convex sub-chain property is satisfied.

Let us now analyze the running time of this algorithm. Sorting the points according to their x coordinate takes $O(n \log n)$ time. Further, each point is pushed on the stack only once, and once it is popped, it is not pushed again. So, each element is popped from the stack at most once. Thus, the total time for stack operations is $O(n)$. Note that the running time is dominated by the time to sort.

7.4.3 Sorting and convex hulls

There is a very close relationship between sorting and convex hulls and we can reduce sorting to convex hull in the following manner. Suppose we want to sort the numbers $x_1, \dots, x_n$ in increasing order. Consider the parabola $y = x^2$, where we map each point x_i to a point $p_i = (x_i, x_i^2)$ on this parabola. Note that all of these points are extreme points and the ordered set of points on the convex hull is same as the points sorted by x_i values. Thus, any algorithm for constructing a convex hull would be required to sort these points.

In fact, it is hardly surprising that almost all sorting algorithms have a counterpart in the world of convex hull algorithms. An algorithm based on a divide-and-conquer paradigm which works by arbitrary partitioning is called **merge hull**. The idea is similar to merge sort. We first partition the points into two arbitrary subsets of equal size. We recursively construct the upper hull of both the subsets. We would now like to merge the two upper hulls in $O(n)$ time. If we could achieve this, then the running time would obey the recurrence

$$T(n) = 2T\left(\frac{n}{2}\right) + O(n),$$

which would imply $O(n \log n)$ running time.

The key step here is to merge the two upper hulls in $O(n)$ time. Note that the two upper hulls are not necessarily separated by a vertical line, and could be intersecting each other. The merge step computes the common tangent, called bridge over line L, of these two upper hulls, as shown in Figure 7.9. For the separated hulls, the merge step computes the common tangent, called *bridge* of these two upper hulls, as shown in Figure 7.9(a). We leave it as an exercise to find this bridge in $O(n)$ time – the idea is similar to merge sort – we scan both the upper hulls in left to right order while computing the merged hull.

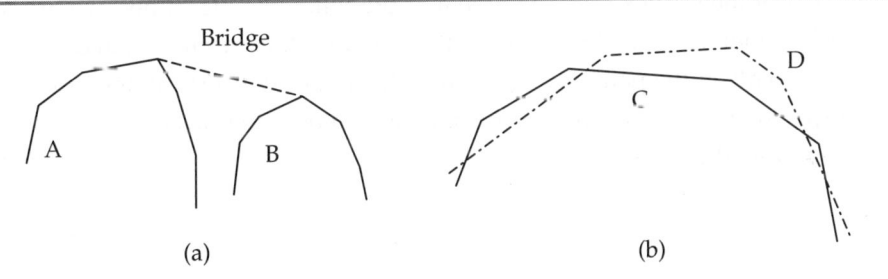

Figure 7.9 *Merging upper hulls – for the separated case in (a) we compute the bridge joining the two hulls. For the non-separable case depicted in (b) the two boundaries could have multiple intersections and can be computed in a sequence.*

7.5 Quickhull Algorithm

We describe an algorithm for convex hull which is based on ideas from the quick-sort algorithm – partitioning points into two disjoint subsets based on some criteria, recursively solving each sub-problem, and then combining the two solutions should be easy. Recall that quick-sort can take $O(n^2)$ time in the worst-case scenario, and one needs to use randomization (or a median-finding algorithm) to ensure that it runs in $O(n \log n)$ time. We will see that similar ideas are needed here as well.

Let S be the set of n points whose convex hull has to be constructed. As before, we compute the convex hull of S by constructing the upper and the lower hulls of S. Let p_l and p_r be the extreme points of S in the x direction. Let S_a (S_b) be the subset of S which lies above (below) the line through p_l and p_r. As we had noted previously, $S_a \cup \{p_l, p_r\}$ and $S_b \cup \{p_l, p_r\}$ determine the upper and the lower convex hulls of S respectively. We will describe the algorithm *Quickhull* to determine the upper hull using $S_a \cup \{p_l, p_r\}$.

We first give some definitions. The slope of the line joining p and q is denoted by $\mathtt{slope}(pq)$. The predicate *left-turn*(x, y, z) is true if the sequence x, y, z has a counter-clockwise orientation, or equivalently, the area of the triangle has a positive sign. Recall that this can be figured out by computing the determinant of a 3×3 matrix.

We first pair the points and pick a random pair. Consider the line joining these two points; we move it parallel to itself (i.e., its slope does not change) till all the points lie below it. Thus, this line will be *supported* by some input point and the rest will lie below it. Let p_m be such a point; we use p_m as a pivot to partition the points into two parts (as in quick-sort). Consider the vertical line containing p_m: points which lie to the left of p_m are in one half and those to the right are in the other half. Clearly, if we could construct the upper convex hulls of the two halves, then the upper hull of the entire set of points can be obtained by just combining these two upper hulls (because p_m is an extreme point and so will be part of the convex hull). In Step 4 of the following algorithm, we prune some points which cannot lie on the boundary of the upper hull. For example, in case (i) of Step 4, if both the points in the pair (p_{2j-1}, p_{2j}) lie to the left of the vertical line below p_m, and the triplet (p_m, p_{2j}, p_{2j-1}) forms a right turn, then we know that p_{2j} cannot be part of the upper hull, and so, discard it.

Algorithm   **Quickhull**(S_a, p_l, p_r)

Input: Given $S_a = \{p_1, p_2, \ldots, p_n\}$ and the leftmost extreme point p_l and the rightmost extreme point p_r. All points of S_a lie above the line $\overline{p_l p_r}$.

Output: Extreme points of the upper hull of $S_a \cup \{p_l, p_r\}$ in clockwise order.

Step 1.   If $S_a = \{p\}$, then return the extreme point $\{p\}$.

Step 2.   Select randomly a pair $\{p_{2i-1}, p_{2i}\}$ from the the pairs $\{p_{2j-1}, p_{2j}\}$, $j = 1, 2, \ldots, \lfloor \frac{n}{2} \rfloor$.

Step 3.   Select the point p_m of S_a which supports a line with slope $(p_{2i-1}p_{2i})^a$. (If there are two or more points on this line then choose a p_m that is distinct from p_l and p_r). Assign $S_a(l) = S_a(r) = \emptyset$).

Step 4.   For each pair $\{p_{2j-1}, p_{2j}\}$, $j = 1, 2, \ldots, \lfloor \frac{n}{2} \rfloor$ do the following (assuming $x[p_{2j-1}] < x[p_{2j}]$)

        Case (i): $x[p_{2j}] < x[p_m]$
           if *left-turn* (p_m, p_{2j}, p_{2j-1}) then $S_a(l) = S_a(l) \cup \{p_{2j-1}, p_{2j}\}$
           else $S_a(l) = S_a(l) \cup \{p_{2j-1}\}$.

        Case (ii): $x[p_m] < x[p_{2j-1}]$
           if *left-turn* (p_m, p_{2j-1}, p_{2j}) then $S_a(r) = S_a(r) \cup \{p_{2j}\}$
           else $S_a(r) = S_b(r) \cup \{p_{2j-1}, p_{2j}\}$.

        Case (iii): $x[p_{2j-1}] < x[p_m] < x[p_{2j}]$
           $S_a(l) = S_a(l) \cup \{p_{2j-1}\}$;
           $S_a(r) = S_a(r) \cup \{p_{2j}\}$.

Step 5.   If $S_a(l) \neq \emptyset$ then $Quickhull(S_a(l), p_l, p_m)$.

        Output p_m.
        If $S_a(r) \neq \emptyset$ then $Quickhull(S_a(r), p_m, p_r)$.

---

[a]In the early versions of quickhull, the point p_m was chosen to be furthest from $\overline{p_l p_r}$. The reader may want to analyze its performance.

7.5.1 Analysis

To get a feel for the convergence of the algorithm *Quickhull*, we must argue that in each recursive call, some progress is achieved. This is complicated by the possibility that one of the endpoints can be repeatedly chosen as p_m. However, if p_m is p_l, then at least one point is eliminated from the pairs whose slopes are larger than the supporting line L through p_l. If

L has the largest slope, then there are no other points on the line supporting p_m (Step 3 of the algorithm). Then, for the pair (p_{2j-1}, p_{2j}), whose slope equals that of L, *left-turn* (p_m, p_{2j}, p_{2j-1}) is true, so p_{2j-1} will be eliminated (Figure 7.10). Hence, it follows that the number of recursive calls is $O(n)$, since each call leads to either an output vertex or to an elimination of at least one point.

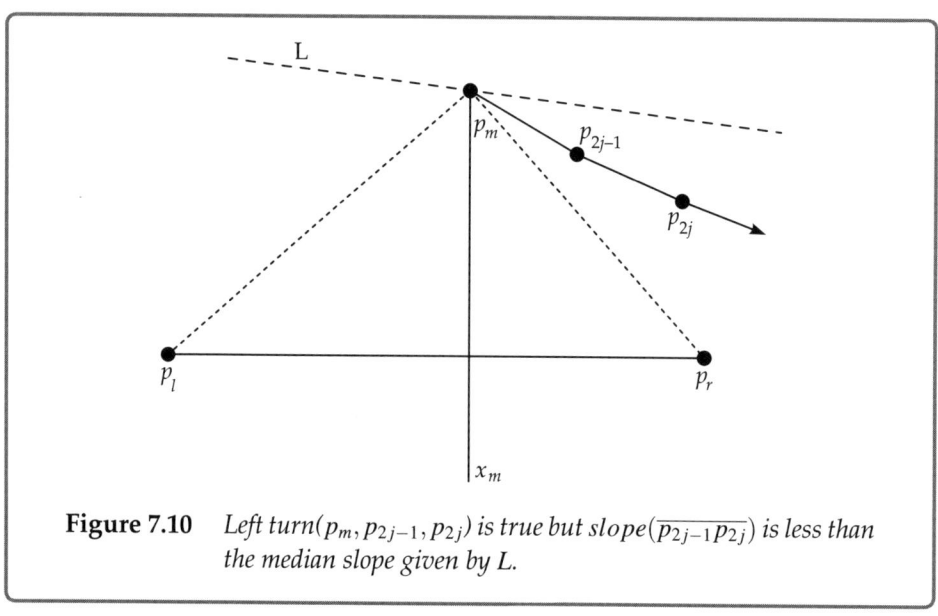

Figure 7.10   *Left turn(p_m, p_{2j-1}, p_{2j}) is true but slope$(\overline{p_{2j-1}p_{2j}})$ is less than the median slope given by L.*

Let N represent the set of $slopes(p_{2j-1}p_{2j})$, $j = 1, 2, \ldots \lfloor \frac{n}{2} \rfloor$. Let k be the rank of the $slope(p_{2i-1}p_{2i})$, selected uniformly at random from N in Step 2 of the algorithm. Let n_l and n_r be the sizes of the sub-problems determined by the extreme point p_m supporting the line with $slope(p_{2i-1}, p_{2i})$. We can show the following.

Observation 7.2 $\max(n_l, n_r) \leq n - \min(\lfloor \frac{n}{2} \rfloor - k, k)$.

Without loss of generality, let us bound the size of the right sub-problem. There are $\lfloor \frac{n}{2} \rfloor - k$ pairs with slopes greater than or equal to $slope(p_{2i-1}p_{2i})$. At most one point out of every such pair can be an output point to the right of p_m.

If we choose the median slope, that is, $k = \frac{n}{4}$, then $n_l, n_r \leq \frac{3}{4}n$. Let h be the number of extreme points of the convex hull and $h_l(h_r)$ be the extreme points of the left (right) sub-problem. We can write the following recurrence for the running time.

$$T(n, h) \leq T(n_l, h_l) + T(n_r, h_r) + O(n)$$

where $n_l + n_r \leq n$, $h_l + h_r \leq h - 1$. Exercise 7.10 requires you to show that the solution of this recurrence relation is $O(n \log h)$. Therefore, this achieves the right balance between Jarvis march and Graham's scan as it scales with the output size at least as well as Jarvis march and is $O(n \log n)$ in the worst case.

7.5.2 Expected running time*

Let $T(n, h)$ be the expected running time of the algorithm randomized *Quickhull* to compute hull vertices of a set of n points, given the extreme points p_l and p_r, where h denotes the number of vertices on the upper hull. Let $p(n_l, n_r)$ be the probability that the algorithm recursively calls two smaller size problems of sizes n_l and n_r containing h_l and h_r extreme vertices respectively. Therefore, we can write

$$T(n, h) \leq \sum_{\forall n_l, n_r \geq 0} p(n_l, n_r)(T(n_l, h_l) + T(n_r, h_r)) + bn \tag{7.5.3}$$

where $n_l, n_r \leq n - 1$, $n_l + n_r \leq n$, $h_l, h_r \leq h - 1$, $h_l + h_r \leq h$, and $b > 0$ is a constant. Here we are assuming that the extreme point p_m is not p_l or p_r. Although, in the *Quickhull* algorithm, we have not explicitly used any safeguards against such a possibility, we can analyze the algorithm without any loss of efficiency.

Lemma 7.1 $T(n, h) \in O(n \log h)$.

Proof: We will use the inductive hypothesis that for $h' < h$ and for all n', there is a fixed constant c, such that $T(n', h') \leq cn' \log h'$. For the case that p_m is not p_l or p_r, from Eq. (7.5.3) we get

$$T(n, h) \leq \sum_{\forall n_l, n_r \geq 0} p(n_l, n_r)(cn_l \log h_l + cn_r \log h_r) + bn$$

Since $n_l + n_r \leq n$ and $h_l, h_r \leq h - 1$,

$$n_l \log h_l + n_r \log h_r \leq n \log(h - 1) \tag{7.5.4}$$

Let $\mathcal{E}$ denote the event that $\max(n_l, n_r) \leq \frac{7}{8}n$ and p denote the probability of $\mathcal{E}$. Note that $p \geq \frac{1}{2}$.

From the law of conditional expectation, we have

$$T(n, h) \leq p \cdot [T(n_l, h_l | \mathcal{E}) + T(n_r, h_r) | \mathcal{E}] + (1 - p) \cdot [T(n_l, h_l + | \bar{\mathcal{E}}) + T(n_r, h_r | \bar{\mathcal{E}})] + bn$$

where $\bar{\mathcal{E}}$ represents the complement of $\mathcal{E}$.

When $\max(n_l, n_r) \leq \frac{7}{8}n$, and $h_l \geq h_r$,

$$n_l \log h_l + n_r \log h_r \leq \frac{7}{8}n \log h_l + \frac{1}{8}n \log h_r \tag{7.5.5}$$

The RHS of Eq. (7.5.5) is maximized when $h_l = \frac{7}{8}(h - 1)$ and $h_r = \frac{1}{8}(h - 1)$. Therefore,

$$n_l \log h_l + n_r \log h_r \leq n \log(h - 1) - tn$$

where $t = \log 8 - \frac{7}{8}\log 7 \geq 0.55$. We get the same bounds when $\max(n_l, n_r) \leq \frac{7}{8}n$ and $h_r \geq h_l$. Therefore,

$$
\begin{aligned}
T(n,h) &\leq p(cn\log(h-1) - tcn) + (1-p)cn\log(h-1) + bn \\
&= pcn\log(h-1) - ptcn + (1-p)cn\log(h-1) + bn \\
&\leq cn\log h - ptcn + bn
\end{aligned}
$$

Hence, from induction, $T(n,h) \leq cn\log h$ for $c \geq \frac{b}{tp}$.

If p_m is an extreme point (say p_l), we cannot apply Eq. (7.5.3) directly, but some points will still be eliminated according to Observation 7.2. This can happen a number of times, say $r \geq 1$, at which point, Eq. (7.5.3) can be applied. We will show that this is actually a better situation, that is, the expected time bound will be less and hence, the previous case dominates the solution of the recurrence.

The rank k of slope$(p_{2i-1}p_{2i})$ is uniformly distributed in $[1, \frac{n}{2}]$, so that the number of points eliminated is also uniformly distributed in the range $[1, \frac{n}{2}]$ from Observation 7.2. (We are ignoring the floor in $\frac{n}{2}$ to avoid special cases for odd values of n – the same bounds can be derived even without this simplification). Let $n_1, n_2, \ldots, n_r$ be the r random variables that represent the sizes of subproblems in r consecutive times; p_m is an extreme point. It can be verified by induction that $E[\sum_{i=1}^{r} n_i] \leq 4n$ and $E[n_r] \leq (3/4)^r n$, where $E[\cdot]$ represents the *expectation* of a random variable. Note that $\sum_{i=1}^{r} b \cdot n_i$ is the *expected* work done in r divide steps. Since $cn\log h \geq 4nb + c(3/4)^r \cdot n\log h$ for $r \geq 1$ (and $\log h \geq 4$), the previous case dominates. □

7.6   Point Location Using Persistent Data Structure

The *point location* problem involves an input planar partition (a planar graph with an embedding on the plane), for which we build a data structure such that given a point, we want to report the region containing the point. This fundamental problem has numerous applications including cartography, GIS, computer vision, etc.

The one-dimensional variant of the problem has a natural solution based on binary search – in $O(\log n)$ time, we can find the interval containing the query point. In two dimensions, we can also consider a closely related problem called *ray shooting*, in which we are given a set of line segments in the plane, we shoot a vertical ray and report the first segment that it hits. In the context of point location problem in a planar partition, this problem is relevant because each edge of the planar graph can be thought of as a line segment. Notice that every segment here borders two regions and we can use the ray shooting problem to report the region below this segment (see also Exercise 7.14). Consider a vertical slab which is criss-crossed by n line segments such that no pair of segments intersect within the slab. Given a query point, we can use binary search to answer a ray shooting query in $O(\log n)$ primitives of the following kind – Is the point below/above a line segment? This strategy works since the line segments are totally ordered within the slab (they may intersect outside).

This is illustrated in Figure 7.11.

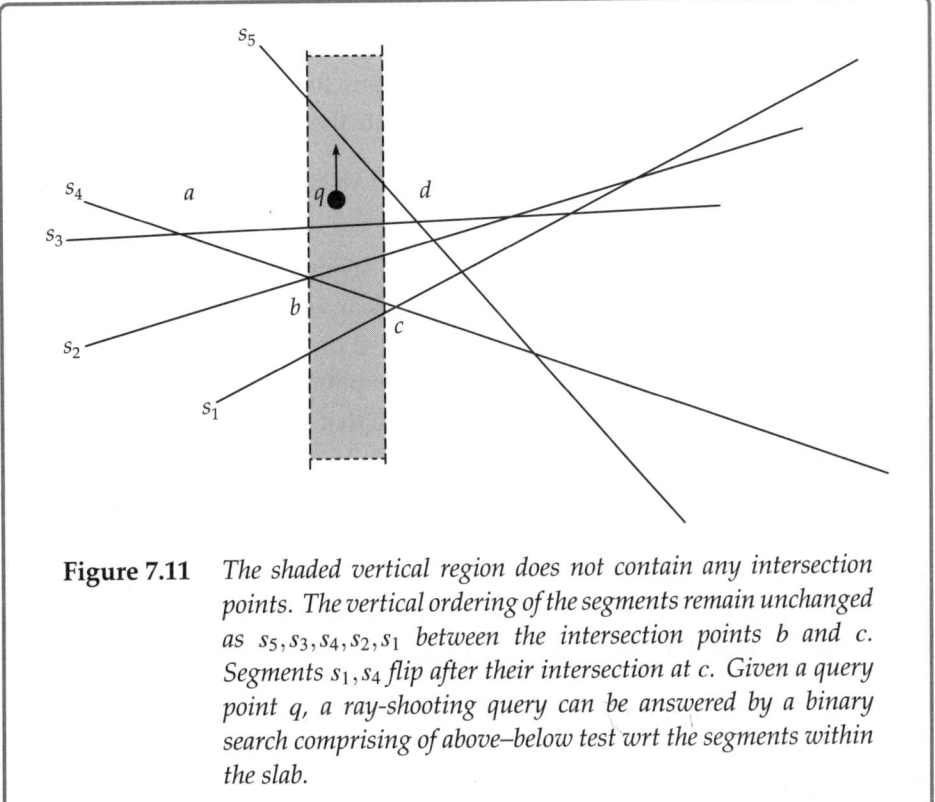

Figure 7.11   *The shaded vertical region does not contain any intersection points. The vertical ordering of the segments remain unchanged as s_5, s_3, s_4, s_2, s_1 between the intersection points b and c. Segments s_1, s_4 flip after their intersection at c. Given a query point q, a ray-shooting query can be answered by a binary search comprising of above–below test wrt the segments within the slab.*

For the planar partition, imagine a vertical line V being swept from left to right and let $V(x)$ represent the intersection of V with the planar partition at an X-coordinate value x. For simplicity, let us assume that no segment is vertical. Further, let us order the line-segments according to $V(x)$ and denote it by $S(x)$. While $V(x)$ is continuously changing as V sweeps across, $S(x)$ remains unchanged between consecutive intersection points. In Figure 7.11, $S(x)$ can be seen as s_5, s_3, s_4, s_2, s_1 where the query point q lies between s_3 and s_5 and the answer to the ray-shooting query is s_5.

Observation 7.3 *Between two consecutive (in X direction) end points of the planar partition, $S(x)$ remains unchanged.*

The region between two consecutive end points is a situation similar to the vertical slab discussed before. So once we determine which vertical slab contains the query point, in an additional $O(\log n)$ above–below tests, we can solve the ray shooting problem. Finding the vertical slab is a one-dimensional problem and can be answered in $O(\log n)$ steps involving a binary search. Therefore, the total query time is $O(\log n)$ but the space bound is not nearly as desirable. If we treat the $(2n-1)$ vertical slabs corresponding to the $2n$ end points, we are required to build $\Omega(n)$ data structures, each of which involves $\Omega(n)$ segments. Figure 7.12 depicts a worst-case scenario in terms of space. A crucial observation is that the two consecutive vertical slabs have almost all the segments in common except for the one whose end points separate the region.

Can we exploit the similarity between two ordered lists of segments and support binary search on both lists efficiently?

In particular, can we avoid storing the duplicate segments and still support $\log n$ step binary searches. Here is the intuitive idea. For each vertical slab, we are storing the segments in it in a balanced binary search tree (based on their order in terms of vertical direction). This will ensure that the search time for a point in this vertical slab is $O(\log n)$. Now, we can consider two adjacent slabs and notice that the set of segments can change as follows: one new segment can enter the set or an existing segment can go away. Let us assume that an element, i.e., a segment, is inserted in the adjacent vertical slab and we would like to maintain both versions of the tree (before and after the insertion). Observe that if a new node leaf v is inserted, then we may perform balancing operations to make the BST balanced again – however all of these operations will only affect the path from v to the root. We use the following idea to maintain both versions of the tree:

Path copying strategy If a node changes, then make a new copy of its parent and also copy the pointers to its children.

Once a parent is copied, it will lead to copying its parent, etc, until the entire root–leaf path is copied. At the root, create a label for the new root. Once we know which root node to start the binary search, we only need to follow pointers and the search proceeds in the normal way that is completely oblivious to fact that there are actually two *implicit* search trees (Fig. 7.13). The search time also remains unchanged at $O(\log n)$. The same strategy works for any number of versions except that to start searching at the correct root node, we may require an additional data structure. In the context of planar point location, we can build a binary search tree that supports one-dimensional search.

The space required is $(path\ length) \cdot (number\ of\ slabs) + n$ which is $O(n \log n)$. This is much smaller than the $O(n^2)$ scheme that stores each tree explicitly.

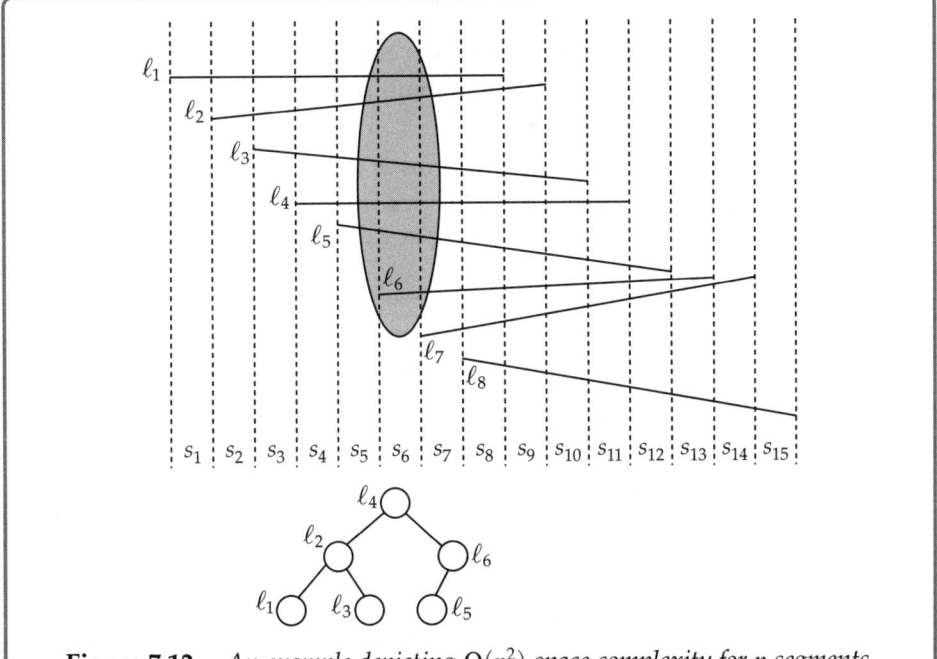

Figure 7.12 *An example depicting $\Omega(n^2)$ space complexity for n segments. The search tree corresponds to the slab s_5 and each node corresponds to an above–below test corresponding to the segment.*

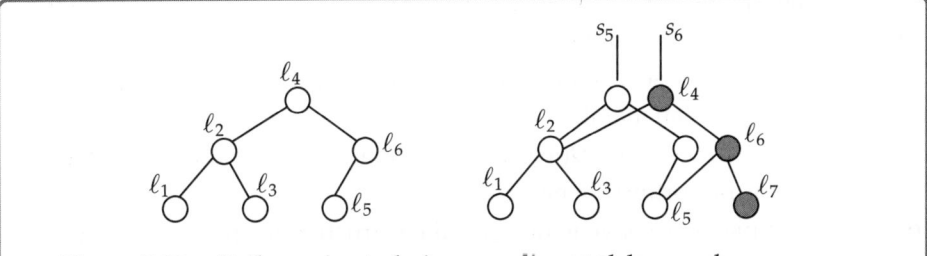

Figure 7.13 *Path copying technique on adjacent slabs s_5 and s_6.*

7.7 Incremental Construction

Given a set $S = \{p_1, p_2, \ldots, p_n\}$ of n points on a plane, we want to find a pair of points $q, r \in S$ such that $d(p,q) = \min_{p_i, p_j \in S} d(p_i, p_j)$, where $d()$ computes the Euclidean distance between two points. The pair (q, r) is known as the closest pair and it may not be unique. Moreover, the closest pair has distance zero if the points are not distinct.[4]

---

[4] So the lower bound for element distinctness would hold for the closest pair problem.

In one dimension, it is easy to compute the closest pair by first sorting the points and choosing an adjacent pair which has minimum separation. A trivial algorithm is to compute all the $\binom{n}{2}$ pairs and choose the minimum separated pair; so we would like to design a significantly faster algorithm.

A general approach to many similar problems is given in Figure 7.14. The idea is to maintain the closest pair in an incremental fashion, so that in the end, we have the required result.

Algorithm 1: Closest pair(S)

1 *Input* $P = \{p_1, p_2 \ldots p_n\}$;
2 $S = \{p_1, p_2\}$; $C = d(p_1, p_2)$; $j = 2$;
3 **while** $j \leq n$ **do**
4 $\quad$ **if** $d(p_j, S) < C$ **then**
5 $\quad\quad$ $C = d(p_j, q)$ where $q = \arg\min_{p \in S} d(p_j, p)$
6 $\quad$ $P \leftarrow P \cup \{p_j\}$ $j \leftarrow j+1$
7 Output C as closest pair distance.

Figure 7.14   *Incremental algorithm for closest pair computation*

While the correctness of the algorithm is obvious, the analysis depends on the test in line 3 and the update time in line 5. For simplicity, let us analyze the running time for points in one dimension. Suppose the distances $d(p_{j+1}, S_j)$ are decreasing where $S_j = \{p_1, p_2, \ldots, p_j\}$. Then the closest pair distance C is updated in every step. To find the closest point from p_{j+1} to S_j, we can maintain S_j as a sorted set and find the closest point from p_{j+1} using a binary search in $O(\log j) = O(\log n)$ time. Overall, the algorithm takes $O(n \log n)$ time which is the same as presorting the points.

For points on a plane, we have to design a data structure to efficiently perform the test in line 3 and update in line 5. Trivially, it can be done in $O(n)$ steps leading to an $O(n^2)$ time algorithm. Instead, we analyze the algorithm for a random ordering of points in S. This will potentially reduce the number of updates required significantly from the worst case bound of $n-2$ updates. Let q_i denote the probability that point p_i causes an update when the points are inserted in a *random order*. A random ordering corresponds to a random permutation of points in P. To avoid extra notations, let us assume that $p_1, p_2, \ldots, p_n$ are numbered according to a randomly chosen permutation.

We can restate our problem as follows.

When $p_1, p_2, \ldots, p_i$ is a random ordering of the set of points $P = \{p_1, p_2, \ldots, p_i\}$, what is the probability that p_i defines the closest pair?

Suppose the closest pair is unique, that is, $C = d(r,s)$ for some $r,s \in \{p_1, p_2, \ldots, p_i\}$. Then, this probability is the same as the event that $p_i = \{r,s\}$. The total number of permutations of i objects is $i!$ and the total number of permutations with r or s as the last element is $2(i-1)!$. So the probability that p_i defines C equals $\frac{2(i-1)!}{i!} = \frac{2}{i}$. In a random permutation of n elements, the previous argument holds for a fixed set of i points. The *law of total probability* states that

$$\Pr[A] = \Pr[A|B_1] \cdot \Pr[B_1] + \Pr[A|B_2] \cdot \Pr[B_2] + \ldots + \Pr[A|B_k] \cdot \Pr[B_k]$$

$$\text{for disjoint events } B_1, B_2 \ldots B_k$$

(7.7.6)

In this situation, B_i represents each of the $\binom{n}{i}$ possible choice of i elements as the first i elements and by symmetry, the probabilities are equal as well as $\sum_i \Pr[B_i] = 1$. Since $\Pr[A|B_i] = \frac{2}{i}$, the unconditional probability of update in the ith step is $\frac{2}{i}$.

This is very encouraging since the expected update cost of the ith step is $\frac{2}{i} \cdot U(i)$, where $U(i)$ is the cost of updating the data structure in the ith step. Therefore, even for $U(i) = O(i \log i)$, the expected update time is $O(\log i) = O(\log n)$.

The situation for the test in line 3 is somewhat different since we will execute this step regardless of whether update is necessary. Given S and a new point p_i, we have to find the closest point from p_i and S (and update if necessary). Suppose the closest pair distance in S is D then consider a $D \times D$ grid of the plane and each point of S is hashed to the appropriate cell. Given the new point $p_i = (x_i, y_i)$, we can compute the cell as $\lceil \frac{x_i}{D} \rceil, \lceil \frac{y_i}{D} \rceil$. It can be seen from Figure 7.15 that if the closest point to p_i is within distance D; then, it must lie in one of the neighboring grid cells, including the one containing p_i. We can exhaustively search each of the nine cells.

Claim 7.1 *None of the cells can contain more than 4 points.*

This implies that we need to do at most $O(1)$ computations. These neighboring cells can be stored in some appropriate search data structure (Exercise 7.25) so that it can be accessed in $O(\log i)$ steps. In line 4, this data structure can be rebuilt in $O(i \log i)$ time, which results in an expected update time of $O(\log i)$. So, the overall expected running time for the randomized incremental construction is $O(n \log n)$.

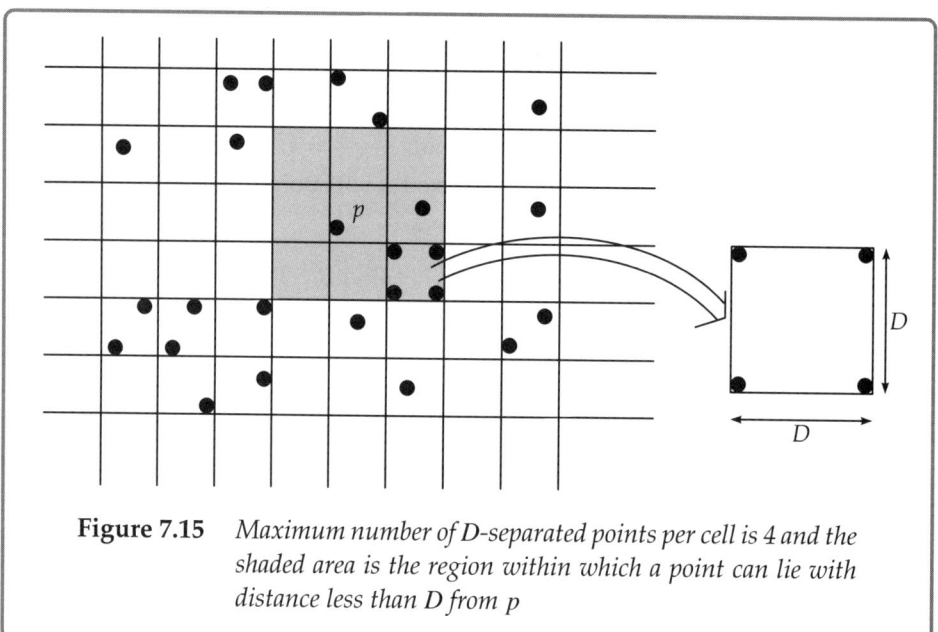

Figure 7.15 *Maximum number of D-separated points per cell is 4 and the shaded area is the region within which a point can lie with distance less than D from p*

Further Reading

The notion of multidimensional searching was motivated by an early paper of Bentley [19] that introduced the *k-d* trees. The closely related classes of nested data structures like range-search trees, interval trees, and segment trees were discovered in the context of natural geometric search problems. The book by Mehlhorn [101] and Preparata and Shamos [118] are excellent sources for these data structure techniques. Planar convex hull was a subject that got a lot of attention during the early days of computational geometry like Graham's scan [59] and gift wrapping method [69, 117] including its relationship to sorting algorithms. Kirkpatrick and Seidel [80] opened up a new direction by introducing the output size as an important parameter in time complexity. Quickhull, originally christened in the textbook [118] to describe an algorithm of Bykat [25] turned out to be not comparable to quicksort from which it gets its name. The description given in the chapter follows that of Bhattacharya and Sen [21] which is a randomized version of Chan, Snoeyink, and Yap [28]. The priority search data structure was first presented by McCreight [99]. The notion of persistent data structures was proposed in the context of planar point location by Sarnak and Tarjan [127]. This was developed further to accommodate *updates in the past* called *fully persistent* by Driscoll et al. [44]. The framework of *randomized incremental construction (ric)* can be found in the works of Clarkson and Shor [32] and Mulmuley [107]. The closest pair algorithm using ric has been adapted from the paper by Khuller and Matias [79]. For readers who are interested in pursuing more in-depth study in geometric algorithms are referred to many excellent textbooks [20, 45, 118]. The book by Edelsbrunner [45] makes many interesting connections between the combinatorial aspects of geometry and the analysis of geometric algorithms.

Exercise Problems

7.1 Given a set of n points in d dimensions, show that the immediate predecessor of a query point can be determined in $O(d + \log n)$. The predecessor is according to the lexicographic ordering described in the beginning of the chapter.

7.2 Show that for any query in the range search tree, the points belong to at most $2 \log n$ subtrees.

7.3 Show how to use threading to solve the range query in a BST without having leaf-based storage.

7.4 How would you modify the data structure for the **counting** version of orthogonal range trees to obtain a polylogarithmic query time?

7.5 Find a solution rigorously for the recurrence given in Eq. (7.1.2).

7.6 Work out the details of performing a three-sided query using a priority search tree and also analyze the running time.

   If the given set of points is sorted by y coordinates, show that the priority search trees can be constructed in $O(n)$ time.

7.7 Given a set of n line segments, design a data structure such that for any query rectangle, the set of line segments intersecting (including those fully contained) can be reported in $O(\log^2 n + k)$ time. Here k represents the size of output.

7.8 Given a set of n horizontal line segments, design a data structure that reports all intersections with a query vertical segment.

   Hint: Use segment trees.

7.9 Given a set of n horizontal and vertical segments, design an algorithm to identify all the *connected components* in $O(n polylog(n))$ time. A connected component is a set of line segments defined as follows:

   Two segments that intersect are connected. A segment intersecting any segment of a connected component belongs to the connected component.

7.10 Complete the solution of the recurrence for running time of the quickhull algorithm to show that $T(n, h)$ is $O(n \log h)$.

7.11 In some early versions of quick-hull, in Step 3, the point p_m was chosen to be furthest from $\overline{p_l p_r}$. The reader may want to analyze its performance carefully and demonstrate some bad inputs that would lead to quadratic running times.

7.12 Given a set S of n points on the plane, the diameter of S is defined as the maximum Euclidean distance between $p_1, p_2 \in S$ over all pairs of points. Design an efficient algorithm to find the diameter of S.

    Hint: Prove that the diameter is defined by a pair on the convex hull of S.

7.13 The **width** of a set S of n points in a plane is defined as the minimum width of a slab defined by a pair of parallel lines, such that S is completely contained in the slab. More formally, let $\Delta(S) = \min_\theta \max_{p \in S}\{d(p, \ell)\}$, where ℓ is a line that makes an angle θ with the x-axis and $d(p, \ell)$ is the Euclidean distance between p and ℓ. The line ℓ that minimizes this distance is called the **median axis** and the width of S is 2Δ.

    Design an $O(n \log n)$ algorithm for finding the width of a point set.

7.14 Design an efficient solution to the ray shooting problem by extending the interval trees.

7.15 Analyze the performance of range trees for reporting orthogonal range queries for dimensions $d \geq 3$. In particular, what are the preprocessing space and query time?

7.16 If we allow for insertion and deletion of points, how does the performance of range trees get affected? In particular, what are the time bounds for the orthogonal range query, insertion, and deletion of points? Discuss the data structure in details.

7.17 Design an efficient algorithm to construct the *intersection* of two convex hulls.

7.18 Design an algorithm to merge two upper hulls in $O(n)$ time where n is the sum of the vertices in the two hulls.

    Further show how to find a bridge between two linearly separable upper hulls in $O(\log n)$ steps using some variation of the binary search. Note that the bridge is incident on the two hulls as a tangent on points p_1 and p_2. Use binary search to locate these points on the two convex hulls by pruning away some fraction of points on at least one of the hulls.

7.19 If we want to maintain a convex hull under arbitrary insertion and deletion of points without recomputing the entire hull, we can use the following approach. Use a balanced BST (binary search tree) to store the current set of points. At each of the internal nodes, store the bridge between the upper hull of the points stored in the left and right subtrees. Use this scheme recursively – the leaves store the original points. In case of any changes to the point set, the bridges may be re-computed using the algorithm in the previous problem. Provide all the missing details to show that this data structure can be maintained in $O(\log^2 n)$ time per update, either insertion or deletion.

7.20 Given a set S of n points in a plane, design a data structure to

    (i) Find the closest point from a query line.

(ii) Find the closest point to the convex hull of S from a query point (if the point is inside the hull, then distance is 0).

7.21 A point $p_1 \succ p_2$, (p_1 *dominates* p_2) iff all the coordinates of p_1 are greater than or equal to p_2. A point p is *maximal* if it is **not** dominated by any other point in an input set S.

(i) Design an $O(n \log n)$ time algorithm to find all maximal points in an input set S of n distinct points on the plane. This is known as the DOMINANCE problem.

(ii) Design an $O(n \log h)$ algorithm for the three-dimensional version of the DOMINANCE problem.

7.22 Design an $O(n \log n)$ algorithm for finding all h maximal points among a set of n points on a plane.

7.23 **Point–line duality** Consider a transform $\mathcal{D}(a,b) = \ell: \ y = 2ax - b$ which maps a point $p = (a,b)$ to a line $\ell: y = 2ax - b$. Moreover, $\mathcal{D}$ is a 1-1 mapping, that is, $\mathcal{D}(\ell: y = mx + c) = (m/2, -c)$, such that $\mathcal{D}(\mathcal{D}(p)) = p$. Note that $\mathcal{D}$ is not defined for $m = \infty$, that is, vertical lines.

Prove the following properties of $\mathcal{D}$, where p is a point and ℓ is a (non-vertical) line.

1. If p is incident on ℓ, then $\mathcal{D}(\ell)$ is incident on $\mathcal{D}(p)$.
2. If p lies below ℓ, then $\mathcal{D}(\ell)$ lies above $\mathcal{D}(p)$ and vice versa.
3. Let p be the intersection point of lines ℓ_1 and ℓ_2. Then the line $\mathcal{D}(p)$ passes through the points $\mathcal{D}(\ell_1)$ and $\mathcal{D}(\ell_2)$.

7.24 **Intersection of half-planes** Given a set of n half-planes $h_i : y = m_i \cdot x + c_i$, we want to compute the intersection of h_is. Note that the intersection of half-spaces in a convex region may be bounded, unbounded, or even empty.

1. If the intersection is non-empty, show that the boundary is the intersection of two convex chains C^+ and C^-, where

   (i) $C^+ = \bigcap_{h_i \in H^+} H_i$ (ii) $C^- = \bigcap_{h_i \in H^-} H_i$

   H^+ (respectively H^-) denotes the set of positive (negative) half-planes, that is, the planes that contain $(0, \infty)$ and $(0, -\infty)$.

2. Design an $O(n \log n)$ algorithm for constructing the intersection of n half-planes by exploiting the dual transform described in the previous problem.

   Hint: There is a 1–1 correspondence between the boundary of the intersection of half-planes and the boundary of the convex hull of $\mathcal{D}(h_i)$. If a point belongs to the lower hull, then there exists a line through the point that contains all the remaining points on the positive half-plane. Likewise, if a half-plane forms the boundary of C^+, then a point on the boundary satisfies $\bigcap_{h_i \in H^+} H_i$.

7.25 Let c_d denote the maximum number points in a d-dimensional unit cube that are mutually separated by at least distance 1. Calculate a tight upper-bound on c_2 and c_3.

Design an efficient data structure based on this observation to determine if the value of the closest pair distance has changed in the algorithm given in Figure 7.14 and also analyze the time to rebuild this data structure if the closest pair distance changes. Compare and contrast tree-based and hashing-based schemes.

7.26 Prove that Graham's scan algorithm correctly outputs the convex hull of a set of points in the plane.

7.27 Consider the quick-hull algorithm and the notation used in its description. Show that the points in $S_a(l)$ which lie below the line joining p_m and p_l cannot be part of the upper hull, and so, can be eliminated from $S_a(l)$. Argue similarly for $S_a(r)$.

7.28 Consider the quick-hull algorithm. In step 3, show that if the pair $\{p_{2i-1}, p_{2i}\}$ satisfies the property that the line containing $\overline{p_{2i-1}p_{2i}}$ does not intersect the line segment $\overline{p_l, p_r}$, then it guarantees that p_{2i-1} or p_{2i} does not lie inside the triangle $\triangle p_l p_{2i} p_r$ or $\triangle p_l p_{2i-1} p_r$ respectively. This could improve the algorithm in practice by eliminating all points within the quadrilateral $p_l, p_{2i-1}, p_{2i}, p_r$.

8

CHAPTER

String Matching and Finger Printing

Given a string P and a text T, where T is usually much longer than P, the (exact) string matching problem is to find all or some of the occurrences of P in T. This may also be viewed as the linear version of the more interesting family of pattern matching problems which is one of the central problems in the area of artificial intelligence and machine learning. Discerning patterns in high dimensional spaces is believed to drive all our cognition processes. Readers familiar with automata theory and formal languages will be familiar with the complexity of recognizing complex patterns in strings that are generated according to some rules. In this chapter, however, we will be mostly concerned with identifying explicit substrings in a long text. Much of the contemporary work in document classification uses keywords identification as a basic routine that feeds into higher level heuristics.

8.1 Rabin–Karp Fingerprinting

We first introduce the notation needed to describe the algorithm. We shall use Σ to denote the set of symbols which form the input string. Let Y denote the text string of length m, where each character belongs to Σ. Let Y_j denote the jth character of Y.

We shall use $Y(j,k)$ to denote the substring $Y_j \cdot Y_{j+1} \ldots Y_{j+k-1}$. In other words, $Y(j,k)$ is the *substring* of length k starting at the jth position. The length of a string Y is denoted by $|Y|$. Similarly, the pattern string X is a length n string with symbols from Σ where $n \leq m$.

The string matching problem is as follows:

Given a pattern string X, $|X| = n$, find an index i such that $X = Y(i, n)$, else report *no match*.

Other common variations include finding the smallest index where a match occurs, or finding all indices where the match occurs. For example, given string $Y = abbaabaababb$ over alphabet $\{a, b\}$ and patterns $A = aaba$ and $B = aaab$, $A = Y(4, 4) = Y(7, 4)$ whereas B does not occur in Y. Note that the two occurrences of A actually overlap.

One obvious and natural way of finding matches is brute-force comparison of each $Y(i, n)$ with X for all i, that could result in $\Omega(nm)$ comparisons. This can actually happen for the input strings $Y = ababab \ldots ab$ and $X = abab \ldots ab$.

Consider the following idea : let $F()$ be a function which maps strings of lengths n to relatively shorter strings. Now, we compute $F(X)$ and compare it with $F(Y(1, n))$, $F(Y(2, n))$, and so on. Clearly, if $F(X) \neq Y(i, n)$, then X and $Y(i, n)$ are distinct strings. The converse may not hold true.

For example, let $F(a) = 1, F(b) = 2$ and for a string $X = X_1 X_2 \ldots X_n$, define $F(X)$ as $\sum_{i=1}^{n} F(X_i)$. Consider strings $S_1 = abbab$, $S_2 = abbaa$, and $S_3 = babab$. Then, $F(S_1) = 8$, $F(S_2) = 7$, and $F(S_3) = 8$. Since $F(S_1) \neq F(S_2)$, $S_1 \neq S_2$. However, the converse is not true, i.e., even though $F(S_1) = F(S_3) = 8$, we know that $S_1 \neq S_3$. In fact, this issue is unavoidable because F is mapping a large set to a smaller set viz., 2^n strings to values in $[n, 2n]$, and so, cannot be a 1-1 function.

The function F is known as the *fingerprinting* function (also called a hash function) and may be defined according to the application. Although the function in the previous example did not give us the desired result, let us try

$$F(X) = (\sum_{i=1}^{n} 2^{n-i} X_i) \mod p$$

where p is a prime number. Here, X is assumed to be a binary pattern (of 0 and 1) and x is the corresponding integral value[1], i.e., if $X = 10110$, then $x = 22$ (in decimal).

To get some intuition about why this is a good idea, let us consider a slightly different, though related, problem. Let x_1, x_2 be two integers and consider the expression $(x_1 - x_2) \mod p$ for some prime p.

Observation 8.1 *If $(x_1 - x_2) \mod p \neq 0$ then $x_1 \neq x_2$.*
If $(x_1 - x_2) \mod p = 0$ then $(x_1 - x_2) = k \cdot p$ for some integer k.

It follows that only when $k = 0$, $x_1 = x_2$. Consider another prime $p' \neq p$. If $(x_1 - x_2) \mod p' \neq 0$, then $x_1 \neq x_2$ from our previous observation. However, it may happen that $(x_1 - x_2) \mod p' = 0$. Does it imply that $x_1 = x_2$? It may if $(x_1 - x_2)$ is simultaneously a multiple of p and p'.

---

[1] We will use the notation for string X and its integer cast x interchangeably when it is clear from the context.

For example, let $x_1 = 24, x_2 = 9, p = 3, p' = 5$, then $15 \mod 3 = 15 \mod 5 = 0$. Consider the natural extension of this strategy given in Fig. 8.1.

Procedure Verifying equality of numbers(x_1, x_2)

1 Input: Integers x_1, x_2, k ;
2 **Repeat** k times ;
3 Choose a random prime $p \in [1, M]$;
4 **if** *If* $x_1 \mod p \neq x_2 \mod p$ **then**
5 $\quad \lfloor$ Return **NO**;
6 Return **YES** COMMENT: *If for all the k iterations* $(x_1 - x_2) \mod p = 0$

Figure 8.1   *Testing equality of two large numbers*

When the test returns *NO*, the answer is correct; to analyze the other case, let us denote by $p_1, p_2, \ldots, p_k$ such that $x_1 \equiv_{p_i} x_2$ or equivalently $(x_1 - x_2) \equiv_{p_i} 0$ for $i = 1, 2, \ldots, k$. The number $x_1 - x_2$ is bounded by $\max\{x_1, x_2\}$, which in turn is bounded by the range r such that $x_1, x_2 \in [1 \ldots r]$. Since p_is divide $(x_1 - x_2)$, we can bound the number of prime factors of x_1 or x_2 by $\log r$. Indeed, if $p_1, \ldots, p_\ell$ are distinct primes dividing x_1, then $\prod_{i=1}^{\ell} p_i \geq 2^\ell$. So, ℓ is $O(\log r)$.

If the number of primes from where we choose a random prime is at least $t \log r$, then the probability that a random prime divides $x_1 - x_2$ is less than $\frac{1}{t}$. So we summarize our analysis as follows.

Claim 8.1 *If the primes are chosen from the range* $[1, M]$, *then the probability that the answer is incorrect is less than* $\left(\frac{\log r}{\pi(M)} \right)^k$, *where* $\pi(M)$ *is the number of primes in the range* $[1, M]$.

This follows easily from the observation that at most $\log r$ out of a possible $\pi(M)$ primes can yield an incorrect answer (YES) and this should happen in all the k iterations independently. The quantity $\pi(M)$ is known to satisfy the following inequality (using results on density of primes)

$$\frac{M}{\ln M} \leq \pi(M) \leq 1.26 \frac{M}{\ln(M)}$$

This gives us a way of reducing the error probability to any $\varepsilon > 0$ by choosing M and k appropriately. If $M \geq 2 \log r \log \log r$, then the error probability is less than $\frac{1}{2}$ and by repeating it k times, it decreases to $\frac{1}{2^k}$. By choosing a much larger range $M = r$, the error probability is $\frac{1}{r^2}$ even for a very small constant k.

To get a better perspective of what we can achieve, let us consider a scenario with two individuals A and B who have the two numbers x_1, x_2 respectively. They want to compare the numbers by exchanging as little information as possible. Since both numbers are in $[1, r]$, any one of the parties can send its individual numbers to the other person and they can be compared easily. This involves sending across $r' = \lceil \log_2 r \rceil$ binary bits. Can we do the same by communicating fewer bits? The strategy given in Fig. 8.1 can be used very effectively by choosing p_is that are much smaller than r and can be represented using fewer bits. If p_is are in the range $[1, \log^2 r]$, then by sending the remainders (modulo p_i) having $O(\log \log r)$ bits, we can compare them with reasonable probability of being error free. By sending across multiple such fingerprints (modulo multiple primes), we can drive down the error probability even further and the total number of bits is bounded by $k \lceil \log \log r \rceil$ which is much smaller than r'.

With this observation, let us now return to the string matching problem and consider the following algorithm in Fig. 8.2. Note that $F_p(i)$ denotes the fingerprint of the string $Y(i, n)$.

Procedure String match(Y, X)

---

1 Input: String $Y = Y_1, Y_2 \ldots Y_m$, Pattern $X = X_1, X_2, \ldots X_n$;
2 Output: $\{j | Y(j, n) = X\}$, else *nil* ;
3 Choose a random prime $p \in [1, n^3]$;
4 Compute hash $F_p(X) = (\sum_{i=1}^{n} 2^{n-i} X_i) \mod p$;
5 Compute the initial hash $F_p(1) = F_p(Y(1, n))$;
6 *Match* $\leftarrow \phi$ (Initialize the match vector) ;
7 **for** $i = 1$ *to* $m - n + 1$ **do**
8 $\quad$ $F_p(i+1) = [2 \cdot F_p(i) + Y_{i+n}] \mod p - Y_i \cdot 2^{n-1} \mod p$;
9 $\quad$ **if** *If* $F_p(i+1) = F_p(X)$ **then**
10 $\quad\quad$ Verify if $Y(i, n) = X$ and add i to *Match* accordingly ;
11 $\quad\quad$ *Match* $= Match \cup \{i\}$;

12 Return *Match* if non-empty else *nil* ;

Figure 8.2   *Karp–Rabin string matching algorithm*

Claim 8.2 *The procedure String match(Y, X) always returns all the correct matches.*

First note that because of the explicit verification step that checks if $Y(i, n) = X$ character by character, we have eliminated the case of false matches when the fingerprints match without an actual match. This introduces some added cost when we analyze the overall algorithm.

Let us now analyze the cost involved in each of the steps. In the logarithmic cost RAM model, we assume that each word has $O(\log m)$ bits and any arithmetic operation on a word can be done in $O(1)$ steps [2]. From Claim 8.1, it follows that the probability of a false match is $\frac{n}{n^3} = \frac{1}{n^2}$ since the strings are binary strings of length n and their values are bounded by 2^n. Therefore, the expected cost of a false match is $\frac{n}{n^2} = \frac{1}{n}$ and the total expected cost of all the false matches (maximum of $m - n$) can be bounded by $m/n \leq m$.

The most important step in the algorithm is line 8 where the fingerprint function $F_p(i)$ is being updated, and $F_p(i) = F_p(Y(i,n))$. Note that $Y(i+1,n) = 2(Y(i,n) - Y_i \cdot 2^{n-1}) + Y_{i+n}$. So,

$$F_p(i+1) = Y(i+1,n) \quad \mod p \tag{8.1.1}$$

$$= [2(Y(i,n) - 2^n \cdot Y_i + Y_{i+n}] \quad \mod p \tag{8.1.2}$$

$$= 2Y(i,n) \quad \mod p - 2^n \cdot Y_i \quad \mod p + Y_{i+n} \quad \mod p \tag{8.1.3}$$

$$= (2F_p(i)) \quad \mod p + Y_{i+n} \quad \mod p - 2^{n-1} \cdot Y_i \quad \mod p \tag{8.1.4}$$

All the terms are modulo p except $2^{n-1} \cdot Y_i$ which can be much larger. However, we can pre-compute it in $O(n)$ time by noting that $2^i \mod p = 2(2^{i-1} \mod p) \mod p$. So as long as p can fit in one (or $O(1)$) word in the memory, this can be done in constant update time.

The actual running time of the algorithm depends on the number of (correct) matches. If there are t matches of X in the string, then, the cost of verification is $O(t \cdot n)$. Since the expected cost of false matches is m/n, we can summarize the overall analysis as follows

Theorem 8.1 *The expected running time of the algorithm in Figure 8.2 is $O((t + m/n) \cdot n)$, where t is the number of matches of the the pattern X in the string Y.*

Clearly, if we are only interested in finding the first match, the algorithm can be tweaked so that it takes linear time. However, for multiple matches, the verification step is expensive. The characterization of the algorithm minus the verification step (line 9) is left as an exercise problem.

8.2 KMP Algorithm

Although the previous technique based on a random fingerprinting function is simple, it can give erroneous results if the verification step is eliminated. Otherwise for multiple matches, the verification could make it behave like a brute force $O(m \cdot n)$ algorithm. The reader may have noted that the main challenge for designing an efficient algorithm is to avoid repeated matching of symbols on the same part of the string. From a partial match, we have already gained information about the string that we can make use of during the next partial match. The pattern is like a moving window of fixed length n on the string; if the windows are disjoint (or nearly so), then we will be in business. Let us illustrate this using an example.

---

[2] More specifically, any operation like integer division involving $O(\log n)$ sized integers can be done in $O(1)$ time.

Consider $Y = aababbabaabb$ and $X = aabb$. We start matching from $Y(1,4)$ and we find a mismatch $Y_4 = a \neq X_4 = b$. In the obvious brute force method, we will try to match $Y(2,4)$ with X. However, we have gained the information about $Y_2Y_3Y_4 = aba$ from the previous partial match. So there is no point in trying a match starting from Y_2 since we will fail at $Y_3 = b \neq X_2 = a$. Similarly, there is no hope for $Y(3,4)$ either. However, $Y_4 = X_1 = a$ and we must try $Y(4,4)$. How do we formalize this heuristic for skipping some search windows of the pattern?

For this, we have to deal with *prefix* and *suffix* of strings. A *prefix* of length i is the first i characters of a string. For example, $X(1,i)$ is a prefix of X. Likewise, the *suffix* of a string of length i is the last i characters of a string. If $|X| = n$, then $X(n-i+1,n)$ is a suffix of X of length i.

Let $\alpha \sqsubset \beta$ denote that α is a *suffix* of β. As in the previous example, we would like to make use of matches done so far before proceeding to the next match. For example, suppose while matching strings X and $Y(k,n)$, we realize that the first i characters of X match with those of $Y(k,n)$, but $X_{i+1} \neq Y_{k+i}$. Now next time we would like to align X with $Y(k+1,n)$ and again check for a match. But note that we have already seen the first i characters of $Y(k+1,n)$ and in fact, $Y(k+1,i-1)$ overlap with $X_2, \ldots, X_i$. Therefore, we should try this alignment only if $X_2, \ldots, X_i$ is equal to $X_1, \ldots, X_{i-1}$. Note that this property can be ascertained without even looking at Y. Similarly, we should try to align X with $Y(k+j,n)$ only if $Y_{k+j}, \ldots, Y_{k+i-1}$ is same as $X_1, \ldots, X_{i-j-1}$, and Y_{k+i} matches with X_{i-j}. Again, we know that $Y_{k+j}, \ldots, Y_{k+i-1}$ is same as $X_j, \ldots, X_{i-1}$. Thus, we again get a property involving X only (except for the character X_{i-j} matching with Y_{k+i}). This discussion can be summarized as follows.

Given a prefix $X(1,i)$ of X and the character Y_{k+i} in the earlier discussion, we would like to find $\arg\max_j \{X(1,j) \sqsubset Y(k,i+1) = X(1,i) \cdot Y_{k+i}\}$. Following this, we try matching X_{j+1} with Y_{k+i+1} which is yet to be read next. As observed before, this property is dependent mostly on the pattern except the last symbol of the string. From this intuitive description, we can define a function of a given pattern X as follows. For every $i, 1 \leq i \leq n$ and character $a \in \Sigma$, define

$$g(i,a) = \begin{cases} \max_j\{X(1,j) \sqsubset X(1,i) \cdot a\} & \text{if such an index } j \text{ exists} \\ 0 & \text{otherwise} \end{cases}$$

The function g for the pattern $X = aabb$ is given by the Table 8.1. Note that the columns represent the extent of partial match of the relevant portion of the text with pattern. This table can be naturally associated with a DFA (deterministic finite automaton) for the pattern that we are trying to find. At any stage, the state of the DFA corresponds to the extent of partial match – it is in state i if the previous i symbols of the text have matched the first i symbols of the pattern. It reaches the final stage iff it has found a match. Given this DFA, we can find all occurrences of an n symbol pattern in an m symbol text in m

steps, where there is a transition for every input symbol of the text. Indeed, suppose we are in state i and the last scanned character in the text is Y_k. We check the table T for the entry (Y_k, i) – say $T(Y_k, i)$ is j. Then, we know that $X_1, \ldots, X_j$ match with $Y_{k-j+1}, \ldots, Y_k$. Now we match X_{j+1} and Y_{k+1}. Two things can happen – (i) If these two characters do not match, we move to state $T(Y_{k+1}, j)$ in the DFA and proceed as before; (ii) If these two characters match, we move to state $T(Y_{k+1}, j) = j + 1$ in the DFA and continue.

Table 8.1   *Finite automaton transition function for the string aabb matching*

	0	1	2	3	4
a	1	2	2	1	1
b	0	0	3	4	0

The size of the DFA is $O(n|\Sigma|)$, where Σ is the alphabet which is optimal if Σ is of constant size. The algorithmic complexity should also include the construction of the DFA, that we will address in the context of the next algorithm.

With some additional ideas, the previous approach can be made to run in $O(n + m)$ steps without dependence on the alphabet size. We modify the definition of g as follows. Let us define the *failure* function of a given pattern X as

$$ f(i) = \begin{cases} \max_{j:j<i}\{X(1,j) \sqsubset X(1,i)\} & \text{if such an index } j \text{ exists} \\ 0 & \text{otherwise} \end{cases} $$

Note that the subtle change in the definition of f makes it purely a function of X and just one index. The failure function for $X = aabb$ is given by $f(1) = 0, f(2) = 1, f(3) = 0, f(4) = 0$. Let us postpone the method for computing the failure function and assume that we have the failure function available.

The overall idea of the algorithm is same as earlier. Let Y_k denote the kth symbol of the text for which we have a partial match up to i symbols of the pattern. We then try to match X_{i+1} with Y_{k+1}. In case of a match, we increase the partial match and if it is n, then we have found a match. Otherwise (if X_{i+1} does not match Y_{k+1}), we try to match $X_{f(i)+1}$ with Y_{k+1} and again if there no match, we try $X_{f(f(i))+1}$ with Y_{k+1} and so on till the partial match becomes 0. The algorithm is described formally in Figure 8.3. Note that this differs from the earlier algorithm (based on the function g) in only one situation – if X_j does not match with Y_i, the earlier algorithm would compute the partial match according to Y_i and proceed to Y_{i+1}. The current algorithm will keep on reducing the partial match j till it gets a match at Y_i. Therefore, it is not immediately clear if the running time of this algorithm is linear since Y_i is being repeatedly compared till partial match becomes 0.

Let us consider an example to illustrate the algorithm in Table 8.2. The first time there is a mismatch between X and Y is at Y_7. Subsequently, we shift the pattern right by $6 - f(6)$

and next again by $4 - f(4)$. The notation $X(+i)$ denotes that the pattern is shifted right by i symbols.

The analysis of this algorithm centers around a situation where the pattern string keeps sliding to the right till it either finds a match or it slides beyond the mismatched symbol. Therefore, either the algorithm progresses right on Y or the pattern moves forward. We can analyze it in many ways – here we will use the technique of potential function.

Procedure Deterministic String Match(Y, X)

1 Input: String $Y = Y_1, Y_2 \ldots Y_m$, Pattern $X = X_1, X_2, \ldots X_n$;
2 Output: $\{j | Y(j, n) = X\}$, else *nil* ;
3 $j \leftarrow 0$, $Match \leftarrow \phi$;
4 **for** $i = 1$ *to* m **do**
5    $j \leftarrow j + 1$;
6   **if** $Y_i = X_j$ **then**
7   **if** $j = n$ **then**
8          $Match \leftarrow Match \cup \{(i - n + 1)\}$ (a match is found starting in Y_{i-n+1}) ;
9          $j \leftarrow f(j)$ (trying the next potential match)
10   **else**
11   **while** $(f(j) \neq 0) \wedge (Y_i \neq X_j)$ **do**
12          $j \leftarrow f(j)$;

13 Return *Match* if non-empty else *nil* ;

Figure 8.3   *Knuth–Morris–Pratt string matching algorithm*

Table 8.2   *Illustration of matching using KMP failure function f for the pattern abababca.*

	1	2	3	4	5	6	7	8
X	a	b	a	b	a	b	c	a
$f(i)$	0	0	1	2	3	4	0	1
Y	a	b	a	b	a	a	b	a
$X(+2)$		a	b	a	b	a	b	
$X(+4)$				a	b	a	b	

8.2.1 Analysis of the KMP algorithm

During the algorithm, we may be comparing any given element of the text Y,[3] a number of times, depending on the failure function. Let us define the potential function as the extent of partial match. In other words, at any point of time t, define $\Phi(t)$ as the value of the index j as described in Figure 8.3. Recall that the amortized cost of an operation at time t is defined as the actual cost plus the change in potential, i.e., $\Phi(t+1) - \Phi(t)$.

We have two distinct cases – one where the symbols in the pattern and string match; otherwise, we use the failure function to shift the pattern appropriately. So the amortized cost works out as follows.

Case: match Here Y_i matches with X_j (as in the algorithm described in Figure 8.3). The algorithm incurs one unit of cost, but the potential also increases by 1. Therefore, the amortized cost of a match is 2.

Case: mismatch or $j = n$ The amortized cost is ≤ 0, since the potential is strictly decreasing as the value of j strictly decreases.

By associating this amortized cost with the index of the string Y, and summing over all indices, the total amortized cost of all the operations is $O(m)$. Since the initial potential is 0, it follows that the running time is $O(m)$.

8.2.2 Pattern analysis

It remains to describe how to construct the failure function f. We use the following crucial observation:

Observation 8.2 *If the failure function $f(i) = j$, $j < i$, it must be true that $X(j-1) \sqsubset X(i-1)$ and $X_i = X_j$.*

This shows that the computation of the failure function is very similar to the KMP algorithm itself and we compute the $f(i)$ incrementally with increasing values of i. The details are left as an exercise for the reader. Therefore, we can summarize as follows.

Theorem 8.2 *The failure function of a pattern X can be computed in $O(|X|)$ comparisons so that the total running time of the algorithm described in Figure 8.3 is $O(|Y|)$, where Y is the string. The number of comparisons is not dependent on the size of the alphabet $|\Sigma|$.*

8.3 Tries and Applications

A *trie* or digital tree data structure is tailor-made for addressing a wide range of problems related to strings including *string matching*. It is particularly useful if the text remains

---

[3] Unlike the DFA construction, we do not always move ahead on Y, which is handled in the inner loop.

fixed, and we need to query it with many different patterns. The text is pre-processed in a manner that each query takes time proportional to the size of the pattern (and independent of the size of the text). This can be very useful if the text is huge and patterns are small strings. We will think of the text as consisting of a set of strings, and a pattern as a single string. Given a pattern, we would like to check if the pattern appears in the set of strings (of course, in real applications, we would also like to know the locations in the text where the pattern appears, but such extensions are easy to achieve here).

We assume that the text consists of a set of strings (e.g., the set of words in a document). As before, we use Σ to denote the alphabet. A trie on the given text can be thought of as a k-ary tree, where $k = |\Sigma|$ with a depth that is dependent on the length of the strings in the text. Each internal node has k children corresponding to each symbol of the alphabet and the edges are labeled accordingly (it is useful to think about edge labels rather than child nodes in this framework). A single string is defined by a path in this tree, where the edge labels on this path corresponds to the string (see Figure 8.4). This simple and intuitive mechanism for storing a string provides many applications including locating exact and closest substrings, finding common substrings, and many applications in genome research.

Let us see how tries can be used for string matching. First note that it is very easy to check if a given pattern X is a prefix of a string stored in a trie. We simply try to see if there a path starting from the root labeled with X. Therefore, if we have all the suffixes $Y(j) = Y_j Y_{j+1} \ldots Y_m$ of a string Y, then we can easily check if X is a prefix of $Y(j)$. In fact, we can easily keep track of all the matches by storing in each node how many strings share a given prefix. This is equal to the number of strings that pass through a given node. To make this even simpler, we append a special character, say $, to denote the end of a string. This ensures that all strings are uniquely identified by a leaf node that stores $ and also that no string can be a prefix of another string. For example, in Figure 8.4, the number of occurrences of the string ca is 2, which is the number of leaf nodes in the subtree rooted in the node with label ca.

Definition 8.1 *A suffix tree is a data structure that stores all the suffixes of a given string in a trie-like storage including the string itself.*

If we use a simple trie, the storage could become $\sum_{j=1}^{m} |Y(j)|$, which could be as much as $\theta(m^2)$ – see Exercise Problem 8.8. There are sophisticated data structures for reducing the storage to a linear structure by associating a substring (more than one symbol) on an edge.

Observation 8.3 *The path corresponding to a string has two distinct spans – the initial span is the longest common prefix with an existing string and then there is the subsequent path leading to a leaf node with no other shared subpaths with another string.*

This follows from the underlying tree structure – that is, once the paths of two strings diverge, they cannot meet again. This also implies that the additional storage required for a string is $O(1)$, that is, the unique second span of a string. We can store the *forking* node with the string. Moreover, the initial span could also split an existing span into two parts. All these changes can be performed in additional $O(1)$ storage. Figure 8.4 shows the different stages of a suffix tree for the string *catca* beginning with the smallest suffix *a*.

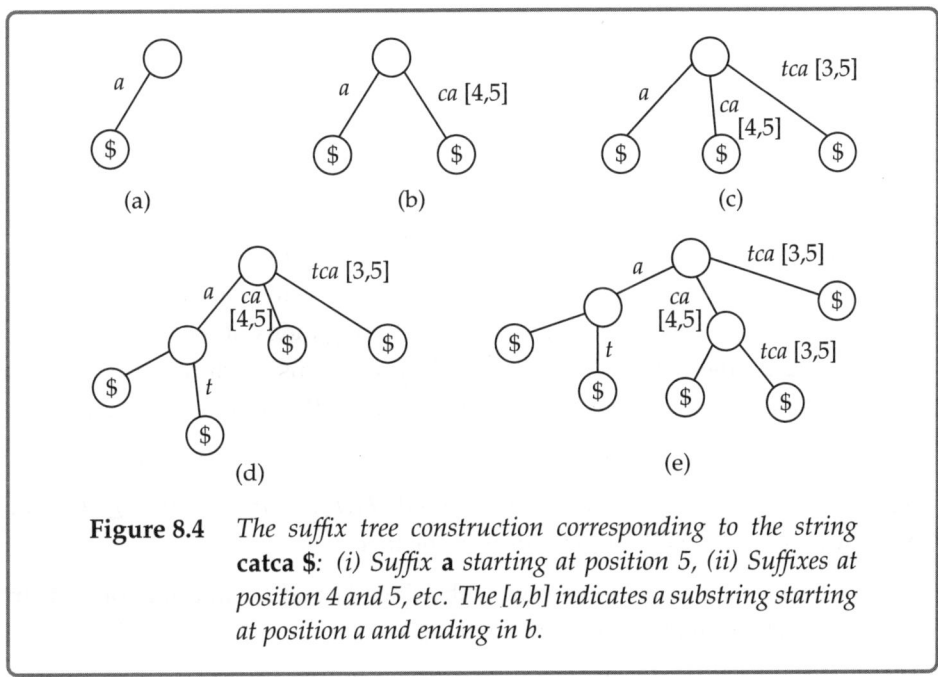

Figure 8.4 *The suffix tree construction corresponding to the string* **catca \$**: *(i) Suffix* **a** *starting at position 5, (ii) Suffixes at position 4 and 5, etc. The [a,b] indicates a substring starting at position a and ending in b.*

The details of these algorithms are somewhat intricate, so we will only provide a high level sketch and illustrate some applications of the suffix trees. Note that an edge labeled with a substring of Y only needs an interval $[a, b]$, where the substring is $Y_a Y_{a+1} \dots Y_b$ such that $1 \leq a \leq b \leq m$. Thus, it takes at most $2 \log m$ bits for any label which is $O(1)$ space in the context of the problem. The path associated with a node is the concatenation of the labels of all the edges comprising the path from the root to the node and is referred to as a *path label of a node*. The known linear time algorithms for constructing suffix trees assume that the alphabet is bounded, that is, some known constant. All the suffixes, $Y(j)$ $1 \leq j \leq (m+1)$ including $Y(m+1) = \$$ are added in a forward or a reverse order, that is, from the longest suffix $Y(1)$ or from the shortest one at $Y(m+1)$. Let us assume that all suffixes $Y(i+1) \dots Y(m+1)$ have been added and we are now trying to add $Y(i)$. Starting from a leaf of $Y(i+1)$, we look for a node with maximum depth that has a path labeled $Y_i \cdot Y(i+1)$. This path could even end in the middle of an edge. The data structure that we maintain at each node contains this information corresponding to every $a \in \Sigma$ and the path denoted by the node. For example, if the path label of a node is a string α, then the node will store the information of all those nodes having path label $a \cdot \alpha$ for all $a \in \Sigma^4$. This data structure must also be updated after every insertion.

Some additional properties of suffix trees are needed to show that the new suffix can modify only a constant number of nodes. The number of levels that must be traversed for a single suffix insertion is not bounded but the amortized number of levels is $O(1)$. This leads to the $O(m)$ bound on the total running time. One interesting and important property of suffix trees can be stated as follows without a formal proof.

Claim 8.3 *In any suffix tree, if there is a node v having label $a \cdot \alpha$, where $a \in \Sigma$ and $\alpha \in \Sigma^*$, then there is a node u having label α.*

The suffix tree data structure has a related cousin called the *suffix array* – this can actually be derived from the suffix trees. It is a sorted array of the suffix strings where its ordering is defined as lexicographic on the strings. The strings are however, not stored explicitly to save space and can be defined by the starting index of the suffix.

We provide a partial list of the plethora of applications of suffix trees to sequence related problems, especially important in bioinformatics. Recall that the length of the text is m.

1. *Keyword set matching* Given a set of keywords $P_1, P_2, \ldots, P_k$ such that $\sum_i |P_i| = n$, find all the occurrences of all the keywords in the text.

2. *Longest common substring* Given two substrings S_1 and S_2, find the longest string s which is a substring (contiguous) of both S_1, S_2.

   Using tries, it is possible to find s in $O(|S_1| + |S_2|)$ comparisons. This is done by building *generalized suffix trees* – a common suffix tree for all the suffixes of S_1 and S_2.

3. *Matching statistics* Given $1 \leq i \leq m$, find the longest substring starting at position i that occurs again in the text, starting from position $j \neq i$.

Further Reading

Along with sorting, string matching is one of the earliest problems in the computer science community that drew a lot of attention. DFA based string matching was a natural algorithm that was improved by the algorithm of Knuth–Morris–Pratt [84], presented as the KMP algorithm. Aho and Corasic [6] generalized it to find all occurrences of a set of keywords. The original algorithm of Rabin and Karp appeared in their paper in 1987 [78]. Tries appeared first in the work of Briandais and Fredkin [41,51]. Linear tree construction of suffix trees was first given by Weiner [152]. Subsequently, McCreight [98] and more

---

[4] If Σ is not bounded, then the space will not be linear.

recently, Ukonnen [143] further simplified and refined the construction. The notion of suffix arrays was proposed by Manber and Myers [97]. The book by Gusfield [60] provides a comprehensive account of string matching algorithms with numerous applications, in particular to computational biology.

Exercise Problems

8.1 Using the following well-known and useful result from number theory, give an alternate proof of the procedure for testing equality of large numbers described in Figure 8.1.

(Chinese Remainder Theorem – CRT) For k numbers $n_1, n_2, ..., n_k$, relatively prime to each other,

$$x \equiv y \bmod n_i \text{ for all } i \Leftrightarrow x \equiv y \bmod n_1 n_2 n_3 ... n_k = M$$

Moreover,

$$y \equiv \sum_{i=1}^{R} c_i d_i y_i$$

where $c_i d_i \equiv 1 \bmod n_i$, $d_i = \Pi_{j=1, j\neq i}^{j=k} n_j$ and $y_i = x \bmod n_i$

Hint: Let k be such that $2^m < M = 2 \times 3 \times ... \times p_k$, that is, the first k primes. From CRT, if $X \neq Y(i, n)$, then for some p in $\{2, 3, ..., p_k\}$, $F_p(X) \neq F_p(Y(i, n))$.

8.2 How does the algorithm in Figure 8.2 behave without the explicit verification step in line 10? In particular, comment about the trade-off between the overall correctness and running time.

8.3 Using the potential function method, show that the failure function can be computed in $O(|X|)$ steps.

8.4 Let y be a string of length m. A substring x of y is called a period of y if $y = (x^k)x'$, where (x^k) is the string x repeated k times and x' is a prefix of x. The period is the shortest period of y. Design an efficient algorithm to determine the period of a string of length n.

Hint: Prove that a string X is a period of a string Y iff Y is a prefix of XY

8.5 If p and q are periods (not the shortest) and $|p| + |q| < m$, then there is a period of length $|p| - |q|$ (assuming p is larger).

(This is the equivalent of Euclid's algorithm for strings).

8.6 Give an example to argue why KMP algorithm cannot handle wild-cards. You may want to extend the definition of failure functions to handle wild-cards.

Hint: Find all occurrences of the pattern $aba * a$ in the text $ababaababa...$

8.7 Geometric pattern Given a set of points on the real line with coordinates $x_1, x_2, \ldots, x_n$, we want to determine if there is a subset $x_i, x_{i+1}, \ldots, x_{i+m-1}$ with separation distances d_i $i \leq m-1$. Design an $O(n)$ algorithm for this problem.

Hint: Use the idea of the Rabin–Karp algorithm with a polynomial having coefficients from x_is.

8.8 Construct a trie for the string $0^m \cdot 1^m$ where each edge is associated with one symbol label. What is the size of the prefix tree?

Using a pair of integers to denote a substring, and using such pairs to denote a path label, show how to reduce the size of the previous trie to $O(m)$.

8.9 How would you use tries to sort s_5 given set of n strings s_i such that $\sum_{i=1}^{n} |s_i| = N$ where $|s_i|$ is the length of string s_i? Analyze your algorithm and compare it to the string sorting algorithm described in Section 3.3.

8.10 Design efficient algorithms to support the following operations on tries
(i) Keyword set matching (ii) Longest common substring (iii) Matching statistics

9

Fast Fourier Transform and Applications

Fast Fourier transform (FFT) is one of the most commonly used algorithms in engineering, and there are dedicated hardware chips which perform this algorithm. Since its development in the 1960s, it has become an indispensable algorithm in many areas of science and engineering. The basic ideas behind the algorithm are rooted in the divide and conquer paradigm, and these ideas are used to design dedicated hardware for FFT as well. We explain FFT by applying it to the problem of multiplying two polynomials. It is easy to show that two degree n polynomials can be multiplied in $O(n^2)$ time if they are stored in suitable data structures. However, FFT allows this computation to be done in $O(n \log n)$ time. As a result, lot of cryptographic algorithms based on polynomial evaluation use FFT as a tool. At the end of this chapter, we will discuss a hardware implementation of FFT and other applications of this algorithm.

9.1 Polynomial Evaluation and Interpolation

A polynomial $\mathcal{P}(x)$ of degree $n-1$ in indeterminate x is a power series with maximum degree $n-1$ and has the general form $a_{n-1}x^{n-1} + a_{n-2}x^{n-2} + \ldots + a_1 x + a_0$, where a_is are coefficients over some field, typically the complex numbers $\mathbb{C}$. One way of storing a polynomial would be to store the coefficients $a_0, \ldots, a_{n-1}$ in an array or a list. This could lead to wastage of space if most of these coefficients are 0. A more suitable data structure is to store the non-zero coefficients (in the order of the degree of the polynomial) in a list. Some of the most common problems involving polynomials are as follows:

Evaluation Given a value for the indeterminate x, say α, we want to compute $\sum_{i=0}^{n-1} a_i \cdot \alpha^i$:
By Horner's rule, the most efficient way to evaluate a polynomial is given by the formula

$$(((a_{n-1}\alpha + a_{n-2})\alpha + a_{n-3})\alpha + \ldots a_0$$

It is easy to check that this expression involves n multiplications and n additions. We are interested in the more general problem of evaluating a polynomial at multiple (distinct) points, say $x_0, x_1, \ldots, x_{n-1}$. If we apply Horner's rule to each of these points, then it will take $\Omega(n^2)$ operations to evaluate the polynomial at these n points. We will see that one can do this much faster if the n points $x_0, \ldots, x_{n-1}$ are chosen in a suitable manner.

Interpolation We are given n pairs $(x_0, y_0), \ldots, (x_{n-1}, y_{n-1})$, where $x_0, \ldots, x_{n-1}$ are distinct, and we need to find a degree $n-1$ polynomial P such that $P(x_i) = y_i$ for $i = 0, \ldots, n-1$.

There is a unique degree $n-1$ polynomial which has this property. This follows from the fundamental theorem of algebra which states that a non-zero polynomial of degree d has at most d roots. Indeed, if there were two such degree $n-1$ polynomials P and P', then $P(x) - P'(x)$ will also be a degree $n-1$ polynomial. But this polynomial has n roots – namely, $x_0, \ldots, x_{n-1}$. It follows that this polynomial must be the zero polynomial, and so, $P = P'$.

To show that there exists a unique polynomial, one can use *Lagrange's formula*, which gives an explicit expression for such a degree $n-1$ polynomial:

$$P(x) = \sum_{k=0}^{n-1} y_k \cdot \frac{\prod_{j \neq k}(x - x_j)}{\prod_{j \neq k}(x_k - x_j)}$$

Claim 9.1 *Lagrange's formula can be used to compute the coefficients a_is in $O(n^2)$ operations.*

The details are left as an exercise problem. One of the consequences of the interpolation is an alternate representation of polynomials as $\{(x_0, y_0), (x_1, y_1) \ldots (x_{n-1}, y_{n-1})\}$ from which the coefficients can be computed (using Lagrange's formula). We will call this representation as the *point-value* representation.

9.1.1 Multiplying polynomials

The product of two polynomials can be easily computed in $O(n^2)$ steps. Consider the polynomials $a_0 + a_1 x + \ldots + a_{n-1}x^{n-1}$ and $b_0 + b_1 x + \ldots + b_{n-1}x^{n-1}$. Then the coefficient of x^i, denoted by c_i, in their product would be $c_i = \sum_{l+p=i} a_l \cdot b_p$ for $0 \leq i \leq 2n - 2$. The coefficients c_i correspond to the *convolution* of the coefficients $(a_0, \ldots, a_{n-1})$ and $(b_0, \ldots, b_{n-1})$.

If the polynomials are given by their point-value with common $x_0, \ldots, x_{n-1}$, then the problem is considerably simpler. Indeed, if $\mathcal{P}_1(x)$ and $\mathcal{P}_2(x)$ are two polynomials, and $\mathcal{P}$ denotes their product, then $\mathcal{P}(x_i) = \mathcal{P}_1(x_i) \cdot \mathcal{P}_2(x_i)$. There is one subtlety here though. The polynomial $\mathcal{P}$ would have degree $2n - 2$, and so, we need to specify its values at $2n - 1$ distinct points. Thus, we would need that the polynomials $\mathcal{P}_1$ and $\mathcal{P}_2$ are specified using the point-value representation at $2n - 1$ common points. The efficiency of many polynomial-related problems depends on how quickly we can perform transformations between the two representations described earlier. We now show that it is possible to do so efficiently assuming that one chooses the points $x_0, \ldots, x_{n-1}$ carefully.

9.2 Cooley–Tukey Algorithm

We now describe the Cooley–Tukey algorithm to evaluate a degree $n - 1$ polynomial at n distinct carefully chosen points in $O(n \log n)$ time. We assume that n is a power of 2 (otherwise, just add zero coefficients, this will at most double the number of terms). Let us choose $x_{n/2} = -x_0, \ x_{n/2+1} = -x_1, \ldots x_{n-1} = -x_{n/2-1}$. You can verify that $\mathcal{P}(x) = \mathcal{P}_E(x^2)$ $+ x\mathcal{P}_O(x^2)$, where

$$\mathcal{P}_E = a_0 + a_2 x + \ldots a_{n-2} x^{n/2-1}$$

$$\mathcal{P}_O = a_1 + a_3 x + \ldots a_{n-1} x^{n/2-1}$$

corresponding to the even and odd coefficients and $\mathcal{P}_E, \mathcal{P}_O$ are polynomials of degree $n/2 - 1$. Then,

$$\mathcal{P}(x_{n/2}) = \mathcal{P}_E(x_{n/2}^2) + x_{n/2}\mathcal{P}_O(x_{n/2}^2) = \mathcal{P}_E(x_0^2) - x_0\mathcal{P}_O(x_0^2)$$

since $x_{n/2} = -x_0$. More generally,

$$\mathcal{P}(x_{n/2+i}) = \mathcal{P}_E(x_{n/2+i}^2) + x_{n/2+i}\mathcal{P}_O(x_{n/2+i}^2) = \mathcal{P}_E(x_i^2) - x_i\mathcal{P}_O(x_i^2), \ 0 \le i \le n/2 - 1$$

since $x_{n/2+i} = -x_i$. Therefore, we have reduced the problem of evaluating a degree $n - 1$ polynomial at n points to that of evaluating two degree $n/2 - 1$ polynomials at $n/2$ points $x_0^2, x_1^2, \ldots, x_{n/2-1}^2$. In addition, we will also need to perform $O(n)$ multiplications and additions to compute the values at the original points. To continue this reduction, we have to choose points such that $x_0^2 = -x_{n/4}^2$ or equivalently $x_{n/4} = \sqrt{-1} \cdot x_0$. This involves complex numbers even if we started with coefficients in $\mathbb{R}^1$. If we continue with this strategy of choosing points, at the jth level of recursion, we require

$$x_i^{2^{j-1}} = -x_{\frac{n}{2^j}+i}^{2^{j-1}} \ 0 \le i \le \frac{n}{2^j} - 1$$

---

[1] Depending on our choice of the field F, we can define ω such that $\omega^2 = -1$.

This leads to $x_1^{2^{\log n-1}} = -x_0^{2^{\log n-1}}$, that is, if we can choose an $\omega \in \mathbb{C}$ such that $\omega^{n/2} = -1$, then the previous conditions can be satisfied by setting $x_i = \omega x_{i-1}$. By letting $x_0 = 1$, the evaluation points are $x_i = \omega^i$, $0 \le i \le n-1$, which are $\{1, \omega, \omega^2 \ldots \omega^{n/2} \ldots \omega^{n-1}\}$. These are usually referred to as the principal nth roots of unity since $\omega^n = 1$.

Algorithm description and analysis

It will be convenient to set $\omega = e^{2\pi i/n}$, which is a primitive nth root of unity. We will evaluate the degree $n-1$ polynomial $\mathcal{P}(x)$ at $\omega^0, \omega^1, \ldots, \omega^{n-1}$. Let $\mathcal{P}_E(x)$ and $\mathcal{P}_O(x)$ be as described earlier. Recall that $\mathcal{P}(x) = \mathcal{P}_E(x^2) + x\mathcal{P}_O(x^2)$ which implies that for $0 \le i < n/2$,

$$\mathcal{P}(\omega^i) = \mathcal{P}_E(\omega^{2i}) + \omega^i \mathcal{P}_O(\omega^{2i}), \text{ and}$$
$$\mathcal{P}(\omega^{i+n/2}) = \mathcal{P}_E(\omega^{2(i+n/2)}) + \omega^{i+n/2}\mathcal{P}_O(\omega^{2(i+n/2)}) = \mathcal{P}_E(\omega^{2i}) - \omega^i \mathcal{P}_O(\omega^{2i})$$

because $\omega^n = 1$ and $\omega^{n/2} = -1$. Since ω^2 is an $(n/2)$th root of unity, we can reduce the problem to evaluating $\mathcal{P}_O$ and $\mathcal{P}_E$ at $\omega'^0, \omega'^1, \ldots, \omega'^{n/2-1}$, where $\omega' = \omega^2$. The algorithm is formally described in Figure 9.1.

Procedure FFT$(a_0, a_1 \ldots a_{n-1}, \omega)$

1 **Input** Coefficients $a_0, a_1, \ldots, a_{n-1}, \omega : \omega^n = 1$;
2 **if** $n = 1$ **then**
3 $\quad$ output a_0 ;
4 Let $\alpha_0, \alpha_1, \ldots, \alpha_{n/2-1} \leftarrow FFT(a_0, a_2, a_4, \ldots, a_{n-2})$, ω^2 ;
5 Let $\beta_0, \beta_1, \ldots, \beta_{n/2-1} \leftarrow FFT(a_1, a_3, a_5, \ldots, a_{n-1})$, ω^2 ;
6 **for** $i = 0$ *to* $n/2 - 1$ **do**
7 $\quad$ $\gamma_i \leftarrow \alpha_i + \omega^i \cdot \beta_i$;
8 $\quad$ $\gamma_{n/2+i} \leftarrow \alpha_i - \omega^i \cdot \beta_i$;
9 Output $(\gamma_0, \gamma_1, \ldots, \gamma_{n-1})$.

Figure 9.1   *FFT computation*

Clearly, the running time follows the recurrence $T(n) = 2T(n/2) + O(n)$, and so, the FFT algorithm takes $O(n \log n)$ time. Let us get back to the problem of multiplying two degree $n-1$ polynomials $\mathcal{P}_1$ and $\mathcal{P}_2$. Let $\mathcal{P}$ denote the product of these two polynomials. Since the degree of $\mathcal{P}$ can be $2n-2$ (and n is a power of 2), we evaluate $\mathcal{P}_1$ and $\mathcal{P}_2$ at $\omega^0, \omega^1, \ldots, \omega^{2n-1}$, where ω is the $(2n)$th primitive root of unity. As explained earlier, this can be achieved in $O(n \log n)$ time. Therefore, we can also find the value of $\mathcal{P}$ at these points in $O(n \log n)$ time. Now, we need to solve the reverse problem – given the value of a polynomial at the roots of unity, we need to construct the coefficients of the polynomial.

Therefore, we consider the problem of interpolation of polynomials, that is, given the values at $1, \omega, \omega^2, \ldots, \omega^{n-1}$, we find the coefficients $a_0, \ldots, a_{n-1}$. Let $y_0, \ldots, y_{n-1}$ denote the value of the polynomial at these points respectively as follows. The process of evaluation of a polynomial functions can be expressed as a matrix vector product.

$$
\begin{bmatrix}
1 & 1 & 1 & \cdots & 1 \\
1 & \omega^1 & \omega^2 & \cdots & \omega^{(n-1)} \\
1 & \omega^2 & \omega^4 & \cdots & \omega^{2(n-1)} \\
\vdots & & & & \\
1 & \omega^{n-1} & \omega^{2(n-1)} & \cdots & \omega^{(n-1)(n-1)}
\end{bmatrix}
\cdot
\begin{bmatrix}
a_0 \\ a_1 \\ a_2 \\ \vdots \\ a_{n-1}
\end{bmatrix}
=
\begin{bmatrix}
y_0 \\ y_1 \\ y_2 \\ \vdots \\ y_{n-1}
\end{bmatrix}
$$

Let us denote this by the matrix equation $A \cdot \bar{a} = \bar{y}$. In this setting, the interpolation problem can be viewed as computing $\bar{a} = A^{-1} \cdot \bar{y}$. Even if we had A^{-1} available, we still have to compute the product which could take $\Omega(n^2)$ steps. However, the good news is that the inverse of A^{-1} is

$$
\frac{1}{n}
\begin{bmatrix}
1 & 1 & 1 & \cdots & 1 \\
1 & \dfrac{1}{\omega^1} & \dfrac{1}{\omega^2} & \cdots & \dfrac{1}{\omega^{(n-1)}} \\
1 & \dfrac{1}{\omega^2} & \dfrac{1}{\omega^4} & \cdots & \dfrac{1}{\omega^{2(n-1)}} \\
\vdots & & & & \\
1 & \dfrac{1}{\omega^{n-1}} & \dfrac{1}{\omega^{2(n-1)}} & \cdots & \dfrac{1}{\omega^{(n-1)(n-1)}}
\end{bmatrix}
$$

which can be verified by multiplication with A. It is a well known fact that

$$
1 + \omega^i + \omega^{2i} + \omega^{3i} + \ldots + w^{i(n-1)} = 0 \quad \text{(Use the identity} \sum_j \omega^{ji} = \frac{\omega^{in} - 1}{\omega^i - 1} = 0 \text{ for } \omega^i \neq 1.)
$$

Moreover, $\omega^{-1}, \omega^{-2}, \ldots, w^{-(n-1)}$ also satisfy the properties of the nth roots of unity. This enables us to use the same algorithm as FFT itself that runs in $O(n \log n)$ operations.

The process of computing the product of two polynomials of degree $n - 1$ by using FFT is often referred to as the **convolution theorem**.

9.3 The Butterfly Network

If we unroll the recursion of an eight point FFT, it looks like Figure 9.2. Let us work through some successive recursive calls. Let $\mathcal{P}(x)$ be $a_0 + a_1 x + a_2 x^2 + \ldots + a_{n-1} x^{n-1}$. For brevity of notation, given the indices $i_0, \ldots, i_k$, we use $\mathcal{P}_{i_0,\ldots,i_k}$ to denote $a_{i_0} + a_{i_1} x + \ldots + a_{i_k} x^k$. Let ω be the primitive 8th root of unity. We shall use ω_i to denote ω^i.

$$
\mathcal{P}_{0,1,..7}(\omega_0) = \mathcal{P}_{0,2,4,6}(\omega_0^2) + \omega_0 \mathcal{P}_{1,3,5,7}(\omega_0^2)
$$

$$
\mathcal{P}_{0,1,..7}(\omega_4) = \mathcal{P}_{0,2,4,6}(\omega_0^2) - \omega_0 \mathcal{P}_{1,3,5,7}(\omega_0^2) \quad \text{since } \omega_4 = -\omega_0
$$

Subsequently, $\mathcal{P}_{0,2,4,6}(\omega_0^2) = \mathcal{P}_{0,4}(\omega_0^4) + \omega_0^2 \mathcal{P}_{2,6}(\omega_0^4)$ and

$$\mathcal{P}_{0,2,4,6}(\omega_2^2) = \mathcal{P}_{0,4}(\omega_0^4) - \omega_0^2 \mathcal{P}_{2,6}(\omega_0^4) \text{ since } \omega_2^2 = -\omega_0^2$$

To calculate $\mathcal{P}_{0,4}(\omega_0^4)$ and $\mathcal{P}_{0,4}(\omega_1^4)$, we compute $\mathcal{P}_{0,4}(\omega_0^4) = \mathcal{P}_0(\omega_0^8) + \omega_0^4 \mathcal{P}_4(\omega_0^8)$ and

$$\mathcal{P}_{0,4}(\omega_1^4) = \mathcal{P}_0(\omega_0^8) - \omega_0^4 \mathcal{P}_4(\omega_0^8)$$

Since $\mathcal{P}_i$ denotes a_i, there is no further recursive call. Notice that in Figure 9.2, a_0 and a_4 are the multipliers on the left-hand side. Note that the indices of the a_i on the input side correspond to the mirror image of the binary representation of i. A *butterfly* operation corresponds to the gadget $\bowtie$ that corresponds to a pair of recursive calls. The black circles correspond to '+' and '−' operations and the appropriate multipliers are indicated on the edges (to avoid cluttering, only a couple of them are indicated).

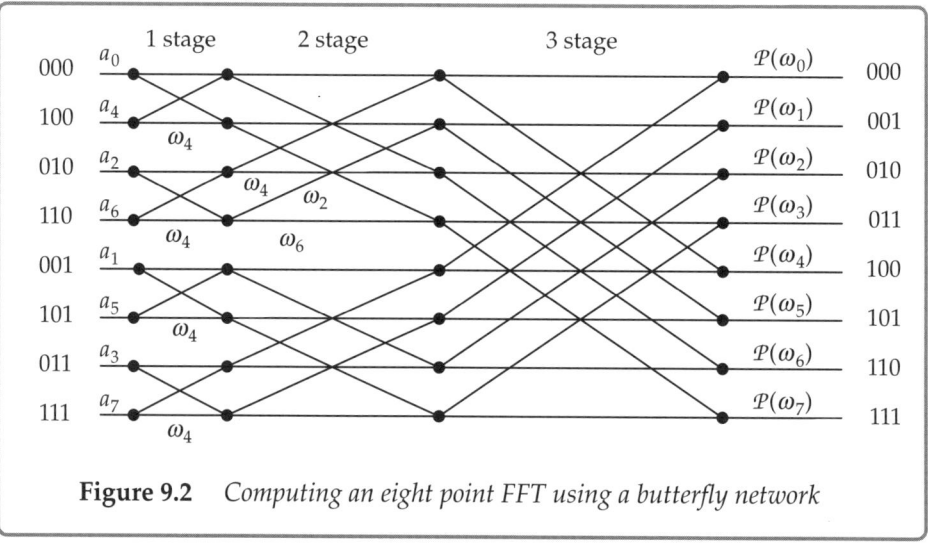

Figure 9.2   *Computing an eight point FFT using a butterfly network*

One advantage of using a network is that the computation in each stage can be carried out in parallel, leading to a total of $\log n$ parallel stages. Thus, FFT is inherently parallel and the butterfly network manages to capture parallelism in a natural manner.

9.4   Schonage and Strassen's Fast Multiplication*

In our analysis of the FFT algorithm, we obtained a time bound of $O(n \log n)$ with respect to multiplication and additions in the appropriate field – implicitly we assumed $\mathbb{C}$, the complex field. This is not consistent with the Boolean model of computation and we should be more careful in specifying the precision used in our computation. Boolean computation is a topic in itself and somewhat out of the scope of the discussion here. In reality, FFT computations are done using limited precision and operations like rounding that inherently result in numerical errors.

In other kinds of applications, like integer multiplication, we choose an appropriate field where we can do exact arithmetic. However, we must ensure that the field contains the nth roots of unity. Modular arithmetic, where computations are done using modulo a prime number, is consistent with the arithmetic done in hardware.

Observation 9.1 *In $\mathbb{Z}_m$, where $m = 2^{tn/2} + 1$ and n is a power of 2, we can use $\omega = 2^t$.*

Since n and m are relatively prime, n has a unique inverse in $\mathbb{Z}_m$ (recall extended Euclid's algorithm). Also,

$$\omega^n = \omega^{n/2} \cdot \omega^{n/2} = (2^t)^{n/2} \cdot (2^t)^{n/2} \equiv (m-1) \cdot (m-1) \bmod m \equiv (-1) \cdot (-1) \bmod m \equiv 1 \bmod m$$

We will actually work in a ring since m may not be prime. However, m and n being relatively prime, it follows that n^{-1} exists as does ω^{-1}.

Claim 9.2 *If the maximum size of a coefficient is b bits, the FFT and its inverse can be computed in time proportional to $O(bn \log n)$.*

Note that the addition of two b bit numbers takes $O(b)$ steps and that multiplications with powers of ω are multiplications by powers of two which can also be done in $O(b)$ steps. From Observation 9.1, if we do an n'-point FFT, and if n' is not a power of 2, then we may choose a $\mathbb{Z}_{m'}$ such that $m' = 2^{tn/2} + 1$, where $n \leq 2n'$ is the closest power of 2. Note that $\frac{m'}{2^{tn'/2}+1} \leq 2^{tn'/2} \times 2^{tn'/2}$, viz., the number of bits may be doubled if n' is not a power of two.

The basic idea of the multiplication algorithm is to extend the idea of polynomial multiplication. Recall, that in Chapter 1, we had divided each number into two parts and subsequently, recursively computed the solution by computing the product of smaller numbers. By extending this strategy, we divide the numbers a and b into k parts $a_{k-1}, a_{k-2}, \ldots, a_0$ and $b_{k-1}, b_{k-2}, \ldots, b_0$.

$$a \times b = \left(a_{k-1} \cdot x^{k-1} + a_{k-2} \cdot x^{k-2} + \ldots a_0 \right) \times \left(b_{k-1} \cdot x^{k-1} + b_{k-2} \cdot x^{k-2} + \ldots b_0 \right)$$

where $x = 2^{n/k}$ – for simplicity, assume n is divisible by k. By multiplying the RHS, and clubbing the coefficients of x^i, we obtain

$$a \times b = a_{k-1}b_{k-1}x^{2(k-1)} + (a_{k-2}b_1 + b_{k-2}a_1)x^{2k-3} + \ldots + a_0 b_0$$

Although in the final product, $x = 2^{n/k}$, we can compute the coefficients using *any* method and perform the necessary multiplications by an appropriate power of two (which is just adding trailing 0s). This is polynomial multiplication and each term is a convolution, so we can invoke FFT-based methods to compute the coefficients. The following recurrence captures the running time

$$T(n) \le P(k, n/k) + O(n)$$

where $P(k, n/k)$ is the time for polynomial multiplication of two degree $k - 1$ polynomials involving coefficients of size n/k. (In a model where the coefficients are not too large, we could have used $O(k \log k)$ as the complexity of polynomial multiplication.) We will have to do *exact* computations for the FFT and for that we can use modular arithmetic. The modulo value must be chosen carefully so that the following conditions are satisfied.

(i) It must be larger than the maximum value of the numbers involved, so that there is no loss of most significant bits.

(ii) It should not be too large, otherwise, operations will be expensive.

Moreover, the polynomial multiplication itself consists of three distinct phases.

1. Forward FFT transform. This takes $O(bk \log k)$ time for b bit operands.

2. Pairwise product of the values of the polynomials at the roots of unity. This will be done *recursively* incurring cost $2k \cdot T(b)$, where $b \ge n/k$.

   The factor two accounts for the number of coefficients of the product of two polynomials of degree $k - 1$.

   Using a technique called *wrapped convolution*, we can avoid this blow-up. The details of wrapped convolution are omitted from this discussion.

3. Inverse FFT, to extract the actual coefficients. This step also takes $O(bk \log k)$, where b is the number of bits in each operand.

So the previous recurrence can be expanded to

$$T(n) \le r \cdot T(b) + O(br \log r) \tag{9.4.1}$$

where $r \cdot b = n$ and we must choose an appropriate value of b. For coefficients of size b, we can argue that the maximum size of numbers during the FFT computation is $2b + \log r$ bits (sum of r numbers where each number is a pairwise multiplication of b bit numbers). Recall that $n = 2^\ell$ is a power of 2, and we will maintain this property in recursive calls. If ℓ is even, then $\ell' = 2$ else $\ell' = \frac{n-1}{2}$, where ℓ' is the new value in the recursive call. The reader may verify that balancing r, b leads to a superior solution of the previous recurrence. So r will be roughly $\sqrt{n/2}$, then $b = \sqrt{2n}$ and we can rewrite the recurrence Eq. (9.4.1) as

$$T(n) \le \sqrt{\frac{n}{2}} \cdot T(2\sqrt{n \log n} + \log n) + O(n \log n) \tag{9.4.2}$$

where we have dropped the factor 2 in the recursive call by invoking wrapped convolution and the number of bits can increase by a factor of 2 for an appropriate choice of $\mathbb{Z}_m$ as noted after Observation 9.1. An underlying assumption in the recurrence is that all the expressions are integral.

To solve the recurrence, we define $T'(n) = T(n)/n$, so that it is transformed to

$$T'(n) \cdot n \leq \sqrt{\frac{n}{2}} \cdot 2\sqrt{2n} \cdot T'(2\sqrt{2n}) + O(n \log n)$$

$$\implies T'(n) \leq 2T'(2\sqrt{2n}) + O(\log n)$$

Using an appropriate terminating condition, this yields a solution $T'(n) = O(\log n \log \log n)$ or equivalently $T(n) = O(n \log n \log \log n)$.

Claim 9.3 *With an appropriate terminating condition, say the $O(n^{\log_2 3})$ time multiplication algorithm, $T(n) \in O(n \log n \log \log n)$.*

A detailed proof of this claim is left to the reader as an exercise problem.

9.5 Generalized String Matching

Very often, we encounter string matching problems where the strings are not represented explicitly. This feature lends versatility to many applications. It gives us a way of compactly representing a set of strings and also dealing with situations when we do not have complete information about strings.[2] One of the fundamental applications is *parsing*, where we have a compact representation of (possibly infinite) strings in the form of a *grammar* and given a query string, we would like to know if the string belongs to the set described by the grammar. For example, consider the regular expression (r.e.) $(aba)^* \cdot b \cdot (ba)^*$. We would like to find all the occurrences of this r.e.[3] in a given text over alphabet $\Sigma = \{a,b\}$. One possible solution is to construct a DFA corresponding to the aforementioned regular expression and mark out the final states so that we can keep track of every occurrence of this r.e. The DFA should be such that the recognition problem can be solved in linear time in a left to right scan of the text which is an input to this automaton. However, the construction of the DFA is quite expensive and would be efficient only if the r.e. is relatively small compared to the text. This is related to the problem of transforming an NFA (non-deterministic finite automaton) to a DFA.

Consider a special case, where we have to deal with *wildcard* symbols. For example, there is a match between the strings $acb * d$ and $a * bed$ by setting the first wild card to e

---

[2] Typical situation in many biological experiments dealing with genetic sequences.
[3] We are assuming that the reader is familiar with automata theory.

and the second one as c. Here a wildcard is a placeholder for exactly one symbol. In other applications, the wildcard may represent an entire substring of arbitrary length. Unfortunately, none of the previous string matching algorithms are able to handle wildcards (see Exercise Problem 8.6).

9.5.1 Convolution based approach

For a start, assume that we are only dealing with binary strings. Given a pattern $X = a_0 a_1 a_2 \ldots a_{n-1}$ and a text $Y = b_0 b_1 b_2 \ldots b_{m-1}$, where $a_i, b_i \in \{0, 1\}$, let us view them as coefficients of polynomials. More specifically, let $\mathcal{P}_A(x) = a_0 x^{n-1} + a_1 x^{n-2} + a_2 x^{n-3} + \ldots + a_{n-1}$ and $\mathcal{P}_B(x) = b_0 + b_1 x + b_2 x^2 + \ldots + x^{m-1}$. Note that the coefficients in the two polynomials are in opposite order of the exponents which will be crucial for our application. The product of $\mathcal{P}_A$ and $\mathcal{P}_B$ can be written as $\sum_{i=0}^{m+n-2} c_i x^i$. It follows that

$$c_{n-1+j} = a_0 \cdot b_j + a_1 \cdot b_{1+j} + a_2 \cdot b_{2+j} + \ldots + a_{n-1} \cdot b_{n-1+j} \quad 0 \le j \le m+n-2$$

which can be interpreted as the dot product of $X = a_0 a_1 \ldots a_n$ and $Y(j,n) = b_j b_{j+1} \ldots b_{j+n-1}$, $0 \le j \le n-1$. Recall the notations for string matching in Chapter 8.

If we replace $\{0, 1\}$ with $\{-1, +1\}$, then we can make the following claim.

Observation 9.2 *There is a match in position j iff $c_{n-1+j} = n$.*

A match occurs iff all the n positions in the string starting from j are identical to the pattern. By taking the index-wise product, we obtain $c_{n-1+j} = \sum_{i=1}^{n} Y_{j+i} \cdot X_i = n$ iff $b_{j+i-1} = a_i \; \forall i$.

This convolution can be easily done using FFT computation in $O(m \log m)$ steps.[4] When wildcard characters are present in the pattern, we can assign them the value 0. If there are w such characters, then we can modify the previous observation by looking for terms that have value $n - w$ (Why?). However, the same may not work if we have wildcards in the text also – try to construct a counterexample.

Wildcard in pattern and text

Assume that the alphabet is $\{1, 2, \ldots s\}$ (zero is not included). We will reserve zero for the wildcard. For every position i of the pattern (assume for now that there are no wildcards in the pattern), we will associate a random number r_i from the set $\{1, 2, \ldots, N\}$ for a sufficiently large N that we will choose later. Let $t = \sum_i r_i a_i \mod N$. We will actually do all computations modulo p for a suitably large prime $p \ge N$. For simplicity of notations, we will omit the modulo notation in the remaining discussion.

Observation 9.3 *For any string $v_1, v_2, \ldots, v_n$, suppose there exists some i for which $a_i \ne v_i$. Then the probability that $\sum_i v_i \cdot r_i = t$ is less than $\frac{1}{N}$.*

---

[4] The numbers involved are small enough so that we can do exact computation using $O(\log n)$ bit integers.

Consider assigning values for all the numbers $r_j, j \neq i$. For any such fixed assignment, $\sum_j v_j r_j$ will be equal to $\sum_j a_j r_j$ only if $(v_i - a_i) r_i = \sum_{j:j \neq i} r_j (a_j - v_j)$. Since $v_i \neq a_i$ and the RHS is some fixed quantity, this condition will be satisfied by at most one value of r_i in the multiplicative modulo prime p field.

We can use this observation to build a string matching algorithm as follows. Compute the quantity t for the pattern X as done earlier. Now for every position of text we compute $\sum_{i=1}^n b_{j+i} r_i$ and check if it is equal to t – if the two are same, we declare this to be a match. Since the probability of a false match is at most $1/N$, we can choose $N \geq n \cdot m$ to ensure that the probability of a false match for any of the locations of text is small.

In the presence of wildcards in the pattern A, we assign $r_i = 0$ iff $a_i =^*$ (instead of a random non-zero number) and the same result holds for positions that do not correspond to wildcards that are blanked out by 0. The number $t = \sum_{j:X_j \neq *} r_j \cdot X_j$ acts like a *fingerprint* or a *hash function* for the pattern.

When the text has wildcards, then the *fingerprint* cannot be fixed and will vary according to the wildcards in the text. The fingerprint t_k at position k of the text can be defined as

$$t_k = \sum_{j=1}^n \delta_{j+k-1} \cdot r_j \cdot a_j$$

where $\delta_i = 0$ if $Y_i = *$ and 1 otherwise where t_k is the fingerprint corresponding to $Y(k,n)$. Recall that $r_j = 0$ if $a_j = *$.

Now we can compute all the t_k values using convolution. Indeed, for the pattern X, we define a string F where $F_j = a_j r_j$, if a_j is not a wildcard, otherwise it is 0. For the text Y, we construct a 0–1 string C which is 1 whenever we have a non-wildcard entry and 0 whenever we have a wildcard entry. Now, we can construct the convolution of these two strings to find all the t_k values. To check for a pattern match, we just need to compute the analogous values t'_k which are the same as t_k with a_j replaced by b_{k+j-1}. Again, this is a convolution of two strings, that can be computed using FFT. There is a possible match at position k of text iff $t_k = t'_k$.

The probability of error (false match) is calculated along similar lines. Thus, the algorithm takes $O(m \log m)$ time. Figure 9.3 illustrates the algorithm on the pattern $X = 321*$.

i	1	2	3	4	5	6	7	8	9	10	11
Y_i	2	2	*	3	*	2	*	3	3	*	2
δ_i	1	1	0	1	0	1	0	1	1	0	1
t_k	9	12	8	12	8	12	8	13	-	-	-
t'_k	3	6	12	13	8	11	0	13	-	-	-

Figure 9.3 *Matching with wildcards: The pattern is $X = 3\ 2\ 1\ *$
and $r_1 = 6, r_2 = 4, r_3 = 11, r_4 = 0$ all chosen from $[1 \ldots 16]$
corresponding to $p = 17$. Since $t_5 = t'_5$ and $t_8 = t'_8$ it can
be verified that X matches with $*\ 2\ *\ 3$ and with $3\ 2\ *\ 2$
respectively and there are no other matches.*

Further Reading

The FFT algorithm was discovered by Cooley and Tukey in 1965 [36], and has found
applications in diverse areas. The use of FFT for pattern matching is due to Fisher and
Patterson [47]. However, because of superlinear running time, it is not the preferred
method for simple string matching for which KMP and Karp–Rabin are more efficient.
The application to string matching with wildcards was shown by Kalai [72].

Exercise Problems

9.1 Show how to compute the polynomial using Lagrange's interpolation formula in $O(n^2)$
operations.

9.2 Describe an efficient algorithm to evaluate a degree n univariate polynomial $P(x)$ at n
arbitrary points $x_1, x_2, \ldots, x_n$ (not necessarily roots of unity). You may assume that polynomial
division takes the same asymptotic time as polynomial multiplication.

   Hint: Use the following observation which is similar to the remainder theorem. Let $D_{i,j} =$
   $\Pi(x - x_i)(x - x_{i+1}) \cdots (x - x_j)$ and let $P(x) = Q_{1,1}(x)D_{1,n} + R_{1,1}(x)$. Then, $P(x_i) = R_{1,1}(x_i)$,
   where the degree of $R_{1,1}$ is less than $D_{1,n}$. To compute $R_{1,1}(x_i)$, we can apply a similar
   decomposition, once with $D_{1,n/2}$ and $D_{n/2+1,n}$ recursively. This defines a tree where at each
   node we do a polynomial division (of degree $n/2^i$ at distance i from the root). At the leaf
   nodes, we have the answers.

9.3 Can a claim similar to Observation 9.2 be proved for the alphabet $\{0, 1\}$ without mapping
them to $\{+1, -1\}$? Explain by giving examples.

9.4 Prove Claim 9.3 rigorously.

Hint: Use appropriate induction hypothesis and bounds like $\frac{\log n}{2} + \log\log n \leq \frac{2\log n}{3}$.

9.5 The RSA cryptosystem described in Section 3.1.2 involves exponentiation of very large numbers. For a k-bit RSA cryptosystem, i.e., $n = p \cdot q$, where n, p, q are roughly k bits, what is the complexity of encryption and decryption if we use the FFT based multiplication algorithm? Recall that we also need to compute the multiplicative inverse but here we are only looking at the complexity of the exponentiation.

9.6 A Toeplitz matrix $A^{n \times n}$ has the property that $A_{i,j} = A_{i+n,j+n}$ for all i, j. In other words, the elements along any diagonal are identical. Design a fast algorithm to compute the matrix vector product $A\bar{v}$ for some given n-vector $\bar{v}$.

Hint: Use a compact representation for the Toeptlitz matrix and reduce this to a convolution problem.

9.7 Consider the set $S_n = \{n+1, 2n+1, \ldots, i \cdot n + 1, \ldots\}$, where n is a power of 2. The following result is due to Dirichlet.

Theorem Any arithmetic progression of the form $a + i \cdot b$, $i \in \mathbb{Z}$ contains infinitely many primes if $\gcd(a, b) = 1$.

Therefore, the set S_n must contain a prime. Consider the smallest such prime $p = kn + 1$. It is known that k is $O(\log n)$. Therefore, p has length $O(\log n)$. In $\mathbb{Z}_p$, let g be a generator, i.e., $g^0, g^1, g^2, \ldots, g^{p-1}$ are exactly all the elements of $\mathbb{Z}_p$ (not necessarily in that order). Then, g^k is a principal nth root of unity. Note that $g^{kn} = g^{p-1} \equiv 1 \bmod p$ from Fermat's theorem.

Use this observation to reduce the size of the ring for computing FFT from roughly n bits to $O(\log n)$ bits.

9.8 Construct an example to illustrate that by taking care of the wild-card characters in the fingerprint function of the pattern there could be incorrect results if the text also contains wild-card characters.

9.9 Suppose we place a charge q_i at the integer coordinate i on the unit line (here q_i could be positive or negative). Also suppose charges are placed at coordinates lying in the range $[0, n]$. Give an $O(n \log n)$ time algorithm to compute the force on each point charge.

(Hint: Frame this problem as computing convolution of two sequences, and then use the FFT algorithm.)

9.10 Let X and Y be two sets containing integer values in the range $[0, M]$. The set $X + Y$ is defined as $\{x + y : x \in X, y \in Y\}$. Given an $O(M \log M)$ time algorithm to compute the size of $X + Y$. Note that $X + Y$ is not a multi-set. For example, if $X = \{1, 2\}, Y = \{3, 4\}$, then $X + Y = \{4, 5, 6\}$.

(Hint: Define degree M polynomials for each of the sets, and use the FFT algorithm.)

Graph Algorithms

Graphs are one of the most versatile and useful data structures that have numerous applications involving representation of relationship between a set of objects. Recall that a graph is represented by a pair $G = (V, E)$, where V denotes the set of vertices and E denotes the set of edges. Depending on whether G is directed or undirected, the pairs may or may not be ordered in E. We assume that the reader is familiar with the data structures used to store a graph. Unless specified otherwise, we assume that graphs are stored using the adjacency list data-structure, where we store the list of neighbors of each vertex (in the case of directed graphs, we separately store the lists of in-neighbors and out-neighbors). We begin by reviewing depth first search (DFS) traversal algorithm and some of its applications. Subsequently, we shall study some of the most widely used graph algorithms – shortest paths, min-cut, and a useful structure called spanners.

10.1 Depth First Search

A number of graph problems use depth first search (DFS) as the starting point. Since it runs in linear time, it is an optimal algorithm. We will explain DFS through an application. Consider the natural problem of sequencing a set of tasks. We are given a set of jobs $J_1, J_2, \ldots, J_n$. Further, a set of precedence constraints are given. A precedence constraint $J_i \prec J_k$ specifies that J_i must be completed before J_k. We want to find a feasible sequence of performing the tasks such that all precedence constraints are satisfied, or determine that such a sequence is not possible.

Example 10.1 *Set of jobs : J_a, J_b, J_c, J_d.*
Precedence constraints : $J_a \prec J_b$, $J_a \prec J_d$, $J_d \prec J_c$, $J_c \prec J_b$.
One possible sequencing is J_a, J_d, J_c, J_b that satisfies all the precedence constraints.

When does a sequence exist? What happens if we change the precedence from $J_a \prec J_b$ to $J_b \prec J_a$? Let us use a graph $G = (V, E)$ to model this problem. The set of vertices correspond to the set of jobs and a directed edge (v_i, v_k) denotes that $J_i \prec J_k$. More formally, we want to define a function $f : V \rightarrow \{1, 2, \ldots, n\}$ such that $\forall i, j \ J_i \prec J_k \Rightarrow f(i) < f(k)$.

Observation 10.1 *There is a feasible schedule if and only if there is no directed cycle in the graph.*

Clearly, there cannot be a consistent ordering if there is a cycle. Suppose there is no cycle. We claim that there must be a vertex of 0 in-degree in the graph. Suppose there is no such vertex. Let us start with a vertex v, and keep following any of the incoming edges. Since this sequence cannot go on without repeating a vertex, it is guaranteed that we will eventually find a directed cycle in the graph. This is a contradiction. Therefore, there is an in-degree 0 vertex v in the graph. We can do the corresponding task first because it does not require any other job as a prerequisite. Now we delete v from the graph and recursively solve the remaining problem. This idea can be implemented in $O(m + n)$ time. However, it turns out that DFS gives a cleaner approach to solving this problem.

We first review DFS. The algorithm is described in Figure 10.1. Whenever DFS visits a vertex, it marks it (initially, all vertices are unmarked). Then it recursively calls DFS on each of the neighbors (or out-neighbors in case of directed graphs) of this vertex which have not been marked. There is also a notion of *time* in DFS. For every vertex x, $start(x)$ is the time at which DFS(x) gets called, and $finish(x)$ is the time at which the procedure ends. Since function calls are nested, it follows that the intervals defined by $[start(x), finish(x)]$ for all the vertices x are also *laminar*, that is, either one is contained inside the other or the two are disjoint. It cannot happen that $start(x) < start(x')$ and $finish(x) < finish(x')$ for two vertices x and x'.

Observation 10.2 *For $u, v \in V$, where DFS(u) is called before DFS(v), either $start(u) < start(v) < finish(v) < finish(u)$, or $start(u) < finish(u) < start(v) < finish(v)$. This is called the bracketing property.*

One can also associate a rooted tree with DFS, often called the DFS tree. This is defined as follows: whenever the DFS(w) gets called during the execution of DFS(v), we make w a child of v – note that w must be a neighbor of v for this to happen. Since DFS(w) is called at most once, it is clear that each node will have exactly one parent, and so, we get a tree structure. Figure 10.2 illustrates the result of running DFS on the given graph.

Consider the starting times $start(v)$. In the DFS tree, the starting time of a node is always less than those of its children. Similarly, the finish time of a node is always more than those of its children. In fact, it is easy to show the following stronger property.

Observation 10.3 *The starting and finishing times correspond to pre-order and post-order numbering respectively of a DFS tree (if we set the starting time of the starting vertex as 1).*

Procedure Depth First Search of a directed graph(G)

---

1 *Input* A directed graph $G = (V,E)$ where $|V| = n$, $|E| = m$;
2 *Output* Starting and Finishing times for every vertex $v \in V$;
3 Initially all vertices are unmarked. A global counter $c = 1$;
4 **while** *Some vertex x is unmarked* **do**
5 $\quad$ $start(x) \leftarrow c$; Increment c ;
6 $\quad$ DFS (x)

Procedure DFS(v)

---

1 Mark v ;
2 **if** *there is an unmarked neighbour y of v* **then**
3 $\quad$ $start(y) \leftarrow c$; Increment c ;
4 $\quad$ DFS (y)
5 **else**
6 $\quad$ $finish(v) \leftarrow c$; Increment c

Figure 10.1 *Algorithm for Depth First Search*

Note that pre-order and post-order numbering are in the range $[1 \ldots n]$, so we will need two distinct counters for starting and finishing times which are incremented appropriately. Alternately, the counts from the global counter can be mapped to the range $\{1, 2, \ldots, n\}$ using integer sorting.

Here is another useful observation about DFS. We use the notation $u \rightsquigarrow v$ to denote the fact that there is a (directed) path from u to v.

Observation 10.4 *If u and v are two vertices with* $start(u) < start(v) < finish(v) < finish(u)$, *then* $u \rightsquigarrow v$.

This observation can be shown using the following argument: for u and v as earlier, $DFS(v)$ must get called after $DFS(u)$ is called, but before $DFS(u)$ ends. So, there must be a sequence of vertices $u = u_1, u_2, \ldots, u_k = v$ such that $DFS(u_i)$ gets called inside $DFS(u_{i-1})$. But this implies that (u_{i-1}, u_i) must be edges in the graph and therefore, $u \rightsquigarrow v$.

One can similarly prove the following observation, which is left as an exercise.

Observation 10.5 *If u and v are two vertices in a graph such that* $start(u) < finish(u) < start(v)$ *then there is no path from u to v.*

Let us now see how DFS can be used to solve the sequencing problem mentioned at the beginning of this section. Recall that we are given a directed graph G which has no cycles (also called a DAG, a directed acyclic graph), and we want to output an ordering $v_1, \ldots, v_n$ of the vertices such that all edges in G are directed from smaller to larger vertices.

Such an ordering is called topological sort of the vertices in the DAG. Note that there may not be a unique ordering here, and any order of vertices satisfying the property is a valid topological sort. It is also important to realize that the DFS also detects the presence of a cycle is the given graph in which case, we cannot do topological sort. In line 2 of the DFS algorithm, all neighbors of v that are marked should have finished DFS, else there is a cycle.

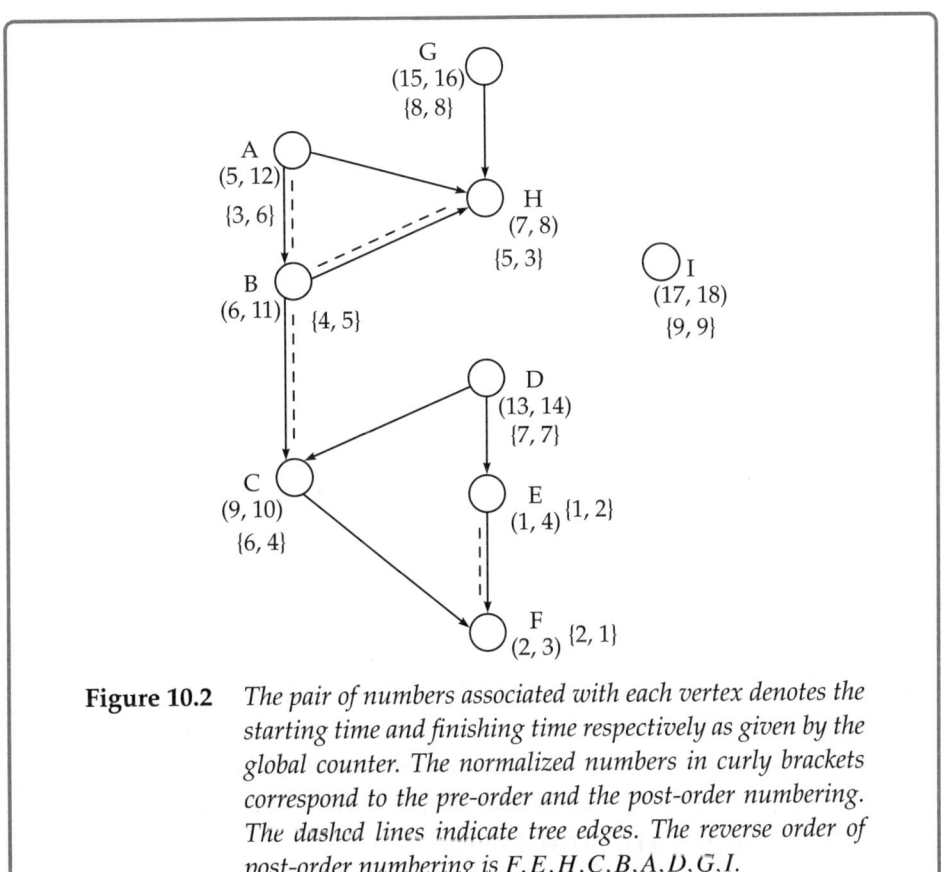

Figure 10.2 *The pair of numbers associated with each vertex denotes the starting time and finishing time respectively as given by the global counter. The normalized numbers in curly brackets correspond to the pre-order and the post-order numbering. The dashed lines indicate tree edges. The reverse order of post-order numbering is F,E,H,C,B,A,D,G,I.*

We first prove a partial converse of Observation 10.4.

Observation 10.6 *If u and v are two vertices in a DAG such that $u \rightsquigarrow v$, then* $finish(u) > finish(v)$.

Consider the two possibilities, either $start(u) < start(v)$ or $start(v) < start(u)$. In the first case, the DFS (u) will terminate before DFS (v) and the result follows. In the second case, using the bracketing property, there are two possibilities: (i) $start(v) < finish(v) < start(u) < finish(u)$, or (ii) $start(v) < start(u) < finish(v) < finish(v)$. The first condition is consistent with the observation whereas the second condition implies that $v \rightsquigarrow u$ from Observation 10.5 which contradicts that the graph is a DAG.

This leads to the following algorithm: Let us run DFS and sort the vertices in decreasing order of finish times which is given by $v_1, \dots, v_n$. We claim that this ordering is consistent with a topological sort. To see this suppose $v_i \rightsquigarrow v_j$ then the previous observation implies that $finish(v_i) > finish(v_j)$, and we have arranged the vertices in decreasing order of their finish times.

Since DFS takes linear time, we have given a linear time algorithm for computing the topological sort of a set of vertices.

10.2 Applications of DFS

A DFS on a directed graph $G = (V, E)$ yields a wealth of information about the structure of the graph. As mentioned earlier, the finish time can be used to yield the topological sort of a DAG. We now give some more non-trivial applications of DFS.

10.2.1 Strongly connected components (SCC)

In a directed graph $G = (V, E)$, two vertices $u, v \in V$ are in the same SCC iff $u \rightsquigarrow v$ and $v \rightsquigarrow u$. It is easy to verify that this is an equivalence relation on the vertices and the equivalence classes correspond to the SCCs of the given graph. We now show how DFS can be used to identify all the SCCs of a graph.

Before doing this, let us understand the structure of the SCC. Let us define a directed graph $\mathcal{G} = (V', E')$ as follows – V' corresponds to the SCCs of G, that is, for every SCC in G, we have a vertex in $\mathcal{G}$. Given two SCCs, c_1 and c_2 in G, we have a directed edge $(c_1, c_2) \in E'$ if there is an edge from some vertex in c_1 to some vertex of c_2. Note that we are abusing notation by using c_1 (or c_2) to denote a vertex in $\mathcal{G}$ and a subset of vertices in G.

It can be shown that $\mathcal{G}$ is a DAG (see Exercise 10.2).

We call a vertex in a DAG to be a *sink* if its out-degree is 0. It is easy to show that every DAG must have at least one sink (e.g., consider the last vertex in a topological sort of the graph). Similarly, define a source vertex as a vertex whose in-degree is 0.

To determine the SCCs of G, notice that if we start a DFS from a vertex of a *sink* component c' of $\mathcal{G}$, then the DFS traversal will visit precisely all the vertices in c'. Let us see why. Say u is a vertex of c and we start DFS from u. We know that v is any other vertex of c, then $u \rightsquigarrow v$, and so, we will visit v while performing DFS starting from u. Conversely, suppose we visit a vertex v while performing DFS from u. We claim that v must be in c as well. Suppose not. Say $v \in c'$, where c' is some other SCC. Since $u \rightsquigarrow v$ (because we visited v), consider a path from u to v. We know that $u \in c$ and $v \notin c$. Therefore, there must be an edge in this path which goes from a vertex in c to a vertex not in c. This contradicts the assumption that c is a sink vertex in G. Thus, DFS from u will reveal exactly the vertices in c, and so, we can identify one SCC of G. To identify another component, we can use the following idea: remove all the vertices of c from G, and repeat the same process. In this manner, we can find all the SCCs of G (Figure 10.3). This strategy works except that we do not know G.

We then have the following property whose proof is left to the reader.

Observation 10.7 *If u and v are two vertices in G belonging to SCC c and c' respectively with $c \leadsto c'$ in G, then $u \leadsto v$ and there is no path from v to u.*

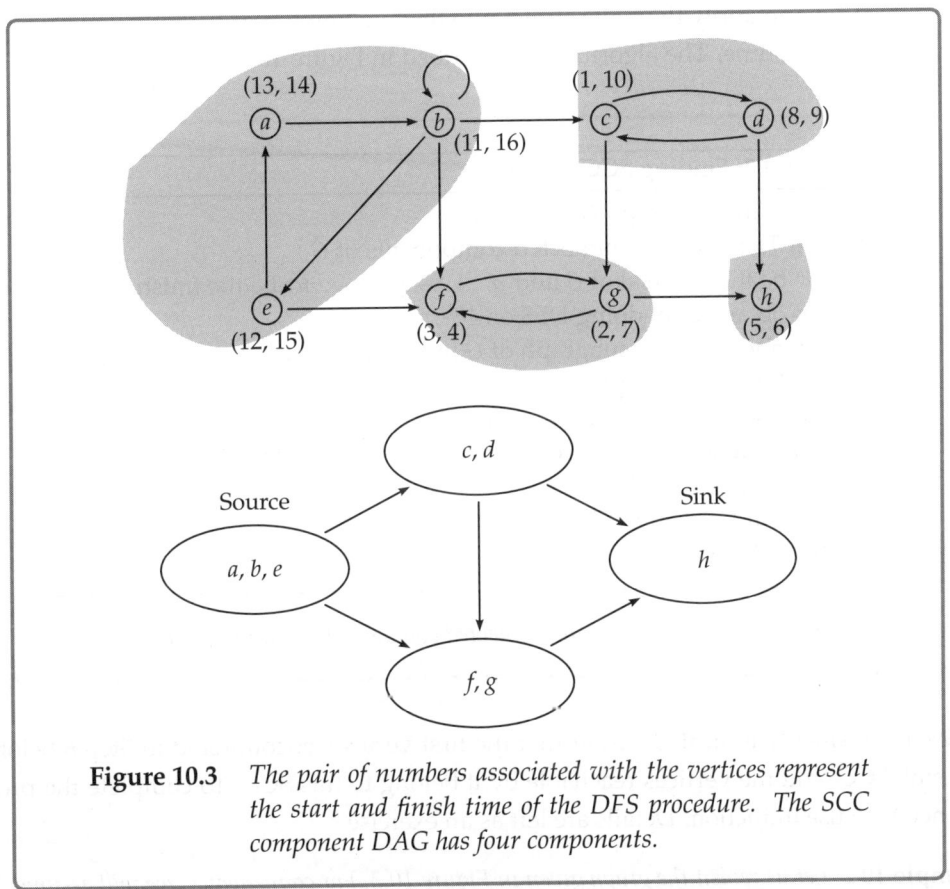

Figure 10.3   *The pair of numbers associated with the vertices represent the start and finish time of the DFS procedure. The SCC component DAG has four components.*

Since G is not explicitly available, we will use the following strategy to determine a sink component of G. First, reverse the edges of G – call it G^R. The SCCs of G^R are the same as those of G, but the sink components and source components of G are interchanged. If we do a DFS in G^R, then the vertex with the largest finish time is in a sink component of G. Let us see why. Let v be the vertex in c with the largest finish time, and suppose it is in a SCC c where c is not a sink component. Let c' be a sink component such that $c \leadsto c'$ in G (check that such a sink component exists). Let u be a vertex in c'. From Observation 10.7, we know that $v \leadsto u$, but there is no path from u to v in G.

Since $finish(v) > finish(u)$, the bracket property implies the following possibilities: (i) $start(u) < finish(u) < start(v) < finish(v)$, or (ii) $start(v) < start(u) < finish(u) < finish(v)$. Note that in G^R, $u \leadsto v$ and so the first property is ruled out because of Observation 10.4. The second property implies that $v \leadsto u$ in G^R, and so, $u \leadsto v$ in G, again a contradiction.

This enables us to output the SCC corresponding to the sink component of G using a DFS in G^R where the vertices are ordered according to the decreasing finish time G^R. Once this component (a sink component) is deleted (where the vertices and the induced edges are deleted), we can apply the same strategy to the remaining graph, that is, start with the next highest finish time. The algorithm is described in Figure 10.4.

Algorithm 2: Finding SCC of (G)

1 *Input* A directed graph $G = (V, E)$;
2 *Output* The strongly connected components of G ;
3 Let G^R be the reversal of G and $p^r : V \rightarrow \{1, 2, \ldots, n\}$ be the finish
    times obtained by doing DFS on G^R ;
4 $W \leftarrow V$ and $G(W)$ be subgraph of G induced by W ;
5 **while** W *is not empty* **do**
6   Choose $v = \arg\max_{w \in W} \{p^r(w)\}$;
7   Let V' denote the vertices of W reachable from v by DFS(v) in
        $G(W)$;
8   Output V' as an SCC of G ;
9      $W \leftarrow W - V'$

Figure 10.4 *Finding strongly connected components using two DFS*

We have already indicated earlier that the first vertex v encountered in Step 6 belongs to a sink SCC, and the vertices reachable by it belong to this SCC. To complete the proof, one needs to use induction. Details are left as an exercise.

Example 10.2 *Let us revisit the graph given in Figure 10.3. For convenience, we will assume that the DFS numbering corresponds to G^R, that is, the original graph has all the edges reversed. The component SCC of the original graph must have the direction of the edges reversed. Since b has the highest value of finish time, we begin DFS from b (in the reverse of the graph). The reachable vertices are a, e which correctly corresponds to a sink component.*

10.2.2 Biconnected components

Biconnected graphs can be defined on the basis of vertex connectivity as well as equivalence classes on edges. Undirected graphs that cannot be disconnected by removing one vertex (along with the incident edges) are *biconnected*. If the removal of a vertex disconnects a graph, then such a vertex is called an *articulation* point.

It will follow from the max-flow min-cut theorem (explained in Chapter 11) that for any biconnected graph, there are at least two vertex disjoint paths between every pair of vertices. The generalization of this observation is known as Whitney's theorem, where a graph is said to be k connected if removal of any subset of $k-1$ vertices does not disconnect the graph.

Theorem 10.1 (Whitney) *A graph is k-connected if and only if there are at least k vertex disjoint paths between any pair of vertices.*

It is clear that if there are k vertex disjoint paths then at least k vertices must be removed to disconnect the graph. However the proof of the converse is non-trivial and is not included here (see Chapter 11). We are interested in 2-connectivity in this section.

As mentioned earlier, the notion of biconnectivity can also be defined in terms of an equivalence relation on edges. We define an equivalence relation on edges as follows.

Definition 10.1 *Two edges belong to the same BCC (biconnected component) iff they belong to a common (simple) cycle. A singleton edge is considered as a biconnected component.*

It is not difficult to show that this is an equivalence relation. Each equivalence class forms a biconnected subgraph, also called, a biconnected component. Therefore, a graph is biconnected if and only if it has only one biconnected component. Although we have defined the BCCs in terms of edges, one can also define this in terms of vertices. A maximal subgraph which is biconnected is called a *block*. One can show that if two different blocks have a common vertex, then this vertex must be an articulation point. In fact, consider the following bipartite graph G – on one side of G, we have all the articulation points, and on the other side we have one vertex for each block in the graph. Further we add an edge between a block B and an articulation point v if v belongs to B. One can show that this graph is a tree (see Exercises). This graph is also called the *component graph*.

Now one can show that the set of edges in a block form a BCC, and conversely, the subgraph consisting of edges in a BCC (and the incident vertices) form a block.

One obvious procedure to check biconnectivity is to test if there is an articulation point in a graph. For every vertex $v \in V$, check if $G - \{v\}$ is connected. This takes $O(n \cdot (m+n))$ time which we will try to improve by using an alternate characterization. Moreover, we will also determine the biconnected components if the graph is not biconnected.

DFS on an undirected graph $G = (V, E)$ partitions the edges into T (tree edges) and B (back edges). Based on the DFS numbering (pre-order numbering or start times) of the vertices, we can direct the edges of T from a lower to a higher number and the edges in B from a higher to a lower number. Let us denote the DFS numbering by a function $d(v)$ $v \in V$.

The basic idea behind the BCC algorithm is to detect articulation points. If there are no articulation points, then the graph is biconnected. Simultaneously, we also determine the BCC. The DFS numbering $d(v)$ helps us in this objective based on the following intuitive observation.

Observation 10.8 *If there are no back-edges out of some subtree of the DFS tree T_u rooted at a vertex u that leads to a vertex w with $d(w) < d(u)$, then u is an articulation point.*

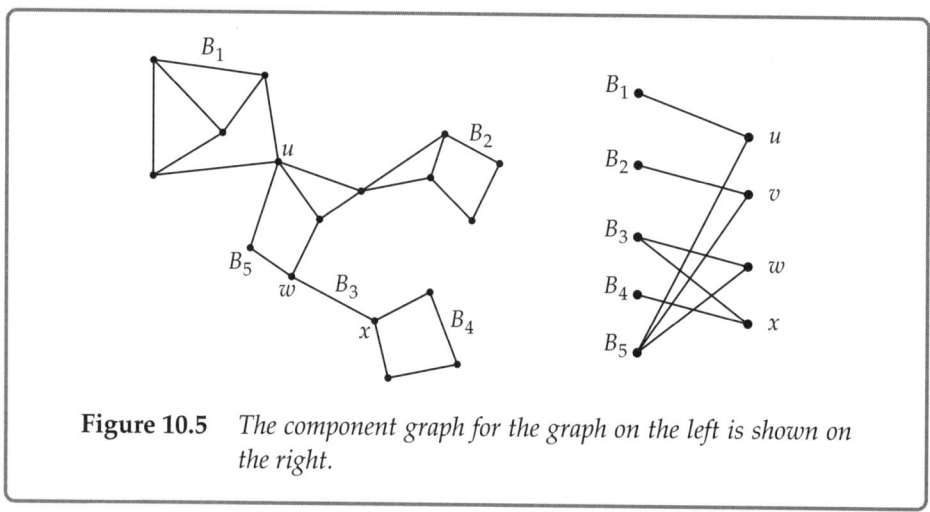

Figure 10.5   *The component graph for the graph on the left is shown on the right.*

This follows because all paths from the subtree to the remaining graph must pass through u making u an articulation point. To detect this condition, we define an additional numbering of the vertices based on DFS numbering. Let $h(v)$ denote the minimum of $d(u)$, where (v,u) is a back-edge, i.e., it is the highest node in the DFS tree which can be reached from u by a back-edge. We also define a quantity $LOW(v)$ with every vertex v as follows: consider the vertices in the sub-tree rooted at v. Then $LOW(v)$ is the minimum over all vertices u in this sub-tree of $h(u)$, i.e., it is the highest node one can reach by following a back-edge from this sub-tree. It is easy to see that $LOW(v)$ can be computed by the following recurrence:

$$LOW(v) = \min_{w|(v,w)\in T} \{LOW(w), h(v)\}$$

Note that $h(v)$ and $LOW(v)$ can be easily computed if v is a leaf node of the DFS tree. Using this as a base case, we can compute the $h(v)$ values and the LOW numbers simultaneously while doing the DFS (Exercise 10.6). Once the LOW numbers are known, we can check if $h(u) \geq d(v)$ for any child u of v. If so, the removal of v would disconnect all the vertices in the subtree rooted at v from the remaining graph and therefore, v is an articulation point. A special case happens if v is the root of the DFS tree, since v does not have any predecessor. In this case, v is an articulation vertex if it has more than one child in the DFS tree since all paths between the children in the subtrees must go through v.

The computation of *LOW* numbers results in an efficient algorithm for testing biconnectivity but it does not yield the biconnected components directly. For this, let us consider the component graph G (recall that this does not contain any cycles). The biconnected component that corresponds to a leaf node of G should be output as we back-up from a subtree w of v such that $LOW(w)$ is not smaller than $d(v)$ (v is an articulation point). After deleting this vertex from G, we consider the leaf components in the remaining G. The edges of a BCC can be kept in stack starting from (v,w) that will be popped out till we reach the edge (v,w).

DFS can start from an arbitrary BCC (biconnected component); this component will be last to be output. An articulation point will be the only way out of a leaf component and all its edges will be output. For a non-leaf component, it will be output after all the neighboring components are output, except for the one through which it first arrived. In other words, the DFS on the component tree has a traversal property similar to the DFS on the vertices of a tree. Formalize this argument into an efficient algorithm that runs in $O(|V|+|E|)$ steps (Exercise 10.12).

10.3 Path Problems

We are given a directed graph $G = (V,E)$ and a weight function $w : E \rightarrow \mathbb{R}$ (may have negative weights also). The natural versions of the shortest path problem are as follows

Distance between a pair Given vertices $x, y \in V$, find the least weighted path starting at x and ending at y.

Single source shortest path (SSSP) Given a vertex $s \in V$, find the least weighted path from s to all vertices in $V - \{s\}$.

All pairs shortest paths (APSP) For every pair of vertices $x, y \in V$, find the least weighted path from x to y.

Although the first problem often arises in practice, there is no specialized algorithm for it. The first problem easily reduces to the SSSP problem. Intuitively, to find the shortest path from x to y, it is difficult to avoid any vertex z since there may be a shorter path from z to y. Indeed, one of the most basic operations used by shortest path algorithms is the *relaxation* step. For any edge (u,v) in the graph, this operation performs the following step where $\Delta(v)$ is an upperbound on the shortest length to v:

    Relax(u, v)
    if $\Delta(v) > \Delta(u) + w(u,v)$ then $\Delta(v) = \Delta(v) + w(u,v)$

Initially, it is set to ∞ but gradually its value decreases till it becomes equal to $\delta(v)$ which is the actual shortest path distance (from a designated source vertex). In other words, the following invariant is maintained for all vertices v: $\Delta(v) \geq \delta(v)$.

The other property that is exploited by all algorithms is as follows.

Observation 10.9 Subpath optimality

Let $s = v_0, v_1, v_2, \ldots, v_i, \ldots, v_j, \ldots, v_\ell$ be the shortest path from v_0 to v_ℓ. Then for any intermediate vertices, v_i, v_j, the subpath $v_i, v_{i+2}, \ldots, v_j$ is also a shortest path between v_i and v_j.

This follows from a simple argument by contradiction. Moreover, any shortest path algorithm using the relaxation step would compute the shortest path to v_i before v_j for $j > i$. In particular, once the shortest path to v_j is successfully computed, viz., $\delta(v_j) = \Delta(v_j)$, then $\delta(v_{j+1}) = \Delta(v_{j+1})$ the next time edge (v_j, v_{j+1}) is relaxed.

10.3.1 Bellman–Ford SSSP algorithm

The Bellman–Ford algorithm is essentially based on the following recurrence

$$\delta(v) = \min_{u \in In(v)} \{\delta(u) + w(u,v)\}$$

where $In(v)$ denotes the set of vertices $u \in V$ such that $(u,v) \in E$. The shortest path to v must have one of the incoming edges into v as the last edge. The algorithm (Figure 10.6) actually, maintains upper bounds $\Delta(v)$ on the distance from the source vertex s to all vertices v – initially $\Delta(v) = \infty$ for all $v \in V - \{s\}$ and $\Delta(s) = 0 = \delta(s)$. The previous recurrence is recast in terms of Δ

$$\Delta(v) = \min_{u \in In(v)} \{\Delta(u) + w(u,v)\}$$

Algorithm 3: SSSP $((V, E, s)$

1 Initialize $\Delta(s) = 0$, $\Delta(v) = \infty$ $v \in V - \{s\}$;
2 **for** $i = 1$ to $|V| - 1$ **do**
3 | **for** *all* $e \in E$ **do**
4 | | Relax (e)
5 Output $\delta(v) = \Delta(v)$ for all $v \in V$.

Figure 10.6   *Bellman–Ford single-source shortest path problem*

that follows from a similar reasoning. Note that if $\Delta(u) = \delta(u)$ for any $u \in In(v)$, then after applying $relax(u, v)$, $\Delta(v) = \delta(v)$.

The correctness of the algorithm follows from the previous discussion and the following key observation.

Observation 10.10 *After i iterations, all vertices v whose shortest paths consist of i edges, satisfy* $\Delta(v) = \delta(v)$.

The proof follows from a simple induction on i.

So, the algorithm finds all the shortest paths consisting of at most $n - 1$ edges within $n - 1$ iterations (Figure 10.7). However, if there is a negative cycle in the graph, then you may require more iterations and in fact, the problem is not well defined any more. We can specify that we will output simple paths (without repeated vertices) but this version is not easy to handle.[1] However, we can use the Bellman–Ford algorithm to detect negative cycles in a given graph (Exercise 10.7). Since each iteration involves $O(|E|)$ relax operations – one for every edge, the total running time is bounded by $O(|V| \cdot |E|)$.

To actually compute the shortest path, we keep track of the *predecessor* of a vertex which is determined by the relaxation step. The shortest path can be constructed by following the predecessor links (Exercise 10.9).

10.3.2 Dijkstra's SSSP algorithm

If the graph does not have negative weight edges, then we can exploit this feature to design a faster algorithm. When we have only non-negative weights, we can actually determine which vertex has its $\Delta(v) = \delta(v)$. In the case of the Bellman–Ford algorithm, at every iteration, at least one vertex had its shortest path computed but we could not identify them. Here, we maintain a partition U and $V - U$ of the set of vertices such that $s \in U$, and the following invariant holds: for any $v \in U$, $\Delta(u) = \delta(u)$. For non-negative weights, we can make the following claim.

Observation 10.11 *The vertex* $v \in V - U$ *for which* $\Delta(v)$ *is minimum, satisfies the property that* $\Delta(v) = \delta(v)$.

We prove this by contradiction. Suppose for some vertex v that has the minimum label after some iteration, $\Delta(v) > \delta(v)$. Consider a shortest path $s \leadsto x \to y \leadsto v$, where $y \notin U$ and all the earlier vertices in the path $s \leadsto x$ are in U. Since $x \in U$, $\Delta(y) \le \delta(x) + w(x, y) = \delta(y)$. Since all edge weights are non-negative, $\delta(y) \le \delta(v) < \Delta(v)$ and therefore, $\Delta(y) = \delta(y)$ is strictly less than $\Delta(v)$ which contradicts the minimality of $\Delta(v)$.

Run the algorithm (Figure 10.8) on the graph given in Figure 10.7 and convince yourself that it does not work. Then make all the weights non-negative and try again.

---

[1] This is equivalent to the longest path problem which is known to be intractable.

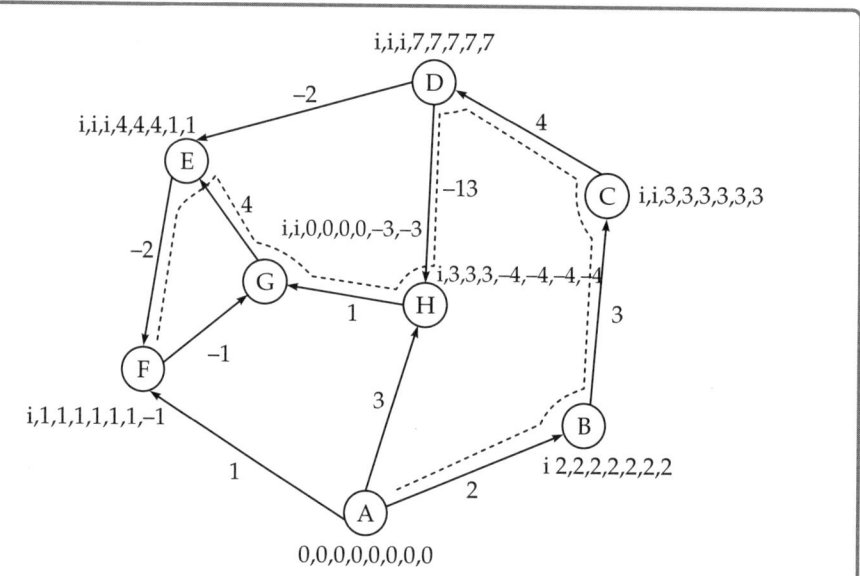

Figure 10.7   *For every vertex, the successive labels over the iterations of the Bellman–Ford algorithm are indicated where i denotes ∞. The dotted line shows the final path computed for vertex F.*

Algorithm 4: SSSP (V, E, s)

1 Initialize $\Delta(s) = 0$, $\Delta(v) = \infty \ v \in V - \{s\}$;
2 $U \leftarrow \{s\}$;
3 **while** $V - U \neq \phi$ **do**
4 $\quad$ $x = \arg\min_{w \in V-U} \Delta(w)$;
5 $\quad$ $\delta(x) = \Delta(x)$; Move x to U ;
6 $\quad$ Relax (x, y) for all edges (x, y) ;

Figure 10.8   *Dijkstra's single source shortest path algorithm*

A crucial property exploited by Dijkstra's algorithm is that along any shortest path $s \rightsquigarrow u$, the shortest path lengths are non-decreasing because of non-negative edge weights. Along similar lines, we can also assert the following

Observation 10.12 *Starting with s, the vertices are inserted into U in a non-decreasing order of their shortest path lengths.*

We can prove this by induction starting with $s - \delta(s) = 0$. Suppose it is true up to iteration i, that is, all vertices $v \in U$ are such that $\delta(v) \leq \delta(x), x \in V - U$. Let $\Delta(u) \in V - U$ be minimum, then we claim that $\Delta(u) = \delta(u)$ (from the previous observation) and $\Delta(u) \leq \Delta(x), x \in V - U$. Suppose $\delta(x) < \delta(u)$, then by an extension of the previous argument, let y be the earliest predecessor of x in $s \rightsquigarrow x$ that is not in U. Then, $\Delta(y) = \delta(y) \leq \delta(x) < \delta(u) \leq \Delta(u)$, thereby violating the minimality of $\Delta(u)$.

To implement this algorithm efficiently, we maintain a min heap on the values of $\Delta(v)$ for $v \in V - U$, so that we can choose the vertex with the smallest value in $O(\log n)$ steps. Each edge is relaxed exactly once since only the edges incident on vertices in U are relaxed – however because of the relax operation, the $\Delta()$ of some vertices may change; these changes need to be updated in the heap. This yields a running time of $((|V| + |E|) \log |V|)$.

10.3.3 All pair shortest paths algorithm

We now consider the version of the shortest path problem where for all pairs $u, w \in V$, we would like to compute $\delta(u, v)$. We can compute this by using the previous algorithm on each of the possible source vertices. The running time will be $O(|V| \cdot |V| \cdot |E|)$, that is, $O(|V|^2 \cdot |E|)$ steps. For a dense graph having close to $|V|^2$ edges, this results in $O(|V|^4)$ running time. We will try to improve this performance significantly.

We start by numbering the vertices arbitrarily with integers $\{1, 2, \ldots, |V| = n\}$. Let us assume that the graph G is represented using an adjacency matrix A_G where $A_G[i, j] = w(i, j)$ which is the weight of the edge (i, j). If there is no edge, then $w(i, j) = \infty$. We define $D_{i,j}^k$ as the length of the shortest path between vertex i to vertex j that does not use any *intermediate* vertex numbered higher than k (i, j are not included among the intermediate vertices). This restricts the paths for a given value of k but since all vertices are numbered $[1..n]$, $\delta(i, j) = D_{i,j}^n$. Moreover, we define $D_{i,j}^0 = w(i, j)$. The following recurrence leads to an efficient dynamic programming based algorithm for all $i, j \in \{1, 2, \ldots, n\}$.

$$D_{i,j}^k = \begin{cases} w(i, j) & \text{if } k = 0 \\ \min\{D_{i,j}^{k-1}, D_{i,k}^{k-1} + D_{k,j}^{k-1}\} & 1 \leq k \leq n \end{cases} \tag{10.3.1}$$

The reasoning is based on comparing $D_{i,j}^k$ and $D_{i,j}^{k-1}$. If the former does not use any vertex numbered k, then $D_{i,j}^k = D_{i,j}^{k-1}$. Otherwise, the shortest path containing k comprises two subpaths – one from i to k (that does not use vertices numbered k) and the remaining path from k to j (again that does not use vertex k). These paths correspond to $D_{i,k}^{k-1}$ and $D_{k,j}^{k-1}$ respectively. The reader may also ponder about why k cannot be visited multiple times in the shortest path between i to j.

The remaining details of refining the recurrence to an algorithm is left as an exercise (Exercise 10.10(b)). We would like to draw attention to computation of the actual paths.

Since each path can be of length $|V| - 1,$[2] the total length of the paths can add up to $\Omega(n^3)$. The reader is encouraged to design such a graph.

Instead, we can exploit the *subpath optimality* to reduce the storage. We will only store the first edge of the path $P_{i,j}$ (the sequence of vertices in the shortest path from i to j). Suppose this is i_1, then we can find the next vertex with the entry $P_{i_1,j}$. If the shortest path $P_{i,j} = i = i_0, i_1, i_2 \ldots i_m = k$, then we will look up the matrix $m - 1$ times which is optimal in terms of the path length.

10.4 Computing Spanners for Weighted Graphs

Given a weighted undirected graph, we can construct a minimal spanning tree which is a subgraph with minimum weight that preserves connectivity. A shortest path tree from a source vertex is a subgraph constructed using the Dijkstra or the Bellman–Ford algorithm that preserves the shortest path distance from the source vertex. What if we want to construct a subgraph that preserves the shortest path between all pairs of vertices and has a significantly smaller size compared to the original graph? Clearly, we cannot delete any edge from the tree since it will disconnect the graph. One would be tempted to conjecture that it could be possible in denser subgraphs, especially when there are $\Omega(n^2)$ edges in an n-node graph.

An immediate counter-example that comes to our mind is a complete bipartite graph. If we remove any edge from this graph, the alternate path must have at least three edges, so even for an unweighted graph, we cannot have a strict subgraph that preserves the distances. In fact, the previous example leads us to the following observation. The *girth* g of a graph is the length of the smallest cycle in a graph. When we remove any edge e from the graph, the alternate path $\Pi(e)$ (if one exists) along with e defines a simple cycle of length at least g, that is, the length of $\Pi(e)$ is at least $g - 1$. Since bipartite graphs have girth four, the previous example is a special case of the dependence on girth of a graph which is 4 for the family of complete bipartite graphs.

This forces us to redefine the problem in terms of allowing more flexibility in the alternate paths provided in a subgraph in terms of allowing longer paths. A t spanner is a subgraph $S = (V, E_S)$ of a given graph $G = (V, E)$ such that for any pair of vertices $u, w \in V$, the shortest path distance between u and v in S, G satisfies $\delta_S(u, w) \geq \delta_G(u, w)$. Here δ denotes the shortest path distance. The bipartite graph example tells us that we cannot have $t < 3$ for some graphs, but it is far from clear if we can get a 3-spanner for any graph and even when the edges have arbitrary real weights. Moreover, can we obtain a much smaller graph of size $o(n^2)$?

The problem of computing, a t spanner involves exploring the trade-offs between t and the size of the spanner. It has real life applications in routing networks, distributed

---

[2] This is again related to the absence of negative cycles.

computing, and in approximating shortest paths within a guaranteed ratio. The following algorithm constructs a 3-spanner having $O(n^{3/2})$ edges.

We shall use the following notations

(i) $w(u,v)$: weight of edge (u,v)

(ii) $N(v,S) = \arg\min_{u \in S} w(v,u)$ and $\mathcal{N}(v,S) = \{x|w(v,x) \leq N(v,S)\}$, i.e., it is the subset of all vertices closer than the closest sampled vertex.

If no neighbor of v is in S, then $\mathcal{N}(v,S) = \{x|(v,x) \in E\}$, that is, all the neighbors of v.

The algorithm in Figure 10.9 has two distinct stages. In stage 1, we are building *clusters* $C(x)$ and in the second stage, we are joining vertices to other clusters (Figure 10.9). A *cluster* $C(x)$ is defined by a sampled vertex x and all other unsampled vertices for which x is the nearest sampled vertex. Figure 10.10 gives an illustration of the two stages. We will first prove that the set of edges output as E_S is a legitimate 3-spanner. For this, we will establish the following property.

Procedure 3-spanner(G (V,E,w))

---

1 Input : Weighted undirected graph $G = (V,E,w)$;
2 Let $R \subseteq V$ be a random sample where each vertex $v \in V$ is
    independently included in R with probability $\frac{1}{\sqrt{n}}$;
3 $E_S \leftarrow \phi$;
4 **Stage 1** ;
5 **for** $v \in V$ **do**
6     $E_S \leftarrow E_S \cup \{(v,u)|u \in \mathcal{N}(v,R)\}$;
7 For every sampled vertex $x \in R$, define a cluster
    $C(x) = \{v|N(v,R) = x\}$;
8 (all vertices v whose nearest sampled vertex is x) ;
9 **Stage 2** ;
10 **for** $v \in V$ **do**
11   **for** *all clusters* $C(x)$, $v \notin C(x)$ **do**
12        $E_S \leftarrow E_S \cup \{(v,y)\}$ where $y = N(v,C(x))$;
13 Output E_S ;

Figure 10.9   *An algorithm for weighted 3-spanner*

Claim 10.1 *For any edge $(u,v) \in E - E_S$, there is an alternate path Π consisting of at most three edges such that the weight of each of those edges is at most $w(u,v)$.*

Once this claim is proved, then the 3-spanner property follows[3] easily. For this, we consider two kinds of edges (u,v) – refer to Figure 10.10 for an illustration.

- *Intracluster missing edges* If $u,v \in C(x)$ for some x, then $(u,x) \in E_S$ and $(v,x) \in E_S$. You can think of x as the center of the cluster $C(x)$. Hence, there exists a path $\{u,x,v\}$ in the spanner between u and v. Since $v \notin \mathcal{N}(u,R)$, $w(u,x) \le w(u,v)$; recall that $\mathcal{N}(u,R)$ consists of all neighbors of u whose weights are smaller than $w(u,x)$. Similarly, $w(v,x) \le w(u,v)$ and there exists a path with two edges between u,v with weight no more than $2w(u,v)$.

- *Intercluster missing edges* Suppose $v \in C(x)$ but $u \notin C(x)$. In stage 2, an edge (u,y) is added where $y = N(u,C(x))$ (the least weight edge from u to the cluster $C(x)$). Clearly, $y \in C(x)$. Consider the path $\{u,y,x,v\}$ – all the three edges belong to E_S. Since $u \notin \mathcal{N}(v,R)$, $w(v,x) \le w(u,v)$. Similarly, $u \notin \mathcal{N}(y,R)$, so $w(y,x) \le w(y,u)$. Moreover, since (u,y) was added in stage 2, $w(u,y) \le w(u,v)$. So

$$w(u,y) + w(y,x) + w(x,v) \le w(u,y) + w(u,y) + w(u,v) \le 3w(u,v)$$

This implies that the weight of the edges in the path $\{u,y,x,v\}$ is no more than $3w(u,v)$.

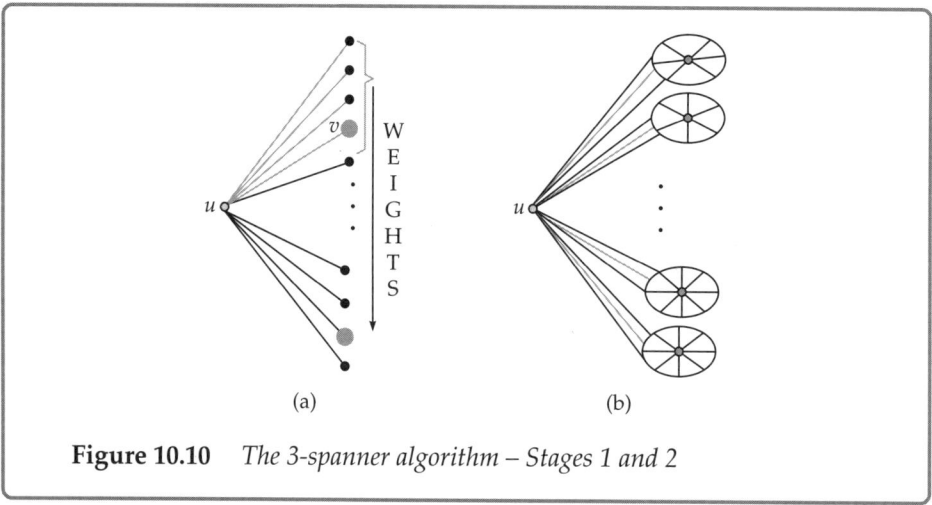

Figure 10.10   *The 3-spanner algorithm – Stages 1 and 2*

We now proceed to prove a bound on the size of the spanner. We will actually bound the expected number of edges output as E_S. Since each vertex is chosen independently with probability $\frac{1}{\sqrt{n}}$, the expected size of $|R|$, which is also the expected number of clusters, is $\sqrt{n}$. Consider the view from each vertex u as shown in Figure 10.10(a). From the vertex u, all the edges before the first sampled vertex v (hypothetically ordered in increasing weights) are

---

[3] Not every 3-spanner needs to satisfy this property.

included in E_S. Let the number of such edges be denoted by X_u which is a random variable taking values between 1 and $\deg(u)$. To estimate $E[X_u]$, note that X_u is equal to k if the first $k-1$ neighbors of u are not chosen and the kth neighbor is selected in R. The probability of this event is exactly $(1-1/\sqrt{n})^{k-1} \cdot 1/\sqrt{n}$. Of course if the degree of u is less than k, then the probability of this event is 0. From this, we see that the expectation of X_u is at most $\sum_k k \cdot (1-1/\sqrt{n})^{k-1} \cdot 1/\sqrt{n}$, which is equal to $\sqrt{n}$. So the total expected number of edges included in stage 1 is $\sum_{u \in V} X_u$ whose expectation is $\mathbb{E}[\sum_{u \in V} X_u] = \sum_{u \in V} \mathbb{E}[X_u] \leq n \cdot \sqrt{n} = O(n^{3/2})$.

In stage 2, each vertex can add 1 edge for each of the $\sqrt{n}$ clusters (expected), that is again bounded easily by $n \cdot \sqrt{n} = O(n^{3/2})$.

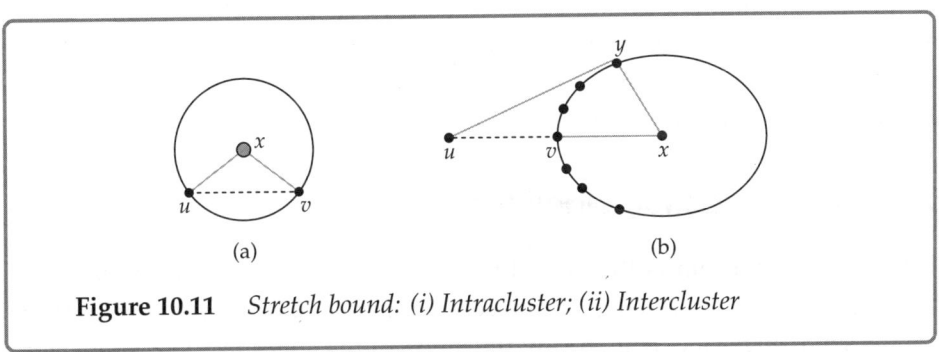

Figure 10.11   *Stretch bound: (i) Intracluster; (ii) Intercluster*

Lemma 10.1 *The expected size of the spanner, output by the previous algorithm is $O(n^{3/2})$ and as stated before, is the best possible.*

To analyze the running time of this algorithm, we note that in stage 1, each vertex u can identify $\mathcal{N}(u,R)$ in time proportional to its degree, $\deg(u)$. This implies an $O(|E|)$ bound. In stage 2, each vertex must identify all neighbors of every cluster and then choose the closest neighbor from each cluster. This can be done by sorting all the vertices on their cluster label (there are $\sqrt{n}$ expected number of clusters) as well as the label u. There are $|E|$ tuples of the form (u,c), where $u \in [1, \dots, |V|]$ and $c \in [1, \dots, \sqrt{V}]$. Using radix sort, this can be done in $O(|V| + |E|)$ steps. Once all the edges to a cluster are adjacent, choosing the closest neighbor can be done in linear time in number of edges.

Theorem 10.2 *A 3-spanner of a weighted undirected graph $G = (V, E)$ of expected size $O(|V|^{3/2})$ can be constructed in $O(|V| + |E|)$ steps.*

10.5   Global Min-cut

A *cut* of a given (connected) graph $G = (V, E)$ is the set of edges which when removed disconnects the graph. In the next chapter, we will consider a related notion, called *s-t cut*, where s and t are two vertices in the graph. Here, an *s-t* cut is a set of edges whose removal disconnects s and t. A *min-cut* is the minimum number of edges that disconnects a graph and is sometimes referred to as *global* min-cut to distinguish it from *s-t* min-cut. The weighted version of the min-cut problem is the natural analog when the edges have non-negative associated weights. A cut can also be represented by a set of vertices S where the cut-edges are the edges connecting S and $V - S$.

It was believed for a long time that the min-cut is a harder problem to solve than the *s-t* min-cut – in fact, the earlier algorithms for min-cuts determined the *s-t* min-cuts for all pairs $s, t \in V$. The *s-t* min-cut can be determined from the *s-t* max-flow flow algorithms and over the years, there have been improved reductions of the global min-cut problem to the *s-t* flow problem, such that it can now be solved in one computation of *s-t* flow.

In a remarkable departure from this line of work, first Karger, followed by Karger and Stein developed faster algorithms (than max-flow) to compute the min-cut with *high probability*. The algorithms produce a cut that is very likely the min-cut.

10.5.1 The contraction algorithm

The basis of the algorithm is the procedure contraction described in this section. The fundamental operation $contract(v_1, v_2)$ replaces vertices v_1 and v_2 by a new vertex v and assigns the set of edges incident on v by the union of the edges incident on v_1 and v_2. If there are edges from v_1 and v_2 to a common vertex w, then we retain all these edges as parallel edges between v and w. Notice that by definition, the edges between v_1 and v_2 disappear.

Procedure Partition(t)

1 Input: A multigraph $G = (V, E)$;
2 Output: A t partition of V ;
3 **while** $|V| > t$ **do**
4 $\quad$ Choose an edge (v_1, v_2) uniformly at random ;
5 $\quad$ **contract**(v_1, v_2)

Procedure contract(u, v)

1 Merge vertices u and v into w such that all neighbours
$\quad$ of u and v are now neighbours of w ;

Figure 10.12 *Algorithm for computing t-partition*

The procedure mentioned here takes a parameter t, and keeps contracting edges till only t vertices remain. At this time, if we remove all the existing edges, it will yield a cut which partitions the graph into k connected components. The procedure *Partition*(2) produces a 2-partition of V which defines a cut. If it is a min-cut, then we are done. There are two issues that must be examined carefully.

1. How likely is it that the cut is a min-cut?

2. How do we know that it is a min-cut?

The second question addresses a more general question, namely, how does one verify the correctness of a *Monte Carlo* randomized algorithm? In most cases, there are no efficient verification procedures and we can only claim the correctness in a probabilistic sense. In our context, we will show that the contraction algorithm will produce a min-cut with probability p, so that, if we run the algorithm $\frac{1}{p}$ times, we expect to see the min-cut at least once. Among all the cuts that are output in $O(\frac{1}{p})$ runs of the algorithm, we choose the one with the minimum cut value. If the minimum cut had been produced in any of the independent runs, we will obtain the min-cut.

10.5.2   Probability of min-cut

We now analyze the algorithm. We will fix a min-cut C (the reader can easily create example graphs where there are multiple min-cuts), and compute the probability that the algorithm outputs C. If C is output, then it means that none of the edges of C has ever been contracted. Let $A(i)$ denote the event that an edge of C is contracted in the ith iteration and let $\mathcal{E}(i)$ denote the event that no edge of C is contracted in any of the first i iterations. Let $\bar{A}(i)$ denote the complement of the event $A(i)$.

If n_i is the number of vertices after i iterations (initially $n_0 = n$), we have $n_i = n - i$. We first argue that the probability that an edge of C is contracted in the first iteration is at most $\frac{2}{n}$. Indeed, let k denote the minimum degree of a vertex. Removing the edges incident to such a vertex will yield a cut, and so, the number of edges in C is at least k. Since the number of edges in the graph is at least $kn/2$ (because the sum of the degrees of vertices is twice the number of edges), it follows that the probability of $A(1)$ is at most $\frac{k}{kn/2} = \frac{2}{n}.$[4]

Thus, we see that $\Pr[\bar{A}(1)] \geq 1 - 2/n$ and similarly, $\Pr[\bar{A}(i)|\mathcal{E}(i-1)] \geq 1 - 2/n_{i-1}$. Then, using the property of conditional probability

$$\Pr[\mathcal{E}(i)] = \Pr[\bar{A}(i) \cap \mathcal{E}(i-1)] = \Pr[\bar{A}(i)|\mathcal{E}(i-1)] \cdot \Pr[\mathcal{E}(i-1)].$$

We can use this equation inductively to obtain

$$\Pr[\mathcal{E}(n-t)] \quad \geq \prod_{i=1}^{n-t} (1 - 2/n_{i-1})$$
$$= \prod_{i=1}^{n-t} \left(1 - \frac{2}{n-i+1}\right)$$
$$\geq \frac{t(t-1)}{n(n-1)}$$

---

[4] We are considering the unweighted version but the proof can be extended to the weighted version using multiset arguments.

Claim 10.2 *The probability that a specific min-cut C survives at the end of Partition(t) is at least* $\frac{t(t-1)}{n(n-1)}$.

Therefore, Partition (2) produces a min-cut with probability $\Omega(\frac{1}{n^2})$. Repeating this algorithm $O(n^2)$ times would ensure that the min-cut is expected to be the output at least once. If each contraction can be performed in $t(n)$ time, then the expected running time is $O(t(n) \cdot n \cdot n^2)$.

By using an adjacency matrix representation, the contraction operation can be performed in $O(n)$ steps. We now address the problem of choosing a random edge using the aforementioned data structure.

Claim 10.3 *An edge E can be chosen uniformly at random at any stage of the algorithm in $O(n)$ steps.*

The selection works as follows.

- Select a vertex v at random with probability $= \dfrac{\deg(v)}{\sum_{u \in V} \deg(u)} = \dfrac{\deg(v)}{2|E|}$

- Select an edge (v, w) at random with probability $= \dfrac{\#E(v,w)}{\sum_{z \in N(v)} \#E(v,z)} = \dfrac{\#E(v,w)}{\deg(v)}$

where $\#E(u,v)$ denotes the number of edges between u and v and $N(v)$ is the set of neighbors of v.

Hence, the probability of choosing any edge (v, w) is given by

$$
= \frac{\#E(v,w)}{\deg(v)} \cdot \frac{\deg(v)}{2|E|} + \frac{\#E(w,v)}{\deg(w)} \cdot \frac{\deg(w)}{2|E|}
$$

$$
= \frac{\#E(v,w)}{|E|}
$$

Therefore, this method picks edges with probability that is proportional to the number of edges between v and w. When there are no multiple edges, all edges are equally likely to be picked. For the case of integer weights, the derivation works directly for weighted sampling. Using an adjacency matrix M for storing the graph, where $M(v, w)$ denotes the number of edges between v and w, allows us to merge vertices v and w in $O(n)$ time. It is left as an exercise (Exercise problem 10.24) to design an efficient method for Partition (2).

Further Reading

There are many excellent textbooks on graph theory (see e.g., [42, 61]). The shortest path problem is one of the most fundamental algorithmic problems in graph theory. Using Fibonacci heaps, Dijkstra's algorithm can be implemented in $O(m + n \log n)$ time [53]. Thorup [141] improved this to an $O(m + n \log \log n)$ time algorithm.

Karger's min-cut algorithm [74] takes $O(n^2 m)$ time in expectation. Karger and Stein [76] extended this result to an $O(n^2 \log^3 n)$ time algorithm. The notion of a graph spanner was formally introduced by Peleg and Schaeffer [115]; the graph spanner has applications to distributed computing. Althöfer et al. [11] described a greedy algorithm based on Dijkstra's algorithm that achieved $2k - 1$ stretch bound, but with a running time of $O(mn^{1+1/k})$. This was improved by Thorup and Zwick [142] to $O(mn^{1/k})$. The algorithm presented here follows Baswana and Sen [15] who developed the first $O(m+n)$ linear time algorithm.

Exercise Problems

10.1 Graph theory

1. Show that in any graph there are at least two vertices of the same degree.
2. Given a degree sequence $d_1, d_2, \ldots, d_n$ such that $\sum_i d_i = 2n - 2$, either construct a tree whose vertices have these degrees, or show that no such tree exists.
3. Show that in a complete graph on six vertices where edges are colored red or blue, there is either a red or a blue triangle (a triangle is a set of three vertices with an edge between every pair of these vertices).

10.2 Prove that the graph G corresponding to strongly connected components in a directed graph (see Section 10.2) is a DAG.

10.3 Based on suitable modifications of the DFS algorithm, design a linear time algorithm for topological sorting or conclude that the given graph is not a DAG.

10.4 Prove rigorously the correctness of the algorithm given in Figure 10.4.

10.5 Consider the following relation on the edges of an undirected graph: two edges e and e' are related if there is a simple cycle containing them; and an edge e is always related to itself. Prove that this relation is an equivalence relation. Show that if a vertex v is incident to two edges belonging to different equivalence classes of this relation, then v is a cut vertex.

10.6 Show how to compute the quantities $LOW(v)$, $v \in V$ along with the DFS numbering in linear time using the recursive definition of LOW. This should be done simultaneously with the depth first search.

10.7 Describe an efficient algorithm to detect if a given directed graph contains a negative weight cycle.

10.8 Describe an efficient algorithm to find the second minimum shortest path between vertices u and v in a weighted graph without negative weights. The second minimum weight path must differ from the shortest path by at least one edge and may have the same weight as the shortest path.

10.9 If there is no negative cycle in a graph, show that the predecessors as defined by the Bellman–Ford algorithm form a tree (which is called the shortest path tree).

10.10 (a) Using the recurrence in Eq. (10.3.1), design and analyze the dynamic programming for computing shortest paths.

(b) How would you modify the algorithm to report the shortest paths?

10.11 Transforming to a non-negative weighted graph

If all the weights $w()$ were non-negative, one could use the Dijkstra algorithm from different vertices to compute APSP that will be faster than using the Bellman–Ford algorithm from each of the vertices. For this, we will use weights on the vertices $g : V \to \mathbb{R}$. Then, the weight of an edge (u, v) is transformed to $w'(u, v) = w(u, v) + g(u) - g(v)$. If one can ensure that $w(u, v) - (g(v) - g(u)) \geq 0$, then $w'() \geq 0$.

(i) For any pair of paths (not just the shortest path), $P = v_s = v_0, v_1 \ldots v_k = v_d$ $P' = v_s = v_0, v'_1, v'_2 \ldots v_m = v_d$. Show that $w(P) \geq w(P') \iff w'(P) \geq w'(P')$, that is, the relative ranking of the paths are preserved.

(ii) Define a function g that satisfies the required property.

(iii) Analyze the overall running time of your algorithm (including the transformation) and compare with the running time of Dijkstra's algorithm where all weights are non-negative.

10.12 Show that the BCC of an undirected graph can be computed in $O(|V| + |E|)$ steps using DFS.

10.13 The **diameter** of an undirected graph $G = (V, E)$ is defined as $\max_{u,v \in V}\{d(u, v)\}$, where $d()$ is the shortest distance between the vertices u, v. A pair that achieves the maximum distance is called a **diametral pair**.

(i) Design an algorithm for finding the diameter of a given graph without computing APSP distances.

(ii) Extend the algorithm to weighted graphs.

Hint: If x is a vertex in the shortest path between a diametral pair, then doing SSSP from x will yield this distance. Choose a random subset of appropriate size.

10.14 Given a directed acyclic graph that has maximal path length k, design an efficient algorithm that partitions the vertices into $k + 1$ sets such that there is no path between any pair of vertices in a set.

10.15 Given an undirected graph, describe an algorithm to determine if it contains an even-length cycle. Can you do the same for odd-length cycle?

10.16 Given an undirected connected graph G, define the biconnected component graph H as follows. For each BCC of G and articulation point of G, there is a vertex in H. There is an edge between vertices x and y in H if x is an articulation point in the BCC.

    1. Prove that H is a tree.

    2. Using H (or otherwise), design an efficient algorithm that adds the minimal number of edges to G to make it biconnected.

10.17 A directed graph is Eulerian if the in-degree equals the out-degree for every vertex. Show that in an Eulerian graph, there is a tour which starts at a vertex v, traverses every edge in the graph exactly once, and ends at v. Design an efficient (linear time) algorithm to find such a tour.

10.18 An n-vertex undirected graph is a scorpion if it has a vertex of degree 1 (the sting) connected to a vertex of degree 2 (the tail) connected to a vertex of degree $n-2$ (the body) which is connected to the remaining $n-3$ vertices (the feet). Some of the feet may be connected among themselves. Give an $O(n)$ algorithm to check if a given $n \times n$ adjacency matrix represents a scorpion.

10.19 Given an undirected graph, we want to orient the edges such that the resulting graph is strongly connected. Design a linear time algorithm which given an undirected graph, either outputs such an orientation or shows that no such orientation is possible.

10.20 Given an undirected graph $G = (V, E)$, create an efficient algorithm to find a maximum size (i.e., number of vertices) subgraph such that the degree of every vertex in this subgraph is at least k (or show that no such subgraph exists).

10.21 Prove Observation 10.7

10.22 Describe an efficient algorithm to find the *girth* of a given undirected graph. The *girth* is defined as the length of the smallest cycle.

10.23 For an unweighted graph, an (α, β) spanner is a subgraph that preserves any path length ρ within distance $\alpha \cdot p + \beta$, where $\alpha \geq 1$ and β is some constant. The t-spanner is a special case of $\alpha = t$ and $\beta = 0$.

Can you modify the construction to yield a $(2, 1)$ spanner?

Hint: For any path, $v_0, v_1, v_2 \ldots$, you can consider the alternate path starting from v_0 and going through $c(v_1)$, followed by v_2, then $c(v_3)$, etc, where $c(v)$ denotes the center of the cluster of v.

10.24 (a) By using an adjacency matrix representation, show that the contraction operation in the min-cut algorithm can be performed in $O(n)$ steps.

    (b) Describe a method to implement **Partition (2)** in $O(m \log n)$ steps. This will be faster for sparse graphs.

Hint: Can you use union-find?

Maximum Flow and Applications

In this chapter, we introduce the maximum flow problem. This is a problem which has numerous applications in many areas of operations research; it is also versatile enough to fit many other problems which may seem unrelated. The problem can be analyzed from the view point of traffic or water flow in a network. Consider a directed graph where edges have capacities – these can be thought of as pipes with capacities being the cross-sectional area; or in case of transportation networks, the edges can be thought of as roads linking two junctions with the capacity of an edge being the maximum rate at which traffic can flow (per unit time) through this edge. There are special 'source' and 'destination' vertices where these flows are supposed to originate and end. At every other vertex, 'flow-conservation' holds, i.e., the total incoming flow equals the total outgoing flow. We now define these concepts formally.

Given a directed graph $G = (V, E)$ and a *capacity* function $c : E \to \mathbb{R}^+$, and two designated vertices s and t (also called 'source' and 'sink' respectively), we want to compute a *flow* function $f : E \to \mathbb{R}^+$ such that

1. **Capacity constraint**

   $$f(e) \leq c(e) \;\; \forall e \in E$$

2. **Flow conservation**

   $$\forall v \in V - \{s,t\}, \;\; \sum_{e \in \text{in}(v)} f(e) = \sum_{e \in \text{out}(v)} f(e)$$

   where $\text{in}(v)$ are the edges directed into vertex v and $\text{out}(v)$ are the edges directed out of v.

As mentioned earlier, we can think of the flow as originating at s and ending into t. The *outflow* of a vertex v is defined as $\sum_{e \in \text{out}(v)} f(e)$ and the *inflow* into v is given by $\sum_{e \in \text{in}(v)} f(e)$. The *net flow* is defined as outflow minus inflow $= \sum_{e \in \text{out}(v)} f(e) - \sum_{e \in \text{in}(v)} f(e)$. From the property of flow conservation, net flow is zero for all vertices except s, t. For vertex s, which is the source, the net flow is positive and for t, the net flow is negative.

Observation 11.1 *The net flow at s and the net flow at t are equal in magnitude.*

To show this, we add the flow conservation constraints for all the vertices except s and t. We get

$$\sum_{v \in V - \{s,t\}} \left(\sum_{e \in \text{out}(v)} f(e) - \sum_{e \in \text{in}(v)} f(e) \right) = 0$$

Let E' be edges that are **not** incident on s, t (either incoming or outgoing).

For an edge $e \in E'$, $f(e)$ is counted once as incoming and once as outgoing, which cancel each other. So

$$\sum_{e \in \text{out}(s)} f(e) - \sum_{e \in \text{in}(s)} f(e) = \sum_{e \in \text{in}(t)} f(e) - \sum_{e \in \text{out}(t)} f(e)$$

Hence, the net outflow at s equals the net inflow at t. We shall denote this as the *value* of the flow f. An *s-t* flow is said to be a maximum *s-t* flow if its value is maximum among all such flows.

Computing maximum flow is a classical problem in combinatorial optimization with numerous applications. Therefore, designing efficient algorithms for maximum flow has been the subject of study for many researchers over many years. Since the constraints and the objective function are linear, we can pose it as a linear program (LP) and use an efficient (polynomial time) algorithm for the LP. However, the algorithms for LP are not known to be *strongly polynomial*[1], and we will explore more efficient algorithms.

Path decomposition of flow

Before we explore algorithms for computing a maximum flow, we will discuss a useful way of decomposing any *s-t* flow into a union of flows along *s-t* paths and cycles where the *value* of flow in a path/cycle is the same for every edge in the path/cycle from the conservation property. In particular, we will prove the following.

Theorem 11.1 *[Path Decomposition] Let f be an s-t flow of value F. Then there exists a set of s-t paths $P_1, \ldots, P_k$ and a set of cycles $C_1, \ldots, C_l$, where $k + l \leq m$, and values $f(P_1), \ldots, f(P_k)$, $f(C_1), \ldots, f(C_l)$ such that for every edge e, $f(e)$ is equal to $\sum_{i : e \in P_i} f(P_i) + \sum_{j : e \in C_j} f(C_j)$.*

---

[1] We define this class more precisely in the next chapter. This may be thought of as an algorithm that scales only with the size of the graph but not the weights of the edges.

Before we prove this theorem, it follows that the value of the flow must be the same as $\sum_{i=1}^{k} f(P_i)$ (see Exercise Problem 11.7). Further, the theorem shows that (ignoring the cycles, which do not add to the value of the flow) one can construct a flow by sending flows along s-t paths.

We prove this theorem by iteratively constructing the paths and the cycles, and then removing an appropriate amount of flow from them. To begin with, let E' be the edges with positive flow, and G' be the sub-graph induced by them. Assume that E' is non-empty, otherwise there is nothing to prove. Let $e = (u, v)$ be an edge in E' having non-zero flow. By flow conservation, either $v = t$ or the out-degree of v must be at least 1 (in G'). Following the out-going edges in this manner, we get a sequence of vertices $v = v_1, v_2, \ldots$ which either ends with t, or repeats a vertex. If a vertex gets repeated, we find a cycle in G'; otherwise, we find a path from v to t in G'. Similarly, following incoming edges from u, we will either find a cycle in G' or a path from s to u. Combining these facts, we see that either (i) G' has a cycle C, or (ii) an s-t path P.

In the former case, let $f_{\min}$ be the minimum of $f(e)$ among all the edges in C. We add the cycle C to our list (of cycles and paths) and define $f(C)$ to be $f_{\min}$. Further, we reduce the flow along all edges in C by $f_{\min}$ (note that they will still be non-negative and satisfy flow conservation). Similarly, in the latter case, define $f_{\min}$ to be the minimum of $f(e)$ among all the edges in P, and $f(P)$ to be equal to $f_{\min}$. We reduce flow along edges in P by $f_{\min}$. We repeat the same process with the new flow till it becomes 0 on all edges. It is clear that when the process stops, we have the desired flow-decomposition property. To see why we will not find more than m paths and cycles, notice that in each iteration, we reduce the flow on at least one edge to 0.

Residual graphs

The path decomposition theorem suggests that one way of finding the maximum flow is by finding appropriate paths and sending flow along them. Unfortunately, a simple greedy strategy which iteratively finds paths from s to t and sends flow along them fails. For example, consider the graph in Figure 11.1 where all edges have capacity 1. Note that the maximum flow from s to t is 2 in this example. But if we send 1 unit of flow from s to t along the path $P = s, v_1, v_4, t$, and remove all the edges with flow 1 on them (the 'saturated' edges), then we can no longer find another s to t path.

To prevent getting stuck with such solutions, we need to introduce the concept of 'un-doing' some of the past mistakes. In the example considered here, this would mean sending 1 unit of flow *back* along the edge (v_4, v_1). One way of capturing this fact is the notion of *residual* graphs. Given a graph G and a flow f in G, the residual graph G_f is defined as follows: the set of vertices is same as that of G. For every edge e in G with flow $f(e)$ and capacity c_e, we consider the following cases (which may not be disjoint): (i) $f(e) < c(e)$: then we add the edge e to G_f as well; the *residual capacity* r_e of e in G' is defined as $c(e) - f(e)$: note that this is the extra flow we can still send on e without violating the capacity constraint, (ii) $f(e) > 0$: we add the edge e' which is the reversal of e to G_f with residual capacity $f(e)$ – sending flow along e' amounts to reducing flow along e, and so, we are allowed to send up to $f(e)$ flow along e' (because we cannot reduce the flow along e by more than $f(e)$). If both the cases happen, then we add two copies of e pointing in opposite directions (see also Figure 11.2). For obvious reasons, the first kind of edges are called forward edges, and the second kind are called backward edges.

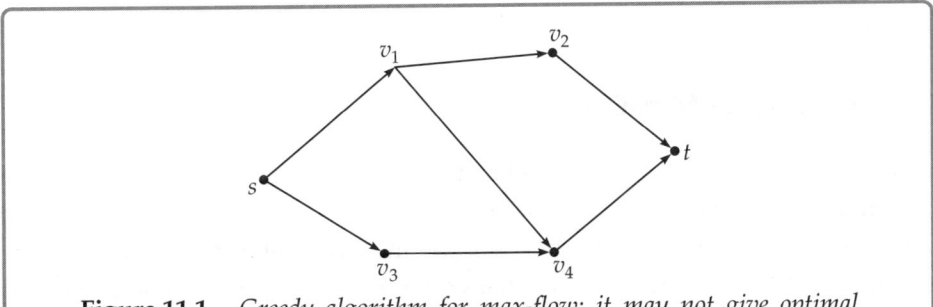

Figure 11.1 *Greedy algorithm for max-flow: it may not give optimal solution.*

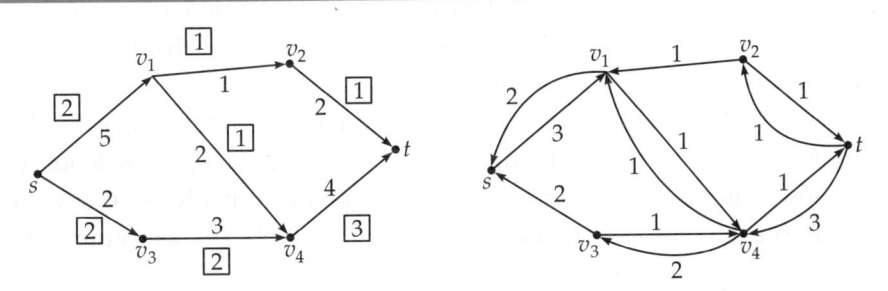

Figure 11.2 *Example of residual graph. On the left side, is a graph with flow shown in boxes, and capacities mentioned along edges. On the right side, the corresponding residual graphs is shown.*

Flow augmentation

Armed with the concept of residual graphs, we can now define a modified greedy-like algorithm for finding max-flow. Initialize the flow f to be 0 on all edges. Let f be the current flow and let G_f be the residual graph with respect to f. Note that as f changes, the graph G_f will also change. In the graph G_f, we find a path from s to t, and let δ be the minimum residual capacity of an edge in this path. We *augment* flow along this path by sending δ units of flow along it. More precisely, for every edge e, we perform the following steps: (i) If e is also present in G, i.e., is a forward edge, then we increase the flow along e by δ units; (ii) If reverse of e, call it e', is present in G, i.e., e is a backward edge, then we decrease the flow along e' by δ units. It is easy to see that we preserve both the capacity constraints and the flow conservation constraints. Such a path is called an *augmentation path*. We now have a new flow whose value is δ more than the previous flow. We repeat the process with this new flow. As long as we can find augmenting paths, we make progress. The algorithm stops when there is no augmenting path in the residual graph (with respect to the current flow). We will establish in the next section that when such a situation arises, the flow must be a maximum flow.

11.0.1   Max-Flow Min-Cut

In this section, we develop a lower bound on the value of a maximum flow. Define a *cut* to be a partition of the set of vertices into two parts such that one part contains s and the other contains t. More formally, an *s-t* cut (we will use the term, cut, for the sake of brevity) is defined as a partition of V into S, T such that $s \in S$ and $t \in T$. We will often use the notation $\bar{S}$ to denote $V - S$ that refers to the complementary partition of the cut S. In the present context, $T = \bar{S}$. The capacity of a cut is defined as $\sum_{(u,v)\in E, u\in S, b\in T} c((u,v))$, i.e., it is the total capacity of edges which leave the set S. We shall use $\text{out}(S)$ to denote edges $e = (u,v)$, where $u \in S, v \in T$, i.e., the edges which leave S. One can define the set $\text{in}(S)$ of edges coming into S similarly. When S consists of a singleton element v, instead of the notation $\in (v)$ and $\text{out}(v)$, we use $\text{in}(S)$ and $\text{out}(S)$ respectively.

It is easy to check that the value of a max-flow cannot exceed the capacity of any cut. Intuitively, even if we saturate all the edges in $\text{out}(S)$ for a cut (S, T), we will not be able to send more flow than the capacity of this cut. To prove this formally, we first write down the flow-conservation constraint for every vertex in S. If $v \in S - \{s\}$, we know that

$$\sum_{e\in\text{in}(v)} f(e) = \sum_{e\in\text{out}(v)} f(e)$$

and for s,

$$\sum_{e\in\text{in}(v)} f(e) + F = \sum_{e\in\text{out}(v)} f(e)$$

where F is the value of the flow. Adding all these constraints, we see that all terms cancel except for those which enter and leave S. In other words, we get

$$F = \sum_{e\in\text{out}(S)} f_e - \sum_{e\in\text{in}(S)} f(e) \qquad (11.0.1)$$

Since $0 \leq f(e) \leq c(e)$, it follows that $F \leq \sum_{e\in\text{out}(S)} c(e)$, which is the capacity of this cut. Since f is an arbitrary flow and (S, T) is an arbitrary cut, we see that the value of maximum flow is at most the minimum capacity of a cut. We now show that the two quantities are in fact equal. In fact, we will show that the aforementioned algorithm finds a flow whose value is equal to the capacity of a cut. Therefore, this flow must be a maximum flow (and the cut must be a min-cut, i.e., a cut of minimum capacity).

Theorem 11.2 (max-flow–min-cut) *The value of the s-t max-flow = s-t min-cut.*

We now prove this result. Recall our algorithm for iteratively finding augmenting paths. Consider a flow f such that there is no augmenting path with respect to it. Let S^* be the set of vertices such that there is an augmenting path from s to $u \in S^*$. By definition, $s \in S^*$ and $t \notin S^*$ and $T^* = V - S^*$. Note that S^* is exactly the set of vertices reachable from s in the residual graph G_f.

Consider an edge $e \in \text{out}(S)$ in the graph G. Such an edge is not present in G_f (by definition of S^*). Therefore, it must be the case that flow on such an edge is equal to its capacity. Similarly, for an edge $e \in \text{in}(S^*)$, it must be the case that $f(e) = 0$; otherwise, the reverse of this edge will be present in G_f, contradicting the definition of S^*. It now follows from Eq. (11.0.1) that the value of this flow must be equal to the capacity of this cut, and so, it must be a maximum flow.

We now discuss some algorithms for finding the maximum flow.

11.0.2 Ford and Fulkerson algorithm

The Ford and Fulkerson strategy for finding max-flow is directly based on the aforementioned result, i.e., we successively find augmenting paths till we cannot find any more paths.

How do we find an augmenting path? We can run a graph traversal algorithm like DFS or BFS in the residual graph. We can find such a path in linear time.

Although the Ford–Fulkerson method converges since the flow increases monotonically, we do not have a bound on the maximum number of iterations that it takes to converge to the max-flow. Bad examples (taking exponential time) can be easily constructed; in fact, for irrational capacities, it converges only in the limit !

However, if all the capacities are integers, then the algorithm does converge. Indeed, consider a graph having n vertices and m edges; let U denote the maximum capacity of any edge (assuming all capacities are positive integers). A trivial upper bound on the value of the maximum flow would be nU. Indeed, the value of a flow is equal to the flow leaving the source vertex, and at most U amount of flow can leave on an edge out of s. In each iteration, the value of the residual capacities will be integers as well, and so, the flow sent along an augmenting path will be an integer as well. Therefore, the value of the flow will increase by at least 1. It follows that we will send flow along augmenting paths atmost nU times. This also allows us to bound the running time of this algorithm. In each iteration, we have to find a path from s to t in the residual graph which will take linear time (using any graph traversal algorithm). Updating the residual graph will also take linear time. Therefore, assuming $m \geq n$, each iteration will take $O(m)$ time, which implies $O(mnU)$ running time for the algorithm. In fact, there exist examples where the running time could be close to this bound (see Exercise Problem 11.3.).

Even though this bound is not very useful in practice, it implies an interesting fact, which is often called *integrality of max-flow*:

Observation 11.2 *Assuming all edge capacities are integers, there exists a maximum flow from s to t which sends integral amount of flow on every edge.*

Observe that not every max-flow has to be integral (even if edge capacities are integers). It is easy to construct such examples, and is left as an exercise. However, the Ford–Fulkerson algorithm shows that there is at least one max-flow which is integral.

Another consequence of this algorithm is that it also allows us to find a min s-t cut (assuming it terminates, which will happen if all quantities are integers). Consider the residual graph G_f when the algorithm terminates. We know that there is no path from s to t in the graph G_f. Let S be the set of vertices which are reachable from s in G_f, i.e., $S = \{u :$ there is a path from s to u in $G_f\}$. We claim that S is a min-cut. First of all, $s \in S$ and as argued earlier, $t \notin S$. Therefore, S is an s-t cut. Now we claim that the value of the flow across this cut is exactly the capacity of the cut. Indeed, consider an edge $e = (u, v) \in \text{out}(S)$. By definition, $u \in S, v \notin S$. We claim that $f_e = c_e$. Suppose not. Then the edge e is present in G_f. Since $u \in S$, there is a path from s to u in G_f. But the fact that $e \in G_f$ implies that v is also reachable from s, which contradicts the fact that $v \notin S$. This proves the claim. One can similarly show that if $e \in \text{in}(S)$, then $f_e = 0$. Now, Eq. (11.0.1) shows that the value of the flow is equal to the capacity of the cut S, and so, S must be a min s-t cut.

11.0.3   Edmond–Karp augmentation strategy

It turns out that if we augment flow along the shortest path (in the unweighted residual network using BFS) between s and t, we can prove much superior bounds. The basic property that enables us to obtain a reasonable bound is the following result. At first, let us define an edge in an augmenting path to be a *bottleneck* edge if it has the minimum residual capacity among all edges in this path.

Claim 11.1 *A fixed edge can become a* bottleneck *edge in at most $n/2$ iterations.*

We will prove the claim shortly. The claim implies that the total number of iterations is $m \cdot n/2$ or $O(|V| \cdot |E|)$, which is polynomial in the input size. Each iteration involves a BFS, yielding an overall running time of $O(n \cdot m^2)$.

11.0.4   Monotonicity lemma and bounding the number of iterations

We now prove Claim 11.1. The idea behind the proof is that every time we augment flow along an edge, the distance of an end-point of this edge from the source s increases (in the residual graph). Since this distance cannot be more than n, we will be able to bound the number of times we augment flow along such an edge. We now prove this claim formally. We begin with some definitions. Let G_k be the residual graph after k iterations of the algorithm. For a vertex v and iteration k of this algorithm, let s_v^k denote the minimum number of edges in a path from s to v in the residual graph G_k.

We claim that

$$s_v^{k+1} \geq s_v^k$$

We will prove it by contradiction. Suppose $s_v^{k+1} < s_v^k$ for some k and vertex v. Further, among all such vertices, we choose v with the smallest s_v^{k+1} value.

Consider the shortest s to v path in G_{k+1}, and let u be the vertex preceding v in this path. Consider the last edge in the path, i.e., the edge (u,v). Then,

$$s_v^{k+1} = s_u^{k+1} + 1 \tag{11.0.2}$$

since u is on the shortest path. By assumption on minimality of violation,

$$s_u^{k+1} \geq s_u^k \tag{11.0.3}$$

From Eq. (11.0.2), it follows that

$$s_v^{k+1} \geq s_u^k + 1 \tag{11.0.4}$$

Consider the flow $f(u,v)$ after k iterations.

Case 1 : $f(u,v) < c(u,v)$ Then there is a forward edge (u,v) in the residual graph and hence, $s_v^k \leq s_u^k + 1$. From Eq. (11.0.4), $s_v^{k+1} \geq s_v^k$, which contradicts our assumption.

Case 2: $f(u,v) = c(u,v)$ Then (v,u) is an edge in the residual graph G_k, and (u,v) is not present in G_k. But we know that (u,v) is present in G_{k+1}. Therefore, the shortest augmenting path during iteration k must contain the edge (v,u). This implies that

$$s_u^k = s_v^k + 1$$

Combining with inequality, Eq. (11.0.4), we obtain $s_v^{k+1} = s_v^k + 2$ that contradicts our assumption.

Let us now bound the number of times an edge (u,v) can be a bottleneck edge for augmentations passing through the edge (u,v) in either direction. Suppose (u,v) is a bottleneck edge during iteration k of the algorithm. It follows that (u,v) lies on a shortest path in G_k from s to another vertex, and so, $s_v^k = s_u^k + 1$. Since (u,v) is a bottleneck edge, the residual graph G_{k+1} will not contain the edge (u,v), but will contain the reverse edge (v,u).

Let $\ell(\geq k+1)$ be the next iteration when an augmenting path passes through $(v,u)^2$. From the monotonicity property, $s_v^\ell \geq s_v^k$, so

$$s_v^\ell \geq s_u^k + 1 \qquad\qquad (11.0.5)$$

Then,

$$s_u^\ell = s_v^\ell + 1 \geq s_u^k + 1 + 1 = s_u^k + 2$$

using inequality, Eq. (11.0.5). Therefore, we can conclude that the distance from u to s increases by at least 2 every time (u,v) becomes a bottleneck and hence, it can become a bottleneck for at most $|V|/2$ augmentations.

11.1 Applications of Max-Flow

In this section, we describe several applications of maximum flow. Although some of these are direct, others require non-trivial ideas.

11.1.1 Disjoint paths

Given a directed graph, we can check if there is a path from a vertex s to a vertex t using DFS. However, we are often interested in more robust versions of this question. Suppose we would like to answer the following question: there is an adversary which can remove any k edges from the graph. Is it true that no matter which edges it removes, there will still be a path from s to t? Consider for example Figure 11.3. It is easy to check that in the graph in Figure 11.3, even after removing any single edge, there still remains a path from vertex 1 to vertex 6. However, if we were to remove 2 edges, we could ensure that there is no such path (for example, we could remove the two edges incident with vertex 1). We would now like to answer the following question – given two vertices s and t, what is the maximum number of edges which can be removed by an adversary such that there still remains a path from s to t.

---

[2] It may not be a bottleneck edge.

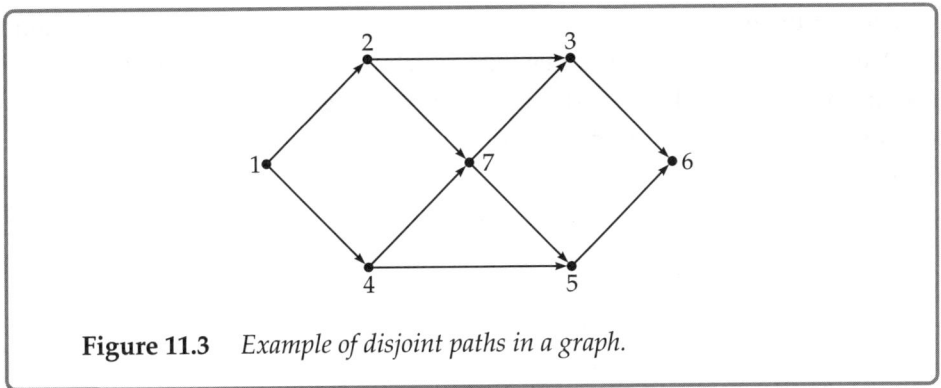

Figure 11.3   *Example of disjoint paths in a graph.*

This question can be easily answered by the techniques we have learned so far. We begin with a definition. Let $\mathcal{P}$ be a set of paths from s to t. We say that these paths are edge disjoint if no two paths in $\mathcal{P}$ share a common edge (they could share a common vertex though). Now suppose there are $k + 1$ edge disjoint paths from s to t (for example, there are two edge disjoint paths from vertex 1 to vertex 6 in Figure 11.3). Now if we remove any k edges, there will still remain a path from which we would not remove any edge. Thus, we see the following observation.

Observation 11.3 *The minimum number of edges which need to be removed to ensure that there is no path from s to t is at least the maximum number of edge disjoint paths from s to t.*

It turns out that the previous equality holds, and is known as Menger's theorem. It is a direct consequence of the max-flow min-cut theorem.

We consider the problem of finding the maximum number of edge disjoint paths from s to t. This can be easily solved by the max-flow algorithm. Indeed, we assign a capacity of 1 to every edge. We now claim that the value of the maximum flow from s to t is equal to the maximum number of edge disjoint paths from s to t. One direction is easy – if there are k edge disjoint paths from s to t, then sending 1 unit of flow along each of these paths yields a flow of value k from s to t. Conversely, suppose there is a flow of value k from s to t. By integrality of maximum flow, we can assume that flow on every edge is 0 or 1. Now proceeding as in the proof of Theorem 11.1, we can find k edge disjoint paths from s to t.

Menger's theorem now follows from the max-flow min-cut theorem applied to the 0-1 flow graph. The proof is left as an exercise.

11.1.2 Bipartite matching

Matching is a classical combinatorial optimization problem in graphs and can be related to a number of natural problems in real life. Given a graph $G = (V, E)$, a matching $\mathcal{M} \subset E$ is a subset of edges that do not have any common end-points in V. A *maximal* matching

M' is such that there is no $e \in E - M'$ such that $M' \cup \{e\}$ is a matching, i.e., M' cannot be augmented. It is easy to see that a maximal matching can be easily constructed using a greedy approach. We will repeatedly choose a new edge (u, v) such that no previously chosen edge is incident on u, v until we cannot pick such an edge.

A *maximum* matching is far more challenging problem – here we want to find a matching which has the highest number of edges.

However, it is easy to show that this problem is a special case of the max-flow problem. Suppose we are given a bipartite graph $G = (V, E)$ with $V = L \cup R$ being the partition of V into the left and right side. From G, we construct a directed graph G' as follows. (See Figure 11.4 for an example.)

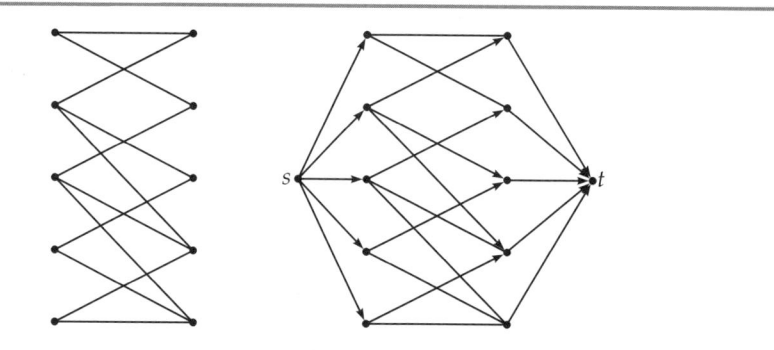

Figure 11.4 *Reduction from a matching instance on the left to a max-flow instance on the right. Note that all edges leaving s and entering t have unit capacity, and the rest have infinite capacity.*

The set of vertices in G' includes V and two new vertices, called s and t. The set of edges include all edges in E directed from L to R. We give infinite capacity to these edges. Now, from s, we add directed edges (s, v) for all $v \in L$, and for t, we add edges (v, t) for all $v \in R$. The capacity of all these edges are 1 (see Figure 11.4 for an example).

Now we argue that the graph G has a maximum matching of size k if and only if the value of maximum flow in G' is k. Let us see the forward implication first. Suppose G has a matching M of size k. Let the edges in this matching be $e_1, \ldots, e_k$, with $e_i = (u_i, v_i), u_i \in L, v_i \in R$. Then, we can find a flow of size k in G' as follows: send 1 unit of flow along each of the paths (s, u_i, v_i, t). Conversely, suppose G' has a max-flow f of value k. By integrality of flow[3], we can assume that f_e is an integer for each edge e. Therefore, on all the edges leaving s and on the edges entering t, the value of the flow is either 0 or 1.

---

[3] Even though some edges have infinite weight, the max-flow value is finite because edges incident to s have unit capacity, so the min-cut is also finite.

Flow conservation implies that if $f_e = 1$ on some edge $e = (s, u)$, then 1 unit of flow must leave u. Again by integrality of flow, this flow can only leave on 1 edge from u, say on edge (u, v). Finally, 1 unit of flow must leave v, and so, there is a unit of flow on the edge (v, t). Thus, we get k disjoint paths from s to t, each containing an edge of the bipartite graph. These edges form a matching in G.

We already know that we can find a maximum flow in G' by the Ford–Fulkerson algorithm. We give a more direct interpretation of the augmenting path algorithm for the case of maximum matching in bipartite graphs. Consider the bipartite graph G, and maintain a matching M. Initially, M is empty, and in each iteration, we increase the size of M by 1 till we are no longer able to increase its size.

An augmenting path with respect to M begins from an unmatched vertex and traces an alternating path of matched and unmatched edges ending with an unmatched vertex – note that this exactly matches with the notion of augmenting path in the max-flow formulation of this problem. If we find such an augmenting path P, then we can increase the size of matching by 1 – just drop the edges in $P \cap M$ and include the edge in $P \setminus M$. What if no such augmenting path exists?

The following claim, analogous to the one in the max-flow problem, forms the basis of all matching algorithms

Claim 11.2 *A matching is maximum (cardinality) iff there is no augmenting path with respect to it.*

The necessary part of the claim is obvious. For the sufficiency, let M be a matching such that there is no augmenting path with respect to it. Let M' be a maximum matching. The following notion of symmetric difference of the two matchings M and M' is useful. Define $M' \oplus M = (M' - M) \cup (M - M')$. We will show in Exercise Problem 11.4 that $M' \oplus M$ consists of a disjoint union of cycles and paths. Suppose M' is maximum; but M is not. There must be a component in $M' \oplus M$ which has more edges of M' than that of M. Any cycle in $M' \oplus M$ will consist of an even number of edges because any cycle in a bipartite graph has even number of edges (this is the only place where we need to use the property that the graph is bipartite). Therefore, there must be path in $M' \oplus M$ which has more edges of M' than those of M. It follows that such a path must be an augmenting path with respect to M (with the starting and the ending vertices belonging to $M' \setminus M$). This proves the claim. (See Figure 11.5 for an example).

It also proves that the augmenting path algorithm described earlier finds a maximum matching. To analyze its running time, note that it will find an augmenting path at most n times – whenever we find an augmenting path, the size of the matching increases by 1. Let us see how we can find an augmenting path in $O(m+n)$ time. This is similar in spirit to constructing the residual graph in case of maximum flow. We construct a directed graph as follows. Consider the directed graph G' as described previously. For every edge in the matching M found so far, we reverse the direction of that edge in G'. It is now easy to check that a path from s to t in this graph yields an augmenting path.

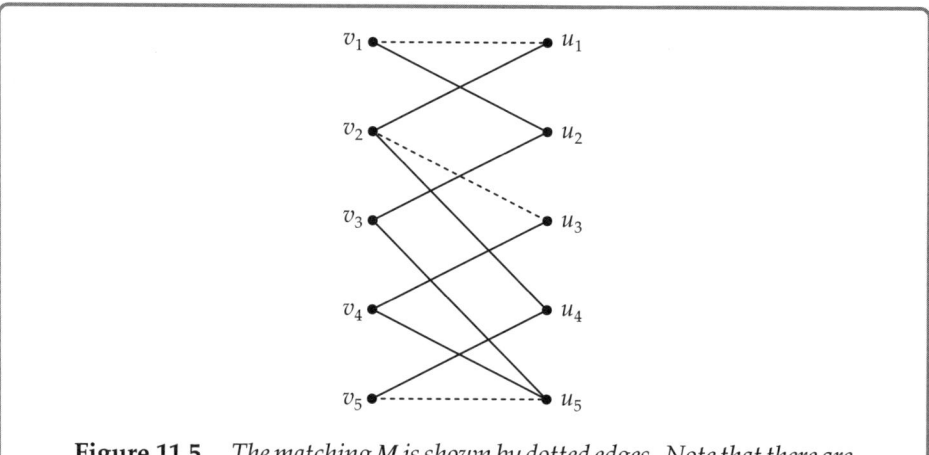

Figure 11.5   *The matching M is shown by dotted edges. Note that there are several augmenting paths with respect to M, e.g., v_4, u_5, v_5, u_4 and v_3, u_2*

It is not difficult to prove that any maximal matching is at least half the size of a maximum cardinality matching (see Section 4.5). There is a useful generalization of this observation using the notion of augmenting paths.

Claim 11.3 *Let M be a matching such that there is no augmenting path of length $\leq 2k-1$. If M' is a maximum matching, then*

$$|M| \geq |M'| \cdot \frac{k}{k+1}$$

From our previous observation, the symmetric difference $M \oplus M'$ consists of a set $\mathcal{P}$ of disjoint alternating paths and cycles (alternating between edges of M and M') such that each path has about half the edges from M. If the shortest augmenting path is of length $2k+1$ (it must have an odd length starting and ending with edges in M'), then there are at least k edges of M in each such augmenting path. It follows that $|M'| - |M| \leq |M' - M| \leq |M' \oplus M| \leq |\mathcal{P}|$. Therefore, $|M'| \leq |M| + |M|/k$ implying the claim[4].

Hall's Theorem

Consider a bipartite graph which has n vertices on both sides. We would like to state simple conditions when the graph has (or does not have) a matching of size n (clearly,

---

[4] A maximal matching has no length 1 augmenting path and hence, it is within a factor 2 of maximum matching.

such a matching is also a maximum matching) – such a matching is also called a *perfect matching*. Consider for example the bipartite graph in Figure 11.6. It is easy to check that it does not have a perfect matching. Here is one way of explaining why it does not have a perfect matching: the vertices A, C, D on the left-hand side can only get matched to vertices E, F on the right. Therefore, there cannot be a matching which matches all of A, C, D. Let us formalize this condition.

Let S be a subset of vertices on the left-hand side. Let $\Gamma(S)$ denote the neighbors of S, i.e., $\{u : (v, u) \in E \text{ for some } v \in S\}$. For example, in Figure 11.7, $\Gamma(\{A, C, D\} = \{E, F\}$. Then, the graph does not have a perfect matching if there exists a set $S \subseteq L$ such that $|\Gamma(S)| < |S|$. The reason is again easy to see – all the vertices in S can only get matched to the vertices in $\Gamma(S)$.

Surprisingly, the converse of this statement is also true. This is the statement of Hall's theorem:

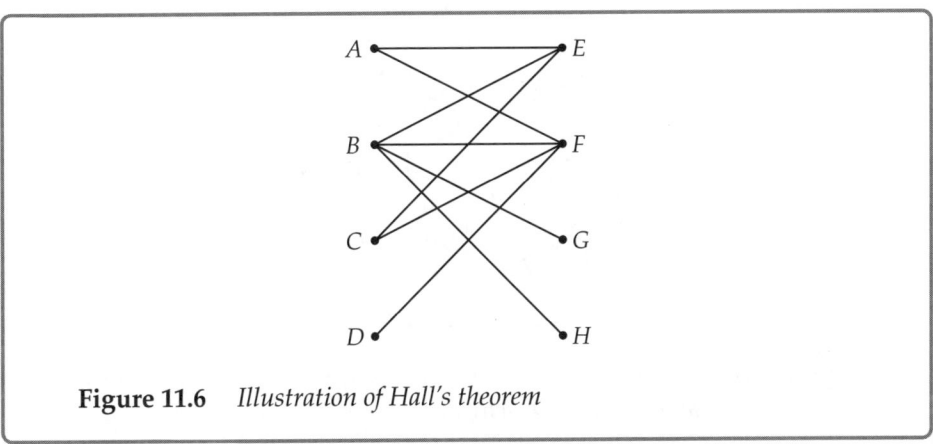

Figure 11.6   *Illustration of Hall's theorem*

Theorem 11.3 *(Hall's theorem) A bipartite graph with equal number of vertices on both sides has a perfect matching iff for every subset $S \subseteq L$,*

$$|\Gamma(S)| \geq |S|$$

We now show that this theorem is a direct consequence of the max-flow min-cut theorem. Note that one direction of Hall's theorem is immediate – if there is a perfect matching, then clearly $|\Gamma(S)| \geq |S|$ for all $S \subseteq L$. The converse is the non-trivial direction. So assume that the graph does not have perfect matching. Now we need to prove that there is a subset S of L for which $|\Gamma(S)| < |S|$.

To prove this, we go back to the directed graph G' which was used in the reduction of the matching problem to the max-flow problem (see Figure 11.4). Assume that the bipartite graph G has n vertices on both sides and it does not have a matching of size n. This implies that the maximum s-t flow in G' is less than n, and therefore, by the max-flow min-cut theorem, there is an *s-t* cut of capacity less than n. Let X denote this *s-t* cut. Note that s belongs to X and t does not belong to X. Also, none of the infinite capacity edges can belong to $\text{out}(X)$; otherwise, the capacity of the cut will not be finite (see Figure 11.7 for an illustration). Let X_L denote the vertices in $X \cap L$, where L denotes the vertices on the left part of the bipartite graph, and $X \cap R$ denotes the ones on the right side. Since no infinite capacity edge belongs to $\text{out}(S)$, it is easy to see that the capacity of the cut is given by $(|L| - |X_L|) + |X_R| = (n - |X_L|) + |X_R|$, where the first term denotes the set of edges in $\text{out}(X)$ which leave s, and the second one denotes the ones entering t. We know that this quantity is less than n, which implies that $|X_L| > |X_R|$. But note that $\Gamma(X_L) \subseteq X_R$ – otherwise there will be an infinite capacity edge in $\text{out}(X)$ to a vertex $w \in R - X_R$. Thus, $|X_L| > |\Gamma(X_L)|$. This proves Hall's theorem.

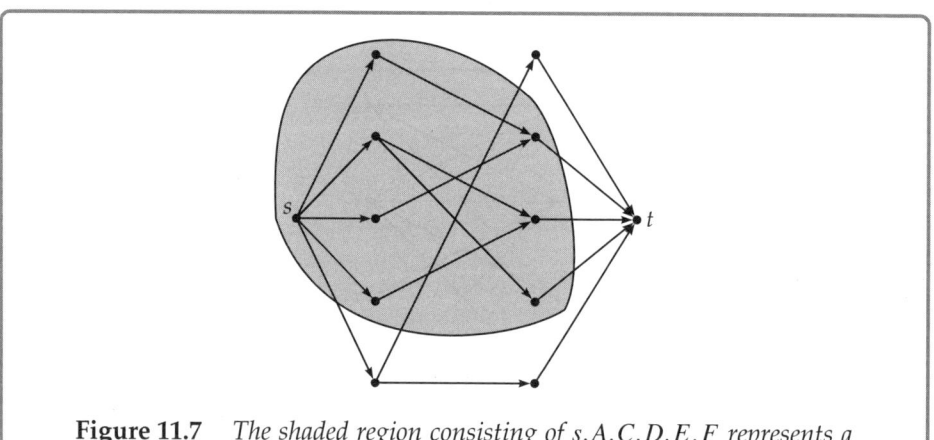

Figure 11.7   *The shaded region consisting of s, A, C, D, E, F represents a min s-t cut of capacity 3.*

11.1.3   Circulation problems

In our maximum flow formulation, there is an upper bound c_e on the amount of flow that can be sent on an edge. We now consider settings where we also have lower bounds l_e on the amount of flow that can be sent on an edge. Thus, each edge e now has a pair of numbers (l_e, c_e) associated with it, where $l_e \leq c_e$. Unlike the previous maximum flow formulation, where setting $f_e = 0$ for all edges e was a *feasible* flow, i.e., it satisfied flow conservation and capacity constraints, it is not clear how to find a feasible flow. In fact, it is easy to show that there may not exist a feasible flow in general. The circulation problem seeks to find a flow which is feasible, i.e., satisfies flow conservation with f_e lying between l_e and u_e for every edge e. Note that we are not trying to maximize any value, and there is no designated s or t vertex here.

It is easy to reduce the maximum flow problem to a circulation problem. Indeed, suppose given an instance of the maximum flow problem (where we only have edge capacities), we would like to check if there is a flow of value at least k from s to t (by trying different k and using a binary search like procedure, we can also find the maximum such k). Then, we can convert it to a circulation problem as follows – we add a new arc e from t to s, and define $l_e = k, u_e = \infty$ on this arc. Now it is easy to show that a circulation in this graph implies a flow from s to t of value at least k, and vice versa.

We now show how to solve a circulation problem by reducing it to a maximum flow problem. Given an instance of the circulation problem, we first send l_e amount of flow on each edge – this may create a surplus or deficit at every node. Define the excess at a node v as

$$e(v) := \sum_{e \in \text{in}(v)} l_e - \sum_{e \in \text{out}(v)} l_e, \text{ where } e(v) \text{ may be negative}$$

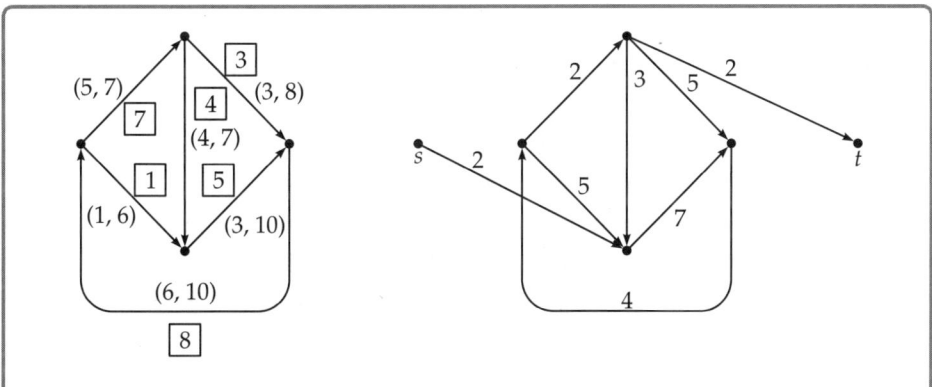

Figure 11.8 *Example of circulation on the left. The numbers in the square boxes represent a feasible flow. The figure on the right shows the reduction to maximum flow problem. The number on the edges on the right show edge capacities. Note that two vertices have 0 excess, and so are not joined to either s or t.*

Let P be the nodes for which excess is positive, and let N be the remaining ones. Note that the total excess of nodes in P is equal to the total deficit (which is just the negative of the excess) of nodes in N – let Δ denote this value. We add two new nodes – a source node s and a sink node t. We add arcs from s to every node in P, and from every node in N to t (see Figure 11.8 for an example). We now set capacities on edges. For an edge e which was present in the original graph, we set its capacity to $u_e - l_e$ – this is the extra flow we can send on this edge. For an edge of the form (s, v) or (v, t), we set the capacity to the absolute value of the excess of v. Now we find a maximum flow from s to t. The claim is that there is a feasible circulation in the original graph if and only if there is a flow of value Δ from s to t. Note that this flow (if it exists) would saturate all edges incident with s or t.

Let us prove this claim. Let $G = (V, E)$ denote the graph in an instance of the circulation problem, and let G' denote the instance of the maximum flow problem as described earlier. Suppose there is a circulation which sends f_e amount of flow on edge e in G. Then we claim that in G', the following is a feasible flow of value Δ – send $f_e - l_e$ flow on all edges e of G, and for edges of the form (s, v) or (v, s), we send flow equal to the absolute value of the excess at this node. Clearly, all capacity constraints are satisfied by this flow in G', because $0 \leq f_e - l_e \leq u_e - l_e$ for every edge e which was present in G. To check flow conservation, note that in G'

$$\sum_{e \in \text{out}(v) \cap E} f_e = \sum_{e \in \text{in}(v) \cap E} f_e$$

for every vertex v in V. Now suppose $e(v)$ is positive. The aforementioned equation can be written as

$$\sum_{e \in \text{out}(v) \cap E} (f_e - l_e) - \sum_{e \in \text{in}(v) \cap E} (f_e - l_e) = e(v)$$

But the RHS is exactly the flow from s to v, and so, G', the total inflow at v is equal to the total out flow. Similar arguments hold if v lies in N.

Let us prove the converse. Suppose there is a flow as stated earlier in G'. Let g_e denote the flow on edge e in G'. In the graph G, define f_e as $g_e + l_e$. Since $g_e \leq u_e - l_e$, it follows that f_e lies between l_e and u_e. To check flow conservation at a vertex v, we first assume that v lies in P (the case when v is in N is similar). Now,

$$\sum_{e \in \text{in}(v)} f_e - \sum_{e \in \text{out}(v)} f_e = e(v) + \sum_{e \in \text{in}(v)} g_e - \sum_{e \in \text{out}(v)} g_e$$

Since g is a feasible flow in G' and the flow on (s, v) is exactly $e(v)$, it follows that RHS is 0, and so, f satisfies flow conservation at v. This proves the equivalence between the two problems.

11.1.4 Project planning

So far we have seen how various problems can be solved using the maximum flow problem. Now we show an application of the min-cut problem. Recall that the algorithms for finding a maximum flow also yield a minimum cut. In the project planning problem, we are given a set of n tasks, and each task i has a profit p_i. The profit could be positive or negative – a positive profit may mean that you gain some amount by completing the task, and a negative profit means that you may have to incur an expenditure in completing the task. Moreover, there are dependencies between the tasks, which is given by a directed acyclic graph (DAG) on these tasks. An arc (i, j) between two tasks indicates that task i must be completed before we can start task j (see Figure 11.9 for an example).

Our goal is now to figure out which tasks to perform so that the overall profit of the completed tasks is maximized (of course, we have to respect the dependencies; if we complete a task, we must complete all tasks which are pre-requisites for it). We show that this problem can be solved by solving a min-cut formulation for a suitable graph.

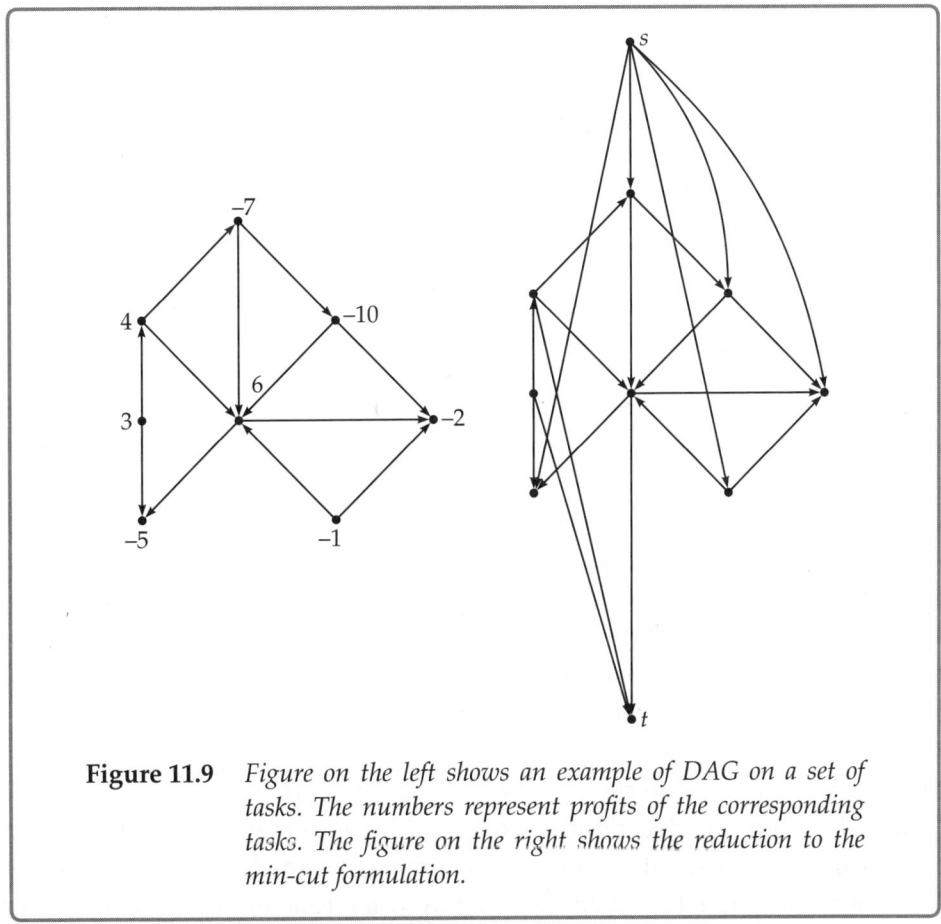

Figure 11.9   *Figure on the left shows an example of DAG on a set of tasks. The numbers represent profits of the corresponding tasks. The figure on the right shows the reduction to the min-cut formulation.*

Let G represent the aforementioned DAG. We would like to convert this into a min-cut problem where the set of tasks in the cut would be the ones which get performed. In order to respect the dependency criteria, if X is a cut in this graph, then there should not be any edge e in the DAG which belongs to $\text{in}(X)$ – otherwise we cannot perform the tasks in X only. If $\bar{X}$ denotes the complement of X, then we can say that there should not be any edge in G which leaves the set $\bar{X}$. With this intuition in mind, we define a graph H as follows. H contains all the vertices and edges in G, and it assigns infinite capacity to all edges in

G. Further, it has a source node s and a sink node t. Let P denote the tasks with positive profit and N denote the nodes with negative profit. For every task $i \in P$, we add an edge (i,t) whose capacity is p_i. For every task $i \in N$, we add an edge (s,i) with capacity $-p_i$ to the graph H. Now let S be an s-t cut with finite capacity (see Figure 11.9). First observe that there is no edge in the DAG H which goes out of S, otherwise the cut capacity will be infinite. Let $\bar{S}$ denote the complement of S (i.e., the tasks which are not in S). The capacity of the cut S is

$$- \sum_{i \in N \cap \bar{S}} p_i + \sum_{i \in S \cap P} p_i = \sum_{i \in P} p_i - \sum_{i \notin S} p_i \text{ by using } P = S \cap P \cup \bar{S} \cap P$$

The first term on the RHS is independent of S, and the second term is the net profit of the tasks in $\bar{S}$. Thus, minimizing the capacity of the cut S is same as maximizing the profit of $\bar{S}$. Therefore, our algorithm for finding the optimal subset of tasks is as follows – find a min s-t cut, and then perform the tasks which belong to the complement of this cut.

Further Reading

Maximum flow is one of the most important topics in operations research, and there are several excellent textbooks on it (see e.g., [8]). We have looked at two elementary algorithms for computing maximum flow. In each iteration of the algorithms discussed in this chapter, we increment flow along a single path. In approaches based on *blocking flows*, we send flows along a maximal set of augmenting paths and make more progress in each iteration. Using blocking flows, one can obtain algorithms with running time $O(mn \log m)$ for maximum flow [112]. Pre-flow push algorithms form another class of algorithms for finding the max-flow, where surplus flow is pushed from the source to intermediate vertices and finally to the sink vertex. State-of-the-art techniques yield $O(mn)$ time strongly polynomial time algorithms. Since m is at most n^2, this implies an $O(n^3)$ time algorithm if we only want to consider the dependence of running time on n. Goldberg and Rao [58] improved this result to $O(\min(n^{2/3}, \sqrt{m}) \log(n^2/m) \log U)$ time algorithm, where U is the largest capacity of an edge (assuming all capacities are integers). For the case of unit capacity graphs, Madry [96] improved the running time to $\tilde{O}(m^{10/7})$, where the $\tilde{O}$ notation hides logarithmic factors.

Menger's theorem dates back to the 1920s, and is a special case of max-flow min-cut theorem. The generalization to the case when we seek edge-disjoint paths between arbitrary pairs of vertices is NP-hard. For directed graphs, finding edge-disjoint paths between two pairs of vertices is an NP-hard problem, whereas in undirected graphs, this can be solved in polynomial time as long as the number of pairs is a constant. The problem of bipartite matching is one of the most fundamental combinatorial optimization problems, and we considered the simplest version in the chapter. One can also consider the case where edges have weights and the goal is to find a perfect matching with the minimum total weight (see e.g. [94]).

Exercise Problems

11.1 Show that even if all capacities are integers, there could be maximum flow which is non-integral.

11.2 Show how you can use maximum flow to find maximum number of edge-disjoint paths in an undirected graph.

11.3 Consider running the Ford–Fulkerson algorithm on the graph shown in Figure 11.10, where L is a large parameter. Show that if we are not careful about choosing the augmenting path, it could take $\Omega(L)$ time.

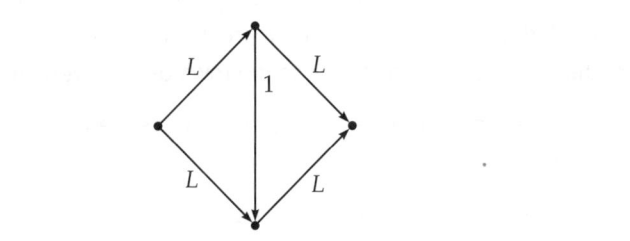

Figure 11.10   *Figure for Exercise 11.3. Numbers denote edge capacities.*

11.4 Let M and M' be two matchings in a graph (which need not be bipartite). Prove that $M' \oplus M$ consists of disjoint alternating cycles and paths.

11.5 Suppose you are given a directed graph with integer edge capacities and a maximum flow from a vertex s to a vertex t in this graph. Now we increase the capacity of an edge e in the graph by 1. Give a linear time algorithm to find the new maximum flow in this graph.

11.6 Suppose you are given a directed graph with integer edge capacities and a maximum flow from a vertex s to a vertex t in this graph. Now we decrease the capacity of an edge e in the graph by 1. Give a linear time algorithm to find the new maximum flow in this graph.

11.7 In the path decomposition theorem (Theorem 11.1), show that the value of the flow is equal to $\sum_{i=1}^{k} f(P_i)$.

11.8 You are given a bipartite graph G and positive integers b_v for every vertex v. A b-matching in G is a subset of edges M such that for every vertex v, at most b_v edges from M are incident with v. Show how you can use maximum flow formulation to efficiently find a b-matching of the largest size.

11.9 Use Hall's theorem to show that any regular bipartite graph, i.e., a bipartite graph where all vertices have the same degree, has a perfect matching.

11.10 A Latin square of size n is an $n \times n$ table where each table entry is filled with one of the numbers $\{1, 2, \ldots, n\}$. Further, no row or column contains a number twice (and so, each number appears exactly once in each row and each column). For example, a Latin square of size 3 is as follows:

2	1	3
3	2	1
1	3	2

You are given a $k \times n$ table, where $k \leq n$. Again, each table entry is a number between 1 and n, and no row or column contains a number more than once. Show that this table can be extended to a Latin square, i.e., there is a Latin square of size n such that the first k rows of the Latin square are same as the rows of the given table.

11.11 Use the max-flow min-cut theorem to prove Menger's theorem. Prove an analogous result for undirected graphs.

11.12 Let G be a directed graph, and s, t, u be three vertices in G. Suppose there are λ edge-disjoint paths from s to t, and λ edge-disjoint paths from t to u. Prove that there are λ edge-disjoint paths from s to u (Hint: use max-flow min-cut theorem).

11.13 You have invited n friends to a party. There are k tables in your home, and each table can accommodate ℓ people. There are s schools in the neighborhood, and each of your friends attends one of these schools. You would like to ensure that at most 2 people from the same school are seated at the same table. Show how you can use maximum flow formulation to find such a seating arrangement (or declare that no such arrangement is possible).

11.14 Consider the same problem as Exercise 11.13, but with the additional constraint that for every table, there must be at least s guests seated at that table, where s is a parameter which is at most ℓ. Show how this problem can be formulated as a circulation problem.

11.15 Consider an instance of the maximum flow problem where the maximum s-t flow is $F^\star$. Prove that there is a path from s to t such that the minimum capacity of any edge on this path is at least $F^\star/m$, where m is the number of edges in the graph.

11.16 Use the max-flow min-cut theorem and the reduction from the circulation problem to the maximum flow problem to prove the following min-max theorem for the circulation problem: consider an instance of the circulation problem where we have a graph $G = (V, E)$ and every edge e has a pair (l_e, c_e) of lower and upper bounds associated with it. Prove that this instance has a feasible flow if and only if for every subset S of vertices, where $S \neq \emptyset, V$,

$$\sum_{e \in \text{out}(S)} c_e \geq \sum_{e \in \text{in}(S)} l_e.$$

11.17 You are given an $n \times n$ matrix X with real positive entries where all row sums and every column sums are integral. You would like to round each entry X_{ij} in the matrix to either $\lfloor X_{ij} \rfloor$ or $\lceil X_{ij} \rceil$ such that the row sums or column sums do not change (i.e., for any row, the sum of the rounded entries is equal to the sum of the actual X_{ij} values in that row, and similarly for any column). Show how you can solve this problem efficiently.

(**Hint:** Use circulation problem)

11.18 There are n teams playing in a tournament. Each pair of teams will play each other exactly k times. So far, p_{ij} games have been played between every pair of teams i and j. Assume that in every game, one of the team wins (draw is not an option). You would like to know if there is any possible scenario in which your favorite team, team 1, can have more wins than any other team at the end of the tournament. In other words, you would like to know if you could decide on the winner of every remaining game; then is it possible that team 1 ends up with more wins than any other team. Show how you can formulate this problem as a maximum flow problem.

11.19 You are running a company with n employees. In the beginning of the year, each employee has specified a subset of days in the year during which he or she is not available. You would like to ensure that on every day at least ℓ employees report to work, and no employee comes to work for more than x days during the year. Show how you can solve this problem efficiently.

11.20 Solve the same problem as Exercise 11.19 with the additional constraint that no employee comes to work for more than 20 days during any month.

11.21 Consider a graph $G = (V, E)$ with edge capacities c_e. Prove that if S and T are any two subset of vertices, then

$$\sum_{e \in \text{out}(S)} c_e + \sum_{e \in \text{out}(T)} c_e \geq \sum_{e \in \text{out}(S \cup T)} c_e + \sum_{e \in \text{out}(S \cap T)} c_e.$$

Use this result to show that if S and T are two min s-t cuts, then $S \cap T$ is also a min s-t cut.

11.22 Given a graph $G = (V, E)$, define the density of a subset S of vertices as the ratio $|e(S)|/|S|$, where $e(S)$ denotes the edges in E which have both the end-points in S. Given a parameter α, we would like to find a subset S whose density is at least α. Show how this can be formulated as a min-cut problem on a suitable graph.

NP Completeness and Approximation Algorithms

In the previous chapters we surveyed many well-known algorithmic techniques and successfully applied them to obtain efficient algorithmic solutions for problems from varied domains. Yet, we cannot claim that there is some general methodology for obtaining efficient algorithms for any given problem. To the contrary any new problem often presents unknown challenges that require new insights and the question that is uppermost in anyone's mind is - what are the problems that are notoriously difficult? We need to first set some target before we can assign a notion of difficulty to a problem. For reasons that will become clear later on, the quest is to design a polynomial time algorithm for any given well-defined problem. This may appear too liberal initially since by definition even n^{100} is a polynomial. However, even this has proved elusive for a large number of problems among which we have come across one in the preceding chapters, namely, 0-1 Knapsack problem. Despite promising starts, we could never claim a truly polynomial time algorithm.

In addition to fixing the limits of practical computation being a polynomial time algorithm, we need to specify the underlying computational model since that will also affect what is achievable in polynomial time. Fortunately, the notion of polynomial time is a robust concept that is not significantly affected by the choice of computational model, except for some constant factors in the exponent of n. We will discuss about a large class of natural and important problems that admits a characterization, that is very intuitive and has resulted in a very interesting theoretical framework. Since the reader is familiar with the Knapsack problem, we will illustrate this framework in the context of the Knapsack problem. Consider a *Prover* , *Verifier* interactive game regarding any given Knapsack instance. The *Prover* is trying to convince the *Verifier* that she has an efficient (polynomial-time) algorithm without actually revealing the technique. For *proof*, given an instance of Knapsack, she provides a list of objects chosen in the optimal solution to the *Verifier* who can easily verify the feasibility and the total profit that can be obtained from this solution. But this is not enough, namely, how do we know that there isn't a superior solution? On the other hand, the *Prover* can easily convince the Verifier that the optimal is at least p by giving out such a solution which can be verified easily. This version of the Knapsack problem is known as the *decision* version of Knapsack and we will henceforth be denoted as decision-Knapsack. The answer is YES/NO depending on whether the optimal profit is $\geq p$ or $< p$. The reader may notice that although the *Prover* can convince the *Verifier* easily if the solution is $\geq p$, it is not obvious how to convince the *Verifier* regarding the contrary. There is some inherent asymmetry in this framework that we will address later in the chapter.

How are the original Knapsack problem and the decision-Knapsack related in terms of computational efficiency?

Clearly, if we can solve the Knapsack in polynomial time, we can also solve the decision-Knapsack easily by comparing the optimal solution O with the threshold p. Conversely, if we can solve the decision version in time T, we can use a binary search like method to compute an optimal solution. Assuming all quantities are integers, the value of the optimal profit lies in the range $[1, n \cdot p_{max}]$, where p_{max} is the maximum profit of any item. Therefore, we can find the optimal profit by calling the decision procedure at most $O(\log n + \log p_{max})$ times, which is linear in the input size (the number of bits required to represent the Knapsack instance). Therefore these two versions are closely related with respect to their computational efficiency. The theory about problems that have been resistant to efficient algorithms (like Knapsack) has been primarily developed around decision problems, i.e., those having YES/NO answers and from our previous discussion, the reader should feel convinced that it is not too restrictive.

Let C be a class of problems characterized by some property - say polynomial time solvability. We are interested in identifying the *hardest* problems in the class, so that if we can find an efficient algorithm for any of these, it would imply fast algorithms for all the problems in C. The class of problems that is considered important is the class $\mathcal{P}$ which is the set of problems for which polynomial time algorithms can be designed. Note that this definition does not preclude problems like Knapsack for which no such algorithm is known today, but there is no proof (more specifically any lower-bound) that such an algorithm cannot be discovered in future using some clever techniques. This subtle distinction makes the definition confusing - so we can think about the problems in $\mathcal{P}$ for which polynomial time algorithms are known today and those that have eluded our efforts so far. One of the latest entrants to the former category is the problem of *primality testing*, i.e., given an integer, an algorithm should answer YES, if it is prime and NO otherwise (see [5]). So, the status of a problem can change from *not known to be in* $\mathcal{P}$ to member of $\mathcal{P}$ when someone discovers a polynomial time algorithm for such a problem. However from our experience with such problems, these can be thought of as *tough nuts*, very few have changed status while others continue to be impregnable.

Over nearly five decades, researchers have tried to find common features of such problems and developed some very interesting theory about a class named as $\mathcal{NP}$, that consists of problems for which *non-deterministic* polynomial time algorithms can be designed. These algorithms can choose among more than one possible actions at any step, and these actions may not depend on any prior information. This additional flexibility can be thought of as *guessing* the next move and there is no cost incurred for guessing [1]. Since it can follow many possible computation trajectories, for the same input there could be a large number of terminating states. These terminating states could be a mix of YES/NO answers often referred to as *accepting / rejecting* states respectively. Even if even one of them is an *accepting* state, the algorithm answers YES to the decision problem (so it ignores the rejecting states). Conversely, the algorithm says NO if all the possible final states are *non-accepting*.

Clearly, this computational paradigm is at least as efficient as the conventional (deterministic) model that does not permit any guesses. However, the obvious question is - does it provide us with any provable advantage? We can relate it to our earlier *Prover /Verifier* game in the following manner. The *Prover* will guess the solution using its magical non-deterministic power and the *Verifier* verifies using conventional computation. While the verifier is limited by polynomial time, there is no such constraint on the *Prover*, so we can think about the class $\mathcal{NP}$ as those problems for which verification of solutions can be done in polynomial time.

Although, $\mathcal{NP}$ doesn't correspond to any realistic computational model, it has been very useful to characterize the important property of efficient verification and has also helped us discover relationship between problems from diverse domains like graphs, number theory, algebra, geometry, etc. In this chapter we will look at many such problems. A deterministic model can emulate this by trying out all possible guesses but it has to pay a huge price for this. Even for two guesses per move, for n moves of the non-deterministic machine, we may have to try 2^n moves. However, this doesn't rule out more efficient emulations or more clever transformations which is one of the central theme of research in this area.

---

[1]The reader should not confuse it with a probabilistic model where a guess is made according to some probability distribution.

More formally, we define the class $\mathcal{P} = \cup_{i \geq 1} C(T^D(n^i))$ where $C(T^D(n^i))$ denotes problems for which $O(n^i)$ time *deterministic algorithms* can be designed. Analogously, $\mathcal{NP} = \cup_{i \geq 1} C(T^N(n^i))$ where $T^N()$ represents *non-deterministic* time. From our previous discussion it is obvious that $\mathcal{P} \subseteq \mathcal{NP}$. However, the Holy Grail of computational complexity is the resolution of the perplexing conundrum $\mathcal{P} = \mathcal{NP}$? or $\mathcal{P} \subset \mathcal{NP}$? (strict subset). For example, if it can be established that Knapsack cannot have a polynomial time algorthm, then the latter holds.

Strong vs Weakly Polynomial Time Algorithms For certain problems, algorithms are classified into weakly and strongly polynomial time algorithms. In such settings, one assumes that all arithmetic operations take $O(1)$ time irrespective of the size of the operands. An algorithm for a problem is said to be *strongly* polynomial if the running time (in this model of computation) is a polynomial in the number of integers in the input. For example, consider the maximum flow problem on a graph containing n vertices and m edges. In this model of computation, an $O(n^3)$-time algorithm will be considered strongly polynomial, whereas an algorithm with running time $O(n^3 \log U)$, where U is the largest capacity of an edge (assuming all edges have integer capacities) will be considered weakly polynomial. Note that the latter running is still polynomial time because the size of the input depends on n and $\log U$.

One of the most well known problems for which a strongly polynomial time algorithm is known as the *linear programming* problem. In the linear programming problem, we are given n variables $x_1, \ldots, x_n$, a vector c of length n, a vector b of length m, and an $m \times n$ matrix A. The objective is to find the value of $\mathbf{x} = (x_1, \ldots, x_n)$ such that $A \cdot \mathbf{x} \leq b$ and $< c, \mathbf{x} >$ is maximized. All known polynomial time algorithms for this problem have running time depending on the number of bits needed to represent A and b, c (even when all arithmetic operations on these numbers are assumed to take $O(1)$ time).

12.1 Classes and Reducibility

The intuitive notion of *reducibility* between two problems is that if we can solve one efficiently, then we can also solve the other efficiently. We will use the notation $P_1 \leq_R P_2$ to denote that problem P_1 is reducible to P_2 using resource R (time or space as the case may be) to problem P_2.

Definition 12.1 *In the context of decision problems, a problem P_1 is **many–one** reducible to P_2 if there is a many-to-one function $g()$ that maps an instance $I_1 \in P_1$ to an instance $I_2 \in P_2$ such that the answer to I_2 is YES iff the answer to I_1 is YES (Fig. 12.1).*

In other words, the many-to-one reducibility function maps YES instances to YES instances and NO instances to NO instances. Note that the mapping need not be 1–1 and therefore, reducibility is not a symmetric relation.

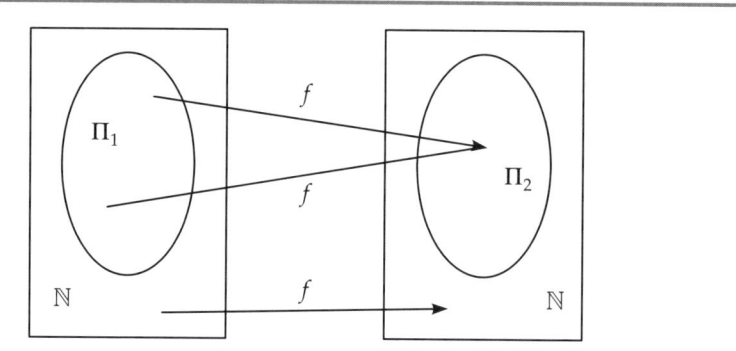

Figure 12.1 *Many-to-one reduction from Π_1 to Π_2 by using a function $f : \mathbb{N} \to \mathbb{N}$. Here, $\Pi_1, \Pi_2 \subset \mathbb{N}$, the set of natural numbers. If f is computable in polynomial time, then it is a polynomial time reduction.*

Definition 12.2 *If the mapping function $g()$ can be computed in polynomial time, then we say that P_1 is* **polynomial time reducible** *to P_2 and is denoted by $P_1 \leq_{poly} P_2$.*

The other important kind of reduction is *logspace* reduction and is denoted by

$$P_1 \leq_{\log} P_2.$$

Claim 12.1 *If $P_1 \leq_{\log} P_2$, then $P_1 \leq_{poly} P_2$.*

This follows from a more general result that any finite computational process that uses space S has a running time bounded by 2^S. A rigorous proof is not difficult but is beyond the scope of this discussion.

Claim 12.2 *The relation $\leq_{poly}$ is transitive, that is, if $P_1 \leq_{poly} P_2$ and $P_2 \leq_{poly} P_3$, then $P_1 \leq_{poly} P_3$.*

From the first assertion, there must exist polynomial time computable reduction functions, say $g()$ and $g'()$ corresponding to the first and second reductions. So we can define a function $g' \circ g$ which is a composition of the two functions and claim that it satisfies the property of a polynomial time reduction function from P_1 to P_3. Let x be an input to P_1, then $g(x) \in P_2$ [2] iff $x \in P_1$. Similarly, $g'(g(x)) \in P_3$ iff $g(x) \in P_2$ implying

---

[2] Note that $g(x)$ may be significantly longer than x.

$g'(g(x)) \in P_3$ iff $x \in P_1$. Moreover, the composition of two polynomials is a polynomial, so $g'(g(x))$ is polynomial time computable.

A similar result on transitivity also holds for logspace reduction, although the proof is more subtle.

Claim 12.3 *Let* $\Pi_1 \leq_{poly} \Pi_2$. *Then*

(i) *If there is a polynomial time algorithm for* Π_2, *there is a polynomial time algorithm for* Π_1.

(ii) *If there is no polynomial time algorithm for* Π_1, *there cannot be a polynomial time algorithm for* Π_2.

Part (ii) is easily proved by contradiction. For part (i), let $p(n)$ denote the running time of the algorithm for Π_2, and $p_1(n)$ denote the running time of the reduction function, where $p(n)$ and $p_1(n)$ are polynomials. Then we have an algorithm for Π_1 which takes at most $p(p_1(n)) + p_1(n)$ steps on an input of size n. This is because the size of the input for Π_2 produced by the reduction function is at most $p_1(n)$, and so, the time to solve the instance of Π_2 is bounded by $p(p_1(n))$. Further, the reduction itself takes $p_1(n)$ steps.

A problem Π is called *NP-hard* under polynomial reduction if for any problem $\Pi' \in \mathcal{NP}$, $\Pi' \leq_{poly} \Pi$.

A problem Π is *NP-complete* (NPC) if it is NP-hard and $\Pi \in \mathcal{NP}$.

Therefore, these are problems that are hardest *within* the class $\mathcal{NP}$. From Claim 12.2, these problems form a kind of equivalent class with respect to polynomial time reductions. However, a crucial question that emerges at this juncture is: *Do NPC problems actually exist ?* A positive answer to this question led to the development of one of the most fascinating areas of theoretical computer science and will be addressed in the next section.

So far, we have only discussed many–one reducibility that hinges on the existence of a many–one polynomial time reduction function. There is another very useful and perhaps more intuitive notion of reducibility, namely, *Turing reducibility*. The many-to-one reduction may be thought of as using *one* subroutine call of P_2 to solve P_1 (when $P_1 \leq_{poly} P_2$) in polynomial time, if P_2 has a polynomial time algorithm. Clearly, we can afford a polynomial number of subroutine calls to the algorithm for P_2 and still get a polynomial time algorithm for P_1. In other words, we say that P_1 is *Turing reducible* to P_2 if a polynomial time algorithm for P_2 implies a polynomial time algorithm for P_1. Moreover, we do not require that P_1, P_2 be decision problems. Although this may seem to be the more natural notion of reducibility, we will rely on the more restrictive definition to derive the results.

12.2 Cook–Levin Theorem

Given a Boolean formula in Boolean variables, the *satisfiability* problem is an assignment of the truth values to the Boolean variables that can make the formula evaluate to TRUE.

For example, $x_1 \lor (\bar{x}_1 \land x_2)$ will be TRUE for the assignment $x_1 =$ TRUE and $x_2 =$ FALSE, whereas the boolean expression $x_1 \land \bar{x}_1$ is always FALSE. A Boolean formula is in a *conjunctive normal form* (CNF) if it is a conjunction of clauses that are a disjunction of literals.[3] A typical CNF formula has the form $(y_1 \lor y_2..) \land (y_i \lor y_j \lor ..) \land \ldots (y_\ell \lor, \ldots, y_n)$, where $y_i \in \{x_1, \bar{x}_1, x_2 \bar{x}_2 \ldots x_n, \bar{x}_n\}$. The satisfiablity problem of the CNF formula is known as CNF-SAT. Further, if we restrict the number of variables in each clause to be exactly k, then it is known as the k-SAT problem. Although any arbitrary Boolean formula can be expressed in an equivalent CNF form, the restriction in the syntax makes it easier and more convenient for many applications. A remarkable result attributed to Cook and Levin says the following.

Theorem 12.1 *The CNF satisfiability problem is NP-complete under polynomial time reductions.*

To appreciate this result, you must realize that there are potentially infinite number of problems in the class $\mathcal{NP}$, so we cannot explicitly design a reduction function for each of them. Other than the definition of $\mathcal{NP}$, we have very little to rely on for a proof of this theorem. A detailed technical proof requires that we define the computing model very precisely – it is beyond the scope of this discussion. Instead, we sketch an intuition behind the proof.

Given an arbitrary problem $\Pi \in \mathcal{NP}$, we want to show that $\Pi \leq_{\text{poly}}$ CNF-SAT. In other words, given any instance of Π, say I_Π, we would like to define a Boolean formula $B(I_\Pi)$ which has a satisfiable assignment iff I_Π is an YES instance. Moreover, the length of $B(I_\Pi)$ should be polynomially bounded by the length of I_Π.

A computing machine is a state transition system that is characterized by the following.

(i) An initial configuration that contains the input

(ii) A final configuration that indicates whether or not the input is an YES or a NO instance

(iii) A sequence of intermediate configurations S_i, where S_{i+1} follows from S_i using a valid transition. In a non-deterministic system, there can be more than one possible transition from a configuration. A non-deterministic machine *accepts* a given input iff there is some valid sequence of configurations that certifies that the input is an YES instance.

All these properties can be expressed as a Boolean formula in a CNF. Using the fact that the number of transitions is polynomial, we can bound the size of this formula by a polynomial. The details are quite laborious and the interested reader can consult a formal proof in the context of the Turing machine model. Just to give the reader a glimpse of the kind of formalism used, consider a situation where we want to write a propositional

---

[3] A literal is a variable x_i or its complement $\bar{x}_i$.

formula to assert that a machine is in exactly one of the k states at any given time i, where $1 \leq i \leq T$. Let us use Boolean variables $x_{1,i}, x_{2,i}, \ldots, x_{k,i}$, where $x_{j,i} = 1$ iff the machine is in state j at time i. We must write a formula that will be a conjunction of two conditions:

(i) At least one variable is true at any time i:

$$(x_{1,i} \lor x_{2,i}, \ldots, \lor x_{k,i})$$

(ii) At most one variable is true at any time i :

$$(x_{1,i} \Rightarrow \bar{x}_{2,i} \land \bar{x}_{3,i} \land \ldots \land \bar{x}_{k,i}) \land (x_{2,i} \Rightarrow \bar{x}_{1,i} \land \bar{x}_{3,i} \land \ldots \land \bar{x}_{k,i}) \ldots \land (x_{k,i} \Rightarrow \bar{x}_{1,i} \land \bar{x}_{2,i} \land \ldots \land \bar{x}_{k-1,i})$$

where the implication $a \Rightarrow b$ is equivalent to $\bar{a} \lor b$ for Boolean expressions a, b.

A conjunction of the aforementioned formula over all $1 \leq i \leq T$ has a satisfiable assignment of $x_{j,i}$ iff the machine is in exactly one state at each of the time instances. The other condition should capture which states can succeed a given state. Note that multiple satisfiable assignments correspond to more than one possible path taken by a non-deterministic machine to reach the terminating state. An equivalent formulation can be in terms of first-order logic using existential quantifiers which can choose successive states that captures the sequence of transitions of the non-deterministic Turing machine.

In this discussion, we sketched a proof that CNF-SAT is NP-hard. Since we can guess an assignment and verify the truth value of the Boolean formula in linear time, we can claim that CNF-SAT is in $\mathcal{NP}$.

12.3 Common NP-Complete Problems

To prove that a given problem Π is NPC, it suffices to establish that

(i) $\Pi \in \mathcal{NP}$: This is usually the easier part.

(ii) CNF-SAT $\leq_{poly} \Pi$: We already know that for any $\Pi' \in \mathcal{NP}$, $\Pi' \leq_{poly}$ CNF-SAT. So, from transitivity (Claim 12.2), $\Pi' \leq_{poly} \Pi$ and therefore, Π is NPC.

Example
Let us show that 3-SAT is NPC given that k-CNF is NPC. [4] Clearly, the 3-CNF formula can be verified in linear time since it is a special case of a k-CNF formula. To reduce k-CNF to 3-CNF, we will do the following. From a given k-CNF formula F_k, we will construct a 3-CNF formula F_3 such that

F_3 is satisfiable if and only if F_k is satisfiable.

---

[4] All clauses have at most k literals. The construction in Cook–Levin theorem actually implies a bounded number of literals in every clause.

This transformation will actually increase the number of variables and clauses but the length of the new formula will be within a polynomial factor of the length of the original formula, thus justifying the definition of polynomial time reducibility. More formally, we have an algorithm that takes F_k as input and produces F_3 as output in polynomial time. We will describe the essence of this algorithm.

The original clauses in F_k are permitted to have $1, 2, \ldots, k$ literals. Only the 3 literal case is compatible with F_3. We will deal with three cases: (i) one literal clause, (ii) two literal clauses, (iii) four or more literal clauses. For each case, we will construct a conjunction of 3 literal clauses that are satisfiable iff the original clause had a satisfiable assignment and for this, we will add variables. To keep these transformations independent, we will introduce a disjoint set of variables so that the conjunction of these clauses is satisfiable iff the original formula was satisfiable.

1. One literal clause: It is a single literal which is a variable or its complement. For a satisfiable assignment of F_k, this literal must be assigned **T** – there is no other option. Suppose the literal is (y). Let us define a set of four 3 literal clauses

$$C_3(y) = (y \vee z_1 \vee z_2) \wedge (y \vee z_1 \vee \bar{z}_2)(y \vee \bar{z}_1 \vee z_2) \wedge (y \vee \bar{z}_1 \vee \bar{z}_2)$$

where z_1, z_2 are new Boolean variables *disjoint* from the given formula and from any other new variables that we may introduce. We want to claim that $y \iff C_3(y)$. Notice that $C_3(y)$ is satisfied by setting $y = $ **T**. Conversely, to satisfy $C_3(y)$, we have to set y to be **T** (the reader may verify this easily). In other words, if there is a satisfiable assignment for F_k, it must have $y = $ **T** and the same assignment would make $C_3(y)$ also **T**. Likewise, if F_3 is satisfiable, then $C_3(y)$ must be **T**, which implies that $y = $ **T**. Since z_1, z_2 are not used anywhere else, it will not interfere with the satisfiability assignment for a similar transformation of other clauses.

2. Two literal clauses: Given $y_1 \vee y_2$, we replace it with

$$C_3(y_1, y_2) = (y_1 \vee y_2 \vee z) \wedge (y_1 \vee y_2 \vee \bar{z})$$

where z is a distinct new variable that is not used elsewhere in the construction of F_3. Along the lines of the previous case, the reader can easily argue that $(y_1 \vee y_2) \iff C_3(y_1, y_2)$.

3. More than 4 literals per clause: Consider $(y_1 \vee y_2 \vee y_3 \vee y_4 \vee y_5 \vee y_6)$ for concreteness, that is, $k = 6$. We will replace it with

$$(y_1 \vee y_2 \vee z_1) \wedge (\bar{z}_1 \vee y_3 \vee z_2) \wedge (\bar{z}_2 \vee y_4 \vee z_3) \wedge (\bar{z}_3 \vee y_5 \vee y_6)$$

where z_1, z_2, z_2 are new Boolean variables disjoint from any other new or original variables. Let us argue that if F_k is satisfiable, then this formula is satisfiable. Without loss of generality, assume y_2 is set to $\mathbf{T}$ in F_k (at least one literal has to be true). Then we can make the aforementioned formula satisfiable by setting $z_1, z_2, z_3 = \mathbf{F}$ which does not affect any other clauses as they do not appear anywhere else. Moreover, the setting of $y_2 \ldots y_6$ also does not affect satisfiability.

Conversely, if this formula is satisfiable, then it must be the case that at least one of $y_1, y_2, \ldots, y_6$ has been set to $\mathbf{T}$ so that F_k can be satisfied. For contradiction, suppose none of the original literals is set to $\mathbf{T}$. Then, $z_1 = \mathbf{T}$ forcing $z_2 = \mathbf{T}$ and $z_3 = \mathbf{T}$. Therefore, the last clause is false contradicting our claims of satisfiability.

The length of F_3 is not much larger compared to the length of F_k. For example, the CNF formula $(x_2 \vee x_3 \vee \bar{x}_4) \wedge (x_1 \vee \bar{x}_3) \wedge (\bar{x}_1 \vee x_2 \vee x_3 \vee x_4)$ gets transformed into the the 3-CNF formula

$$(x_2 \vee x_3 \vee \bar{x}_4) \wedge (x_1 \vee \bar{x}_3 \vee y_{2,1}) \wedge (x_1 \vee \bar{x}_3 \vee \bar{y}_{2,1}) \wedge (\bar{x}_1 \vee x_2 \vee y_{3,1}) \wedge (\bar{y}_{3,1} \vee x_3 \vee x_4)$$

where $y_{2,1}, y_{3,1}$ are new Boolean variables.

Claim 12.4 *If F_k has m clauses in n variables, then F_3 has at most $\max\{4, k-2\} \cdot m$ clauses and $n + \max\{2, k-3\} \cdot m$ variables.*

Therefore, if the original formula has length L, then the length of the equivalent 3-CNF formula is $O(kL)$, which is $O(L)$ assuming k is fixed.

The proof of this claim is left as an exercise. The reader may want to convert this transformation into an actual algorithm that takes in the formula F_k and outputs F_3.

The second step of the reduction can be generalized by reducing any *known* NPC to the new problem P. Having proved the previous result, we can reduce 3-SAT to a given problem to establish NP completeness; 3-SAT turns out to be one of the most useful NPC candidates for reduction because of its simple structure. Some of the earliest problems that were proved NPC include (besides CNF-SAT) the following.

- *Three coloring of graphs*: Given an undirected graph $G = (V, E)$, we want to define a mapping $\chi : V \rightarrow \{1, 2, 3\}$ such that for any pair of vertices $u, w \in V$ such that $(u, v) \in E$ $\chi(u) \neq \chi(w)$, that is, they cannot be mapped to the same value (often referred to as colors). The general problem for k possible values is known as the k-coloring problem.

- *Equal partition of integers*: Given a set of n integers $S = \{x_1, x_2, \ldots, x_n\}$, we want to partion S into S_1 and $S_2 = S - S_1$ such that

$$\sum_{y \in S_1} y = \sum_{z \in S_2} z$$

- *Independent set*: Given an undirected graph $G = (V, E)$ and a positive integer $k \leq |V|$, is there a subset $W \subseteq V$ such that for all pairs $(u, w) \in W$, $(u, w) \notin E$ and $|W| = k$.

  In other words, there is no edge between any of the vertices in W which contains k vertices. In a complete graph, W can have size at most 1.

  A related problem is the *clique* problem where we want to find a $W \in V$ such that every pair in W has an edge between them and W has size k.

  A complete graph has a clique of size $|V|$ by definition.

- *Hamilton cycle problem*: Given an undirected graph $G = (V, E)$, where $V = \{1, 2, \ldots, n\}$, we want to determine if there exists a cycle of length n starting from vertex 1 that visits every vertex j exactly once.

  Note that this is not the same as the better known traveling salesman problem (TSP) that is about finding the shortest such cycle in a weighted graph (which is not a decision problem).

- *Set cover*: Given a ground set $S = \{x_1, x_2, \ldots, x_n\}$, an integer k, and a family $\mathcal{F}$ of subsets of S, that is, $\mathcal{F} \subseteq 2^S$, we want to determine if there is a sub-family $\mathcal{E}$ of k subsets of $\mathcal{F}$ whose union is S.

  A related problem is known as the *hitting set* problem where, given a family of subsets $\mathcal{F}$, we want to determine if there is a subset $S' \subseteq S$ of k elements such that for all $f \in \mathcal{F}$ $S' \cap f \neq \phi$, that is, S' has non-empty intersection with every member of $\mathcal{F}$.

Two outstanding problems whose status is open with regard to NPC are as follows.

- *Graph isomorphism* Given two graphs $G_1 = (V_1, E_1)$ and $G_2 = (V_2, E_2)$, we want to determine if there is a 1–1 mapping $g : V_1 \to V_2$ such that $(u, v) \in E_1 \iff (g(u), g(v)) \in E_2$.

  It is easy to verify this in polynomial time using g but no polynomial time algorithm is known for this problem; the problem is not known to be NPC either.

- Factorization: Although it is not a decision problem, it continues to be elusive in terms of its intractability and has huge ramifications since the security of the RSA is dependent on its hardness.

12.4 Proving NP Completeness

In this section, we consider some of the classic NPC problems and describe reductions from 3-SAT.

12.4.1   Vertex cover and related problems

Given an undirected graph $G = (V, E)$ and an integer k, we want to determine if there is a subset $W \subseteq V$ of k vertices such that for any edge $(u, w) \in E$, either u or v (or both) are in W. The *vertex cover* problem is a special case of the set cover problem. To see this, consider the ground set as E and the family of subsets as $S_v = \{(v, u) | (v, u) \in E\}$. Then it is easy to verify that a subset of k vertices cover all the edges iff the corresponding set cover problem also has a solution. If vertex cover is NPC, then it immediately implies that the set cover problem is also NPC (note that the other direction is not established by this construction).

The vertex cover problem is in $\mathcal{NP}$ since it is easy to verify that a given set of k vertices covers all the edges. To establish the NP-hardness of vertex cover, we will reduce an arbitrary instance of 3-SAT to an instance of vertex cover. More specifically, we will start from a 3-CNF formula F and map it to a graph $G(F) = (V, E)$ such that there is a vertex cover of size $k(F)$ in $G(F)$ iff F is satisfiable. Note that both the graph and the integer k are functions of the given 3-CNF formula F.[5]

Consider a Boolean 3-CNF formula

$$F : (y_{1,1} \vee y_{1,2} \vee y_{1,3}) \wedge (y_{2,1} \vee y_{2,2} \vee y_{2,3}) \wedge \ldots \wedge (y_{m,1} \vee y_{m,2} \vee y_{m,3})$$

where $y_{i,j} \in \{x_1, \bar{x}_1, x_2, \bar{x}_2, \ldots, x_n, \bar{x}_n\}$, that is, it has m clauses in n Boolean variables. We define $G(F)$ on the vertex set $\{y_{1,1}, y_{1,2}, y_{1,3}, y_{2,1}, \ldots, y_{m,3}\}$, that is, $3m$ vertices. The edges are of two types – for all i, we have edges $E_i = \{(y_{i,1}, y_{i,2}), (y_{i,2}, y_{i,3}), (y_{i,3}, y_{i,1})\}$ that define some triangles. The union of all such edges number $3m$. In addition, we have edges $E' = \{(y_{j,a}, y_{k,b}) | j \neq k, y_{j,a} \neq \overline{y_{k,b}}\}$, i.e., E' consists of those pairs of literals which are inverses of each other. The integer $k(F)$ is set to $2m$ (the number of clauses). Figure 12.2 shows an illustration of this construction.

Claim 12.5 *The graph $G(F)$ has a vertex cover of size $k(F) = 2m$ iff F is satisfiable.*

For the proof, we first consider that F has a satisfable assignment – consider one such assignment. Every clause in F has at least one true literal – we choose the other two vertices (recall that every literal is mapped to a vertex in $G(F)$) in the cover. In case there is more than one literal, which is true, we can choose one of them aribitrarily and pick the other two in the cover. Overall, we have chosen 2 vertices from each triangle and it is clear that the vertices picked cover all the edges of E_i for all i. For the sake of contradiction, suppose some edge from E' is not covered by the chosen vertices. It implies that neither of the two literals were set to **F** in the satisfiable assignment. Note that any literal that is set to **F** is picked in the cover. But this is not possible since the two end points of an edge in E' are mapped from complemented literals, so one of them must be **F** and it will be picked in the cover. Thus, there is a cover of size $2m$.

---

[5]We will not be as pedantic in our subsequent proofs but only use G instead of $G(F)$ etc.

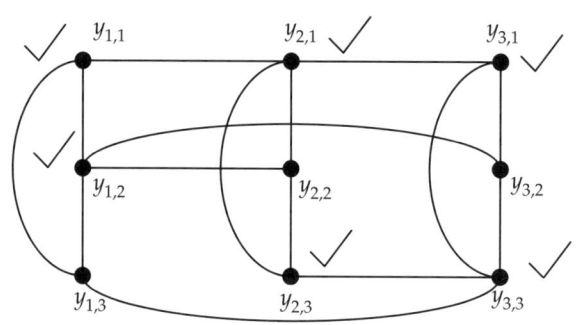

Figure 12.2 *Graph illustrating the reduction for the 3-CNF formula*
$(x_1 \vee \bar{x}_2 \vee x_3) \wedge (\bar{x}_1 \vee x_2 \vee x_3) \wedge (x_1 \vee x_2 \vee \bar{x}_3)$. *Here,* $n = 4, m =$
3, $k = 2 \times 3 = 6$.

*The checked vertices form a cover that defines the truth
assignment* $x_3 = T x_2 = T$ *and* x_3 *can be assigned arbitrarily
for the formula to be satisfiable.*

For the other direction of the claim, suppose there is a vertex cover W of size $2m$. Any vertex cover must pick two vertices from each triangle and since $|W| = 2m$, exactly two vertices are picked from each triangle. For the truth assignment in F, we set the literal not picked in each triangle to be **T**. If some variable has not been assigned any value this way, then we can choose them in any consistent way. Thus, there is a literal set to **T** in every clause, but we must establish that the truth assignment is consistent, that is, both the literal and its complement cannot be set to **T**. For the sake of contradiction, suppose it is so, which implies that there is an edge connecting the two vertices, say (u, v) by construction. Can both be assigned **T**? Since, at least one of the end points u or v must be in the cover, both cannot be assigned **T**.

One can now use the NP completeness of the vertex cover problem to show that the independent set problem is also NP complete. This follows from the observation that a set I is an independent set if and only if $V - I$ is a vertex cover.

12.4.2 Three coloring problem

Given a graph, it is easy to determine if it is two-colorable from the elementary property of *bipartiteness*. This can be done in linear time. Somewhat surprisingly, three coloring turns out to be far more challenging, viz., no polynomial time algorithm is known.[6] It is easy to see that if a graph has a clique of size k, then at least k colors are required but the converse

---

[6] A similar phenomenon is also known in the case of 2-SAT vs 3-SAT.

is not true. So, even if a graph may not have a large clique, it could still require many colors. That makes the coloring problem very interesting and challenging.

We will now formally establish that the 3-coloring problem is NPC. Given a coloring of a graph, it is easy to verify that it is legal and uses only three colors; so the problem is in $\mathcal{NP}$.

We will sketch a reduction from the 3-SAT problem – some of the details are left as exercise problems. Given a 3-CNF formula ϕ that has m clauses over n variables $x_1, x_2, \ldots, x_n$, we define a graph over a vertex set

$$\{x_1, \bar{x}_1, x_2, \bar{x}_2, \ldots, x_n, \bar{x}_n\} \cup \bigcup_{i=1}^{m} \{a_i, b_i, c_i, d_i, e_i, f_i\} \cup \{T, F, N\}$$

This sums to $2n + 6m + 3$ vertices. The edges are defined according to the subgraphs depicted in Fig. 12.3. Broadly speaking, $\{T, F, N\}$ represents the three colors with a natural association of F to **F** and T to **T**; N can be thought of as *neutral*. The second triangle in the figure ensures that the two complementary literals for each variable get distinct colors, which are also distinct from N. The subgraph (a) in the figure is the more critical one that enforces satisfiable assignments in ϕ. There is an inner triangle a_i, b_i, c_i and an outer layer d_i, e_i, f_i connected to the inner triangle enforcing some coloring constraints. Here is the crucial observation about the subgraph.

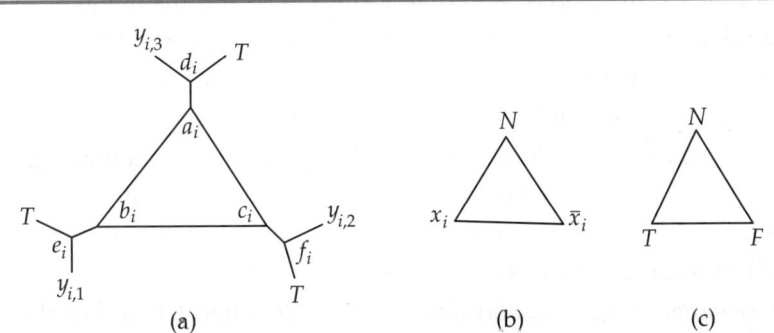

Figure 12.3 *A three coloring of the above subgraph (a) captures the satisfiability of 3-CNF clause $(y_{i,1} \lor y_{i,2} \lor y_{i,3})$. A 3-coloring of subgraph (b) captures the constraint that for every variable x_i, its complement literal must have consistent truth assignment, i.e., if $x_i = $ **T** then $\bar{x}_i = $ **F** and vice versa (as shown in (c)). The third color is the neutral colour N.*

Claim 12.6 *The subgraph is three colorable iff at least one of the literals $y_{i,1}, y_{i,2}, y_{i,3}$ is colored the same as T.*

A formal proof of this claim is left as an exercise problem. However, the reader can see that if $y_{i,1}$ is not colored the same as T then, it must be colored the same as F (it cannot be colored as N because of triangle in Figure 12.3(b)). This forces the color of N on e_i implying that b_i cannot be N. Exactly one of the vertices a_i, b_i, c_i must be colored the same as N but this cannot happen if all the three literals are colored the same as F. From the claim, it follows that the graph defined by the union of the 3-CNF clauses in ϕ is three-colorable iff ϕ is satisfiable. A formal proof is left as an exercise problem.

12.4.3 Knapsack and related problems

We now show that many number theoretic problems like decision-knapsack are NP-complete. We begin with a more elementary problem, called Subset Sum. In this problem, an instance consists of a set of positive integers $\{x_1, \ldots, x_n\}$ and a parameter B. The problem is to decide if there is a subset of $\{x_1, \ldots, x_n\}$ which adds up to exactly B. It is easy to check that this problem can be solved in $O(nB)$ time using dynamic programming. However, this is not polynomial time because the input size only depends on logarithms of these integers. It also suggests that any NP-completeness proof for the Subset Sum problem must use instances where the number of bits needed to represent B depends linearly (or a higher order polynomial) on n.

Proving that Subset Sum is in $\mathcal{NP}$ is easy. A solution just needs to specify the subset of $\{s_1, \ldots, s_n\}$ which adds up to B. A verifier just needs to add the numbers in this subset. Note that addition of a set of k numbers can be done in time proportional to kb, where b is the number of bits needed to represent these numbers. Therefore, a verifier can check, in time proportional to the size of the input, if the solution is valid or not.

We now prove that this problem is NP-complete. Our goal is as follows: given a 3-CNF formula ϕ, we need to produce an instance I of the Subset Sum problem such that the formula ϕ is satisfiable if and only if the instance I has a solution (i.e., there is a subset of $\{s_1, \ldots, s_n\}$ in this instance which adds up to B).

Let ϕ have n variables $x_1, \ldots, x_n$ and m clauses $C_1, \ldots, C_m$. Recall that each clause C_j can be written as $y_{j,1} \vee y_{j,2} \vee y_{j,3}$, where each of the literals $y_{j,l}$ is either one of the variables $x_1, \ldots, x_n$ or its negation. In the instance I, we will have $k := 2n + 2m$ numbers. We will write each of these numbers in decimal, though one could write them in binary by just writing each of the decimal digits in its corresponding binary representation. This would only blow up the size of each number by a constant factor. Writing in decimal would be convenient because our numbers will be such that if we add any subset of them, we would never cause any carry over.

We now give details of the construction. Each number will have $n + m$ digits. These digits will be labeled by the n variables $x_1, \ldots, x_n$ and the m clauses $C_1, \ldots, C_m$. For every variable x_i, we have two numbers s_i and $\bar{s}_i$. The intuition would be that any solution for

I would pick exactly one of s_i and $\bar{s}_i$ – if a satisfying assignment for ϕ sets x_i to **T**, one would pick s_i, else $\bar{s}_i$. So, s_i would correspond to the literal x_i and $\bar{s}_i$ would correspond to the literal $\bar{x}_i$. Both of these numbers will have '1' in the digit corresponding to x_i, and will have 0 in all other digits corresponding to variables x_j, $j \neq i$. Further, if the variables x_i appears (as a literal) in clauses $C_{i_1}, \ldots, C_{i_s}$, then the number s_i will have '1' in the clause digits corresponding to $C_{i_1}, \ldots, C_{i_s}$, and '0' in all other clause digits. The clause digits for $\bar{s}_i$ are defined similarly – if $\bar{x}_i$ appears in a clause C_j, then the corresponding digit for $\bar{s}_i$ is '1', else it is 0.

Let us first see an example. Suppose there are 4 variables x_1, x_2, x_3, x_4 and 3 clauses, $C_1 = x_1 \lor \bar{x}_3 \lor x_4$, $C_2 = \bar{x}_1 \lor x_2 \lor \bar{x}_4$, and $C_3 = x_1 \lor x_3 \lor x_4$. Then, the 8 numbers are as given in Table 12.1.

Table 12.1 *Creating an instance of decision-knapsack from a given instance of 3-SAT. The dummy variables d_j^1, d_j^2 are not shown here. The capacity B can be seen as [1 1 1 1 3 3 3].*

	x_1	x_2	x_3	x_4	C_1	C_2	C_3
s_1	1	0	0	0	1	0	1
$\bar{s}_1$	1	0	0	0	0	1	0
s_2	0	1	0	0	0	1	0
$\bar{s}_2$	0	1	0	0	0	0	0
s_3	0	0	1	0	0	0	1
$\bar{s}_3$	0	0	1	0	1	0	0
s_4	0	0	0	1	1	0	1
$\bar{s}_4$	0	0	0	1	0	1	0

Before we specify the remaining numbers in this instance, let us see what we would like the target number B to be. For each of the variable digits x_i, B would have "1". Since we would never create any carry over, this would ensure that every solution must pick exactly one of $s_i, \bar{s}_i$ for every $i = 1, \ldots, n$. Moreover, we would want such a selection to correspond to a satisfying assignment. In other words, we would like that for every clause C_j, we pick at least one of the literals contained in it (i.e., for the clause C_1 mentioned earlier, we should pick at least one of $s_1, \bar{s}_3, s_4$). So in the digit corresponding to C_j, we would want that B should have a number greater than 0. But this is a problematic issue – a satisfying assignment may choose to set 1 or 2 or 3 literals in C_j to true. And so, we cannot tell in advance if the corresponding clause digit in B should be 1 or 2 or 3. To get around this issue, we add two more 'dummy' numbers for each clause C_j; call these d_j^1 and d_j^2 leading to an additional $2m$ numbers. Both of these dummy numbers have 0 in all the digits, except for the digit corresponding to C_j, where both have 1. The target B is as follows: for each of the variable digits x_i, it has '1', and for each of the clause digits C_j, it has '3'. This completes the description of the input I for the Subset Sum problem.

We now prove that the reduction has the desired properties. First of all, it is easy to check that even if we add all the $2n + 2m$ numbers, there is no carry over. Now suppose there is a satisfying assignment for ϕ – call this assignment α. We now show that there is a solution for I as well. We construct a subset S of the input numbers in I as follows. If α sets x_i to **T**, we add s_i to S, else we add $\bar{s}_i$ to S. This ensures that for each of the variable digits, the sum of the numbers added to S is 1. Now consider a clause C_j. Since α sets at least one literal in C_j to true, there is at least one number in S for which the digit C_j is 1. Let us now see how many such numbers are there in S. Since C_j has 3 literals, there would be at most 3 such numbers in S. If S contains exactly 3 such numbers, we have ensured that adding all the numbers in S would result in digit C_j being 3. If there are only 2 such numbers, we add one of the dummy numbers d_j^1, d_j^2 to S. If S contains exactly one such number, then we add both these dummy numbers to S leading to an additional $3m$ variables. Thus, we have ensured that when we add all the numbers in S, we get '3' in each of the clause digits. Thus, the sum of the numbers in S is exactly B.

The argument for the converse is very similar. Suppose there is a subset S of numbers in the input I which add up to B. From this, we will construct a satisfying assignment α for ϕ. As argued earlier, for each variable x_i, S must contain exactly one of s_i and $\bar{s}_i$. If S contains s_i, α sets x_i to **T**, otherwise it sets x_i to **F**. We claim that this is a satisfying assignment. Indeed, consider the digit corresponding to clause C_j. Even if we pick both the dummy numbers d_j^1, d_j^2 in S, we must have a non-dummy number in S for which this digit is '1'. Otherwise, we will not get a '3' in this digit when we add all the numbers in S. But this implies that α is setting at least one of the literals in C_j to **T**.

Thus, we have shown that Subset Sum is NP-complete. Starting with the Subset Sum, we can show that the decision-knapsack problem is also NP-complete. Recall that in the decision-knapsack problem, we are given a knapsack of size B, and a set of items $I_1, \ldots, I_n$. Each item I_i has a size s_i and a profit p_i. Given a target profit P, we would like to know if there is a subset of items whose total size is at most B, and whose total profit is at least P. We show that this problem is NP-complete by reduction from Subset Sum (proving that the Knapsack problem in $\mathcal{NP}$ is again trivial).

Consider an instance I of Subset Sum consisting of numbers $s_1, \ldots, s_n$ and parameter B. We construct an instance I' of the decision-knapsack problem as follows: the knapsack has capacity B; the profit is also B. For each number s_i in I, we create an item I_i of size s_i and profit s_i as well. It is now easy to check that the Subset Sum problem I has a solution if and only if the corresponding decision-knapsack instance I' has a solution.

12.5 Other Important Complexity Classes

While the classes $\mathcal{P}$ and $\mathcal{NP}$ get the maximum limelight in complexity theory, there are many other related classes which are important in their own right.

- co-$\mathcal{NP}$: Given an NP-complete problem P, the answer to every input is either YES or NO. In fact, we had defined a problem to be the set of inputs for which the answer is YES. A problem P is said to be in the class co-$\mathcal{NP}$ if its complement is in $\mathcal{NP}$, that is, the problem defined by the set of inputs for which P outputs NO is in $\mathcal{NP}$. For example, the complement of 3-SAT, defined as the set of 3-CNF formulas which do not have a satisfying assignment, is in co-$\mathcal{NP}$. Recall that for a problem in $\mathcal{NP}$, there is a short proof which certifies that the answer for an input is YES. It is not clear if the same fact is true for a problem in co-$\mathcal{NP}$. For example, in the 3-SAT problem, a verifier just needs to specify the true/false assignment to each variable, and an efficient verifier can easily check if this solution is correct. However in the complement of 3-SAT, it is not at all clear if one can give a short proof which certifies that a given 3-CNF formula is *not* satisfiable (i.e., all assignments make it false).

  It is not known if $\mathcal{NP}$ is equal to co-$\mathcal{NP}$ or not. It is widely believed that they are different. Note that any problem in $\mathcal{P}$ can be solved in polynomial time, and so, we can also solve its complement in polynomial time. Therefore, $\mathcal{P} \subseteq \mathcal{NP} \cap$ co-$\mathcal{NP}$. Again, it is not known if this containment is strict, but it is widely believed to be so.

- PSpace: So far we have measured the resource consumed by an algorithm in terms of the time taken by it. But we could also look at its space complexity. The class PSpace consists of those problems for which there are algorithms that use polynomial space. Any polynomial time algorithm (deterministic or non-deterministic) in a Turing machine model can only modify a polynomial number of memory locations. Therefore, $\mathcal{NP} \subseteq$ PSpace [7], but it is not known if the containment is strict (though it is conjectured to be so).

  As in the case of $\mathcal{NP}$, we can define problems which are complete for the class PSpace under polynomial time reductions (and so, may not be in $\mathcal{NP}$ as $\mathcal{NP}$ could be strictly contained in PSpace). Here is one example of such a problem. A *quantified Boolean formula* is a Boolean formula where each variable is quantified using either a universal or an existential quantifier. For example, $\exists x_1 \forall x_2 \exists x_3 P(x_1, x_2, x_3)$, where $P(x_1, x_2, x_3)$ is a Boolean propositional formula. One can solve this problem in polynomial space by trying all possible (i.e., 2^n) assignments of Boolean values to the Boolean variables, and evaluating the proposition for each of these assignments. Note that this algorithm would, however, take exponential time. Many counting

---

[7]From Savitch's theorem, PSpace and non-deterministic polynomial space are the same.

problems associated with problems in $\mathcal{NP}$ are in PSpace, for example, counting the number of satisfying assignments to a 3-CNF formula, or counting the number of vertex covers of size at most k in a graph.

- **Randomized classes** Depending on the type of randomized algorithms (mainly Las Vegas or Monte Carlo), we have the following important classes (also refer to Chapter 1, Section 1.4)

    - $\mathcal{RP}$: Randomized polynomial class of problems are characterized by (Monte Carlo) randomized algorithms A such that

        $$x \in L \Rightarrow \Pr[A \text{ accepts } x] \geq 1/2$$

        $$x \notin L \Rightarrow \Pr[A \text{ accepts } x] = 0$$

        These algorithms can err on one side only. The constant $1/2$ in the case $x \in L$ can be replaced by any constant. Indeed, if we have a Monte Carlo algorithm for a language L with the probability of accepting an input $x \in L$ being ε, then we can boost this probability to any constant (close enough to 1) by repeating the same algorithm multiple times, and accepting x if its gets accepted in any of these runs (see Exercise Problems). A famous problem which lies in this class, but is not known to belong to $\mathcal{P}$ is the *polynomial identity testing* problem. An input to this problem is given by a polynomial on n variables. In general, such a polynomial can have exponential number of monomials, and so even writing it down may take exponential space (in n). We assume that the polynomial is given in a compact form by a short Boolean circuit which takes as input n variables (corresponding to $x_1, \ldots, x_n$), and outputs the value of the polynomial (e.g., the polynomial could be $(x_1 + 1) \cdot (x_2 + 1) \cdots (x_n + 1)$). The language consists of those polynomials (in fact, circuits) which are not identically zero. A simple randomized algorithm is as follows: pick $x_1, \ldots, x_n$ uniformly at random (one needs to specify the range, but we ignore this issue). We evaluate the polynomial at $(x_1, \ldots, x_n)$ and check if it evaluates to 0. If it does, we declare that it is identically zero. If the polynomial is identically equal to 0, then the algorithm will not make a mistake. If it is not zero, then the probability of it evaluating to 0 on a random input can be shown to be small. It is a major open problem to come up with a deterministic polynomial time algorithm for this problem. It is easy to check that $\mathcal{RP}$ is contained in $\mathcal{NP}$.

    - $\mathcal{BPP}$: When a randomized algorithm is allowed to err on both sides

        $$x \in L \Rightarrow \Pr[A \text{ accepts } x] \geq 1/2 + \varepsilon$$

        $$x \notin L \Rightarrow \Pr[A \text{ accepts } x] \leq 1/2 - \varepsilon$$

where ε is a fixed nonzero constant. Again, the parameter ε is arbitrary – we can make the gap between the two probabilities larger (e.g., 0.99 and 0.01), by repeating the algorithm multiple times and taking the majority vote (see Exercise Problems). It is not known if $\mathcal{BPP}$ is contained in $\mathcal{NP}$ (in fact some people believe that $\mathcal{P} = \mathcal{BPP}$.

- $\mathcal{ZPP}$ (zero error probabilistic polynomial time): This complexity class corresponds to Las Vegas algorithms, which do not make any error, but take polynomial time in expectation only (i.e., could take more time in worst case).

12.6 Combating Hardness with Approximation

Since the discovery of NP-complete problems in the early 1970s, algorithm designers have been wary of spending efforts on designing algorithms for these problems as it is considered to be a rather hopeless situation without a definite resolution of the $\mathcal{P} = \mathcal{NP}$ question. Unfortunately, a large number of interesting problems fall under this category and so ignoring these problems is also not an acceptable attitude. Many researchers have pursued non-exact methods based on heuristics and empirical results to tackle these problems.[8] Some of the well-known heuristics are *simulated annealing*, *neural network* based learning methods, and *genetic algorithms*. You will have to be an optimist to use these techniques for any critical application.

The accepted paradigm over the last decade has been to design polynomial time algorithms that guarantee *near-optimal* solutions to an optimization problem. For a maximization problem, we would like to obtain a solution that is at least $f \cdot OPT$, where OPT is the value of the optimal solution and $f \leq 1$ is the *approximation factor* for the *worst case* input. Likewise, for a minimization problem, we would like a solution no more than a factor $f \geq 1$ larger than OPT. Clearly, the closer f is to 1, the better is the algorithm. Such algorithms are referred to as *approximation* algorithms. Even though most optimization problems solved in practice are $\mathcal{NP}$-hard, they behave very differently when it comes to approximations. In fact, assuming $\mathcal{P} \neq \mathcal{NP}$, there is a very rich complexity theory of such problems, which tells us that the extent to which they can be approximated can vary widely from problem to problem. In fact, just because we can reduce one NP-complete problem to another NP-complete problem, it does not necessarily follow that the optimization versions of these problems will also have similar approximation algorithms. For example, there is a simple relation between the minimum vertex cover problem and the maximum independent set problem – the complement of an optimal vertex cover is a maximum independent set. While we know of

---

[8] The reader must realize that our inability to compute the actual solutions makes it difficult to evaluate these methods in a general situation.

2-approximation algorithms for the vertex cover problem, no such small approximation ratios are known for the maximum independent set problem.

We now give a brief description of the range of values that the approximation ratio can have (although this is not exhaustive).

- PTAS (polynomial time approximation scheme): For problems falling in this class, we can design polynomial time algorithms with $f = 1 + \varepsilon$, where ε is any user defined constant (the running time may depend on $1/\varepsilon$, and so will get worse as ε approaches 0). Further, if the algorithm is polynomial in $1/\varepsilon$, then it is called FPTAS (fully PTAS). This is in some sense the best we can hope for in an $\mathcal{NP}$-hard optimization problem. The theory of *hardness of approximation* has yielded lower bounds (for minimization and upper bounds for maximization problems) on the approximation factors for many important optimization problems. One example of a problem for which there exists a PTAS is the knapsack problem. Recall that we are given a knapsack of capacity B, and a set of items, where each item has a size and a profit. We would like to find a subset of items whose total size is at most B, and whose profit is maximized. Since Knapsack is NP-complete, this optimization problem is NP-hard and we had discussed how one is reducible to the other in the beginning of this chapter. Recall that one can solve this problem by dynamic programming, but this algorithm takes time exponential in the number of bits needed to specify the input. It turns out that there is a PTAS for this problem.

  There are very few problems for which one can get a PTAS. It is now known that unless $\mathcal{P} = \mathcal{NP}$, we cannot have a PTAS for many problems. This includes problems like minimum vertex cover, minimum set cover, maximum independent set, max-cut.

- Constant factor: If we cannot get a PTAS for a problem, the next best thing is to try get a constant factor approximation (i.e., f is a constant independent of the size of the problem instance). We can get such results for many problems, including minimum vertex cover, max 3-SAT problem, where the max 3-SAT problem involves finding the maximum number of satisfiable clauses in a 3-CNF expression.

- Logarithmic factor: For some NP-hard problems, we cannot hope to get a constant factor approximation algorithm (unless $\mathcal{P} = \mathcal{NP}$). The most well-known of such problems is the minimum set cover problem, for which we only have a $\log n$ approximation algorithm, where n is the number of elements in an instance of the set cover problem (and it is also known that one cannot improve on this bound if $\mathcal{P} \neq \mathcal{NP}$).

- Even harder problems: There are many problems for which we cannot even get a logarithmic approximation factor in polynomial time. One such problem is the maximum independent set problem, for which we cannot even get $f = n^{1-\varepsilon}$ for any constant $\varepsilon > 0$ (assuming $\mathcal{P} \neq \mathcal{NP}$).

There are several problems for which we do not know the right approximability ratio. One such problem is (the optimization version of) the 3-coloring problem. In this problem, we are given a graph G which is known to be 3-colorable. The problem is to color the vertices with as few colors as possible in polynomial time. The best known polynomial time algorithm uses about $n^{0.2}$ colors, even though the possibility of a polynomial time algorithm using a constant number of colors is not ruled out.

In the next section, we give several illustrative approximation algorithms. One of the main challenges in the analysis is that even without explicit knowledge of the optimum solutions, we can still prove guarantees about the quality of the solution of the algorithm.

12.6.1 Maximum knapsack problem

Given n items of sizes $s_1, \ldots, s_n$ and profits $p_1, \ldots, p_n$, and a knapsack of capacity B, we would like to find a subset of items of maximum profit which can be packed in the knapsack.

Recall our earlier approach in Section 5.1 using dynamic programmng: we maintain a table $S(j,r)$, where $0 \le r \le B$, and $1 \le j \le n$. This table entry would store the maximum profit that can be obtained from the first j items if we are given a knapsack of size r. We can write the following recurrence to compute $S(j,r)$ (the base case when $j = 1$ is easy to write):

$$S(j,r) = \max\{S(j-1,r), p_j + S(j-1,r-s_j)\}$$

Although this dynamic programming only computes the value of the optimal solution, it is easy to modify it to compute the subset of items which yields this optimal profit. The running time of this algorithm is $O(nB)$, which is not polynomial time because it is exponential in the number of bits needed to write B. One idea for obtaining a PTAS would be to round the item sizes so that B becomes polynomially bounded. However, this can be tricky. For example, consider an input where the optimal solution consists of 2 items of different sizes and equal profit. If we are not careful in rounding the sizes, it is possible that in the rounded instance, both may not fit in the knapsack, and so, we may not be able to get the profit of both.

Here is another dynamic program formulation which is less intuitive, but works with the profits of the items rather than their sizes. Let $p_{\max}$ denote the maximum profit of any item, that is, $\max_j p_j$. Let P denote $n \cdot p_{\max}$ – this is the maximum profit any solution can have. We have a table $T(j,r), 1 \le j \le n, 0 \le r \le P$, which stores the minimum size of the knapsack such that we can pack a subset of the first j items to get a profit of at least r (this entry is infinity if r happens to be more than the total profit of the first j items or if the minimum size is $> B$). The recurrence for $T(j,r)$ can easily be written:

$$T(j,r) = \min(T(j-1,r), T(j-1,r-p_j) + s_j)$$

The optimal solution can be found in the last row of the table, i.e., $j = n$ which fetches the maximum profit having volume $\leq B$. The running time of this algorithm is $O(Pn)$, which again could be bad if P is very large. The idea now is to round down the values p_j of items j. More formally, let M denote $\varepsilon p_{\max}/n$, where ε is an arbitrary positive constant. Let I denote the original input. Define a new input I', which has the same set of items as that of I, and the sizes of these items are also the same as those in I. However, the profit of item j now becomes $\bar{p}_j = \lfloor p_j/M \rfloor$. Note that the maximum profit in I' is at most n/ε and so we can run the aforementioned dynamic program in polynomial time using an $n \times \frac{n^2}{\varepsilon}$ table. Let S denote the set of items picked by the dynamic program. Let O denote the set of items picked by the optimal solution for I. As far as input I' is concerned, S is at least as good as O. Therefore, we get

$$\sum_{j \in S} \bar{p}_j \geq \sum_{j \in O} \bar{p}_j$$

Clearly, $\frac{p_j}{M} - 1 \leq \bar{p}_j \leq \frac{p_j}{M}$. Therefore, from this inequality, we get

$$\sum_{j \in S} \frac{p_j}{M} \geq \sum_{j \in O} \left(\frac{p_j}{M} - 1 \right) \Rightarrow \sum_{j \in S} p_j \geq \sum_{j \in O} p_j - M|O| \geq \sum_{j \in O} p_j - Mn$$

Let $p(S)$ and $p(O)$ denote the profit of S and O respectively. From this, we get

$$p(S) \geq p(O) - Mn = p(O) - \varepsilon p_{\max} \geq p(O)(1 - \varepsilon)$$

where the last inequality follows from the fact that the optimal profit is at least $p_{\max}$. Thus, we get a PTAS for this problem.

12.6.2   Minimum set cover

We are given a ground set $S = \{x_1, x_2, \ldots, x_n\}$ and a family of subsets $S_1, S_2, \ldots, S_m, \ S_i \subset S$. Each of these subsets S_i also has a cost $C(S_i)$. We want to find a set cover, that is, a collection of these subsets such that their union is S, of minimum total cost. Intuitively, we would like to pick sets which are cheap and cover lots of elements. This motivates the following greedy algorithm: we will pick sets iteratively. Let $V \subset S$ be the set of elements covered by the sets picked so far. At the next step, we pick the set U for which the ratio of its cost to the number of new elements covered by it, that is, $\frac{C(U)}{|U-V|}$, is minimized. We will denote this ratio as the *cost-effectiveness* of U at this time. We do this repeatedly till all elements are covered. We now analyze the approximation ratio of this algorithm.

Let us number the elements of S in the order they were covered by the greedy algorithm (without loss of generality, we can re-number them such that they are $x_1, x_2, \ldots$). We will apportion the cost of covering an element $e \in S$ as $w(e) = \frac{C(U)}{U-V}$, where e is covered for the first time by U. The total cost of the cover is $= \sum_i w(x_i)$.

Claim 12.7

$$w(x_i) \leq \frac{C_o}{n-i+1}$$

where C_o is the cost of an optimum set cover.

In the iteration when x_i is considered, the number of uncovered elements is at least $n - i + 1$. The greedy choice is more cost effective than any left over set of the optimal cover. Suppose the cost-effectiveness of the best set in the optimal cover is C'/U', that is, $C'/U' = \min\left\{ \frac{C(S_{i_1})}{S_{i_1}-S'}, \frac{C(S_{i_2})}{S_{i_2}-S'} \cdots \frac{C(S_{i_k})}{S_{i_k}-S'} \right\}$, where $S_{i_1}, S_{i_2}, \ldots, S_{i_k}$ belong to minimum set cover and S' is the set of covered elements in iteration i. Then

$$C'/U' \leq \frac{C(S_{i_1}) + C(S_{i_2}) + \ldots C(S_{i_k})}{(S_{i_1} - S') + (S_{i_2} - S') + \ldots (S_{i_k} - S')} \leq \frac{C_o}{n-i+1}$$

since the numerator is bounded by C_o and the denominator is more than $n - i + 1$, it follows that $w(x_i) \leq \frac{C_o}{n-i+1}$.

Thus, the cost of the greedy cover is $\sum_i \frac{C_o}{n-i+1}$, which is bounded by $C_o \cdot H_n$. Here $H_n = \frac{1}{n} + \frac{1}{n-1} + \ldots + 1$.

12.6.3 The metric TSP problem

The traveling salesman problem (TSP) is as follows: we are given an undirected graph G with edge lengths. Our goal is to find a tour of minimum total length which starts at a vertex s and comes back to s after visiting every vertex in G (recall that a tour is allowed to traverse an edge multiple times). This problem is NP-hard even if all edge lengths are 0–1. An approximation algorithm for this problem is as follows: let T be a minimum spanning tree of the graph G. We define G_T to be the graph obtained from T by replacing each in T by two parallel copies. So, the total cost of edges in G_T is twice that of T. In G_T, the degree of every vertex is even (because every edge has another parallel copy). So, G_T is Eulerian, and there is a tour in G_T which visits every edge exactly once, and hence, visits every vertex. The cost of this tour is exactly the cost of edges in G_T. We now show that the optimal cost is at least the cost of T, and so, we get a 2-approximation algorithm. This is also easy to see – let E' be the set of edges used (at least once) by the optimal tour. Then, (V, E') is a connected subgraph of G, and so, its cost is at least the cost of T.

It may be noted that there is a much superior approximation algorithm, in fact a PTAS, known for the special case of points on the Euclidean plane.

12.6.4 Three coloring

In this problem, we are given an undirected graph $G = (V, E)$, which is guaranteed to be 3-colorable. We describe a polynomial time algorithm which colors the vertices using $\sqrt{n}$ colors.

We will rely on the following simple observations:

(i) Let $\Gamma(v)$ denote the set of neighbors of a vertex v. Then, $\Gamma(v)$ is 2-colorable. Indeed, otherwise the subgraph induced by $\{v\} \cup \Gamma(v)$ will need more than 3 colors. Since we can color a 2-colorable graph (i.e., bipartite graph) using 2-colors efficiently, it follows that the subgraph induced by $\{v\} \cup \Gamma(v)$ can be colored with 3 colors efficiently.

(ii) Let Δ be the maximum degree of a vertex in G. Then, it can be colored using $\Delta + 1$ colors by a simple greedy algorithm – order the vertices in any manner. When considering a vertex, assign it a color from $\{1, \ldots, \Delta + 1\}$ which is not assigned to any of its neighbors.

We can now describe the coloring algorithm. While there is a vertex v of degree at least $\sqrt{n}$ in the graph G, we color $\{v\} \cup \Gamma(v)$ using a set of 3 new colors (as mentioned in the observation (i)). We now remove v and $\Gamma(v)$ from G, and iterate. Note that in each such iteration, we remove at least $\sqrt{n}$ vertices from G, and so, this process can go on for at most $\sqrt{n}$ steps. Thus, we would use at most $3\sqrt{n}$ colors. When this process ends, every vertex has degree less than $\sqrt{n}$, and so, by the observation (ii), can be efficiently colored using at most $\sqrt{n}$ colors. Thus, our algorithm uses at most $4\sqrt{n}$ colors.

It is a rather poor approximation since we have used significantly more colors than three, but even the best known algorithm uses n^c colors, for some constant $c > 0$.

12.6.5 Max-cut problem

In the max-cut problem, we are given an undirected graph G with edges having weights. We want to partition the vertices into sets $U, V - U$ such that the total weight of edges across U and $V - U$ is maximized.

We have designed a polynomial time algorithm for min-cut but the max-cut problem is an NP-hard problem. Consider the following simple idea: independently assign each vertex uniformly at random to one of the two sets in the partition. Let us now estimate the expected cost of the solution. For any fixed edge $(u, v) \in E$, let X_e be a random variable which is 1 if the end points of e belong to different sets in the partition. If Y denotes the total weight of edges crossing the partition, then it is easy to see that $Y = \sum_{e \in E} w_e X_e$. Linearity of expectation now implies that $\mathbb{E}[Y] = \sum_e w_e \mathbb{E}[X_e]$. Note that $\mathbb{E}[X_e]$ is just the probability that end points of e belong to two different sets in the partition, and so, is equal to $1/2$. This shows that $\mathbb{E}[Y]$ is half of the total weight of all the edges in G. Since the optimal solution can be at most the total weight of all edges, we see that this is a 2-approximation algorithm.

Further Reading

The class of NP-complete problems and the existence of a natural NP-complete problem was given in a classic paper by Cook [35]. Later, it was also attributed to Levin [91] as an independent discovery and now it is known as the Cook–Levin theorem. Shortly following Cook's paper, Karp [77] strengthened the field of NP-completeness by demonstrating a number of very fundamental decision problems like set cover, clique, partition that are also NP-complete. It soon became the holy grail of the CS theory community to resolve the $\mathcal{P} = \mathcal{NP}$ puzzle and it continues to be elusive till this date. Garey and Johnson [57] maintained a compendium of a large number of known NP-complete problems from the literature and to this date, the compendium remains a very reliable repository of such problems. Levin [92] developed the theory further to define the notion of *average NP completeness* which is technically more complex but more relevant in areas like security and cryptography.

There are some excellent textbooks [63, 93] that deal with the formalism of the NP-completeness theory using the Turing machine model of computation. A more recent textbook by Arora and Barak [13] presents many important and interesting results in the area of *complexity theory* that researchers have pursued to solve this long-standing open problem.

The area of approximation algorithms got a big boost with the result on hardness of approximation [14] that rules out efficient algorithms for many classical problems. The books by Vazirani [148] and a relatively recent book by Williamson and Shmoys [154] describe many interesting techniques for approximation algorithm design and different ways of parameterization that help us to understand the deeper issues about the complexity of a problem.

Exercise Problems

12.1 Prove the following

(i) If $P \in \mathcal{P}$, then complement of P is also in $\mathcal{P}$.

(ii) If $P_1, P_2 \in \mathcal{P}$, then $P_1 \cup P_2 \in \mathcal{P}$ and $P_1 \cap P_2 \in \mathcal{P}$.

12.2 If problems A and B are NPC, then $A \leq_{\text{poly}} B$ and $B \leq_{\text{poly}} A$.

12.3 Show that the complement of an NPC problem is complete for the class co-$\mathcal{NP}$ under polynomial time reduction.

12.4 Show that any arbitrary Boolean function of k variables can be expressed by a CNF formula of atmost 2^k clauses.

Note: When k is constant, the formula also has constant size, and therefore, the CNF formula in the proof of Cook–Levin theorem expressing the transition function of the NDTM is of bounded size as it only involves 4 cells.

12.5 Can you design an efficient algorithm that satisfies at least 50% of the clauses in a 3-CNF Boolean formula?

How about 66%? You may want to use a randomized strategy.

12.6 Show that if a Boolean formula in CNF contains at most one un-negated literal, then the satisfiability problem can be solved in polynomial time.

Note: Such clauses are called Horn clause.

12.7 What would it imply if an NPC problem P and its complement $\bar{P}$ are polynomial time reducible to each other?

12.8 Prove Claim 12.4.

12.9 Formulate the vertex cover problem as an instance of a set cover problem.

Analyze the approximation factor achieved by the following algorithm. Construct a maximal matching of the given graph and consider the union C of the end points of the matched edges. Prove that C is a vertex cover and the size of the optimal cover is at least $C/2$. So the approximation factor achieved is better than the general set cover.

12.10 Using the NP completeness of the vertex cover problem, show that the independent set problem on graphs is NPC.

Further show that the clique problem is NPC. For this, you may want to use the notion of complement of a graph. The complement of a graph $G = (V, E)$ is a graph $G' = (V, V \times V - E)$.

12.11 Prove Claim 12.6 and use it to show that the three-coloring problem on a given graph is NPC.

Give a bound on the size of the graph in terms of the given 3-SAT formula.

12.12 Consider the following special case of the `Subset Sum` problem (also called the 'PARTITION' problem). Given n positive integers $s_1, \ldots, s_n$, we would like to know if there is a partition of these numbers into two disjoint subsets such that the sum of the numbers in each of the subsets are equal. Prove that this problem is NP-complete.

12.13 Given an undirected graph G and a parameter k, consider the problem of deciding whether G has a clique of size k and an independent set of size k. Prove that this problem is NP-complete.

12.14 A set of vertices S in an undirected graph G is said to form a near-clique if there is an edge between every pair of vertices in S, except perhaps for one pair of vertices in S (so a clique is also a near-clique). Prove that the problem of deciding whether a graph G has a near-clique of size k is NP-complete.

12.15 Consider a randomized algorithm $\mathcal{A}$ that belongs to the class $\mathcal{BPP}$ which means that it can output erroneous answers in both directions. So, unlike algorithms in $\mathcal{RP}$ where the probability of error can be driven down by repeated runs of the algorithm, how can we interpret an answer output by $\mathcal{A}$? One possibility to run the algorithm multiple times and hope that the mojority answer is correct. Given that the parameter ε bounds the error probability from 1/2, can you use Chernoff bounds to show that by taking the majority of sufficiently large number of independent runs, we can conclude the answer holds with high probability.

12.16 You are given 2 machines and a set of n jobs. Each job j has a size p_j. You want to assign each job to one of the machines. The load on a machine is the total size of all the jobs assigned to it. Prove that the problem of minimizing the maximum load on a machine is NP-hard.

12.17 In the Hamiltonian cycle problem, we are given an undirected graph G and would like to know if there is a simple cycle in it which contains all the vertices (no repetitions allowed). In the Hamiltonian path problem, we are given an undirected graph and we would like to know whether there is a path containing all the vertices. Show that the Hamiltonian cycle problem is polynomial time reducible to the Hamiltonian path problem. Conversely, show that the Hamiltonian path problem is polynomial time reducible to the Hamiltonian cycle problem.

12.18 Consider the Hamiltonian cycle problem described in Exericise Problem 12.18. Suppose you are given a black box algorithm A which, given an undirected graph G, outputs 'yes' if G has a Hamiltonian cycle, otherwise it outputs 'no'. Give a polynomial time algorithm for finding a Hamiltonian cycle in a graph (assuming it has such a cycle). Each call to the black box algorithm A counts as 1 unit of time.

12.19 The path selection problem can be defined as follows: given a directed graph G, a set of directed paths $P_1, \ldots, P_r$ in the graph G, and a number k, is it possible to select at least k of these paths so that no two of the selected paths share any vertices? Prove that the independent set problem is polynomial time reducible to the path selection problem. Note that the parameters r, k are not constant.

12.20 A kite is a graph on an even number of vertices, say $2k$, in which k of the vertices form a clique and the remaining k vertices are connected in a 'tail' that consists of a path joined to one of the vertices of the clique. Given a graph G and a number k, the KITE problem asks whether G has a kite of size $2k$ as a subgraph (i.e., a set of $2k$ vertices $a_1, a_2, \ldots, a_{2k}$ such that $a_1, \ldots, a_k$ form a clique, and there are edges $(a_k, a_{k+1}), (a_{k+1}, a_{k+2}), \ldots, (a_{2k-1}, a_{2k})$ in the graph). Prove that KITE is NP-complete.

12.21 The zero weight cycle problem can be defined as follows: you are given a directed graph $G = (V, E)$, with weights w_e on the edges $e \in E$. The weights can be negative or positive integers. You need to decide if there is a cycle in G so that the sum of the edge weights on this cycle is exactly 0 (i.e., say YES if there is such a cycle, NO otherwise). Prove that the subset sum problem is polynomial time reducible to the zero weight cycle problem. Contrast this result with the problem of negative cycle.

13

Dimensionality Reduction*

There are many applications where we deal with points lying in a very high dimensional Euclidean space. Storing n points in a d-dimensional space takes $O(nd)$ space, and even a linear time algorithm for processing such an input can be impractical. Many algorithms depend only on the pair-wise distance between these points. For example, the nearest-neighbor problem seeks to find the closest input point (in terms of Euclidean distance) to a query point. There is a trivial linear time algorithm to solve this problem, which just looks at every input point and computes its distance to the query point. Since the solution to this problem only depends on the distance of the query point to these n points, we ask the following question: can the points be mapped to a low-dimensional space which preserves all pair-wise distances? It is clear that d can be made at most n (just restrict it to the affine space spanned by the n points), and in general, one cannot do better.

For example, Exercise Problem 13.1 shows that even in trivial settings, it is not possible to reduce the dimensionality of a set of points without distorting pair-wise distances. What if we are willing to incur a small amount of distortion in the pair-wise distances? This is often an acceptable option because in many practical applications, the actual embedding of points in d dimensions is based on some rough estimates. Since the data already has some inherent noise, it should be acceptable to distort the pair-wise distances slightly.

Let us make these ideas more formal. We are given a set V of n points in a d-dimensional Euclidean space. Let f be a mapping of these points to a k-dimensional Euclidean space. We say that this mapping (or embedding) has distortion $\alpha > 1$ if the following condition holds for every pair of distinct points $p_i, p_j \in V$:

$$\frac{1}{\alpha} \cdot ||p_i - p_j||^2 \leq ||f(p_i) - f(p_j)||^2 \leq || \leq \alpha \cdot ||p_i - p_j||^2$$

Note that $||v||^2$ denotes $v \cdot v$ which is the square of the Euclidean distance – this turns out to be much easier to work with than Euclidean distances.

13.1 Random Projections and the Johnson–Lindenstrauss Lemma

The Johnson–Lindenstrauss lemma states that for any small constant $\varepsilon > 0$, there is a linear map f from a Euclidean space containing n points into a Euclidean space of dimension $O(\log n/\varepsilon^2)$ such that the distortion is at most $(1 + \varepsilon)$. In fact, the map f turns out to be quite simple. They show that f just needs to be the projection of the points on a random subspace of appropriate dimension.

As an example, consider the case when we have two points p and q in the two-dimensional plane, and suppose we try to project the points on a 1-dimensional line through the origin. We pick a suitable line L through the origin, and for a point p, define $f(p)$ as the projection of p on L. A moment's thought shows that in general such an embedding can have very high distortion. For example, suppose there are two points p and q such that the line joining them is (nearly) perpendicular to L. In this case, $f(p)$ and $f(q)$ will be very close, even though $||p - q||$ could be large. We can avoid such a situation by picking L to be a line along a *random* direction – we can easily do this by picking a random number θ in the range $[0, \pi)$ and then drawing L as the line which makes angle θ with one of the coordinate axes. Can we now compute the probability with which the distance between $f(p)$ and $f(q)$ is (nearly) the same as that between p and q (see Problem Exercises)?

The aforementioned exercise shows that this probability is quite small. How can we increase this probability to close to 1? One natural idea is to take several such lines, and think of the projection along each line as giving one coordinate of $f(p)$. This does not make much sense when the points are already in 2-dimensions, but can lead to significant savings if the number of such lines is much less than the dimension d.

So suppose we have a set V of n points in a d-dimensional Euclidean space. We pick k lines through the origin along random directions – call these lines $L_1, \ldots, L_k$. We now define $f(p)$ as a k-dimensional vector, where the ith coordinate is the length of the projection of p along L_i. The first non-trivial issue is how to pick a random direction in a d-dimensional Euclidean space. The trick is to pick a distribution whose density does not depend on a particular direction.

Recall that the normal distribution with 0 mean and variance 1, denoted by $N(0,1)$, has density

$$\phi(x) = \frac{1}{\sqrt{2\pi}} e^{-x^2/2}$$

We define a multi-dimensional normal distribution $X = (x_1, \ldots, x_d)$, where the variables x_i are independent and each of them has distribution $N(0,1)$ (such a set of variables are called i.i.d. $N(0,1)$ random variables). The joint distribution of X is given by

$$\phi(X) = \frac{1}{(2\pi)^{d/2}} e^{-(x_1^2 + \ldots + x_d^2)/2} = \frac{1}{(2\pi)^{d/2}} e^{-||X||^2/2}$$

Note that this distribution only depends on the length of X and is independent of the direction of X. In other words, here is how we pick a line along a random direction: sample d i.i.d. $N(0,1)$ random variables $x_1, \ldots, x_d$. Consider the line through the origin and the vector $(x_1, \ldots, x_d)$.

Having resolved the issue of how to pick a line along a uniformly random direction, we can now define what the embedding f does. Recall that f needs to project a point along k such lines. Thus, if p is a point with coordinates $\mathbf{p} = (p_1, \ldots, p_d)$, then $f(p) = R \cdot \mathbf{p}$, where R is a $k \times d$ matrix with entries being i.i.d. $N(0,1)$ random variables. Note that each row of R gives a line along a random direction, and each coordinate of $R \cdot \mathbf{p}$ is proportional to the projection of p along the corresponding line. To understand the properties of this embedding, we first need to understand some basic facts about the normal distribution. We use $N(\mu, \sigma^2)$ to denote a normal distribution with mean μ and variance σ^2. Recall that the distribution of $N(\mu, \sigma^2)$ is given by

$$\phi(x) = \frac{1}{\sqrt{2\pi}\sigma} e^{-(x-u)^2/2\sigma^2}$$

Exercise Problem 13.2 shows that the projection of a vector a along a uniformly random direction also has normal distribution. Using this fact, we can now calculate the expected length of $f(p)$ for a point p. Indeed, each coordinate of $f(p)$ is the projection of p along a random direction (given by row i of R, denoted by R_i). Therefore, using the results of this exercise, and substituting $a_i = p_i$, we get

$$E[||f(p)||^2] = \sum_{i=1}^{k} E[(R_i \cdot p)^2] = k \cdot ||p||^2$$

We would like to normalize $f(p)$ such that $E[||f(P)||^2]$ is the same as that of $||p||^2$. Therefore, we re-define $f(p)$ as $\frac{1}{\sqrt{k}} \cdot R \cdot \mathbf{p}$. Now, the earlier calculations show that $E[||f(p)||^2] = ||p||^2$. We would now like to prove that $||f(p)||^2$ is closely concentrated around its mean with high probability. More precisely, we want to show that given an error parameter $\varepsilon > 0$ (which should be thought of as a small constant),

$$\Pr[||f(p)||^2 \notin (1 \pm \varepsilon)||p||^2] \leq 1/n^3$$

Once we show this, we will be done. We can replace p by $p_i - p_j$ for all distinct pair of points p_i, p_j in V. Thus, for any distinct pair of points p_i, p_j, the distance between them gets distorted by more than $(1+\varepsilon)$-factor with probability at most $1/n^3$. But now, notice that there are at most n^2 such pairs we need to worry about. So, using union bound, the probability that there exists a pair p_i, p_j in V for which $||f(p_i) - f(p_j)||^2$ is not in the range $(1 \pm \varepsilon)||p_i - p_j||^2$ is at most

$$n^2 \cdot 1/n^3 = 1/n$$

Thus, the embedding has distortion at most $(1 + \varepsilon)$ with probability at least $1 - 1/n$ (in particular, this shows that there *exists* such an embedding).

Let us now prove that the length of $f(p)$ is tightly concentrated around its mean. First observe that $||f(p)||^2$ is the sum of k independent random variables, namely, $(R_1 \cdot p)^2, \ldots,$ $(R_k \cdot p)^2$, each of which has mean $||p||^2$. Therefore, as it happens in Chernoff–Hoeffding bounds, we should expect the sum to be tightly concentrated around its mean. However, in the setting of Chernoff–Hoeffding bounds, each of these random variables have a bounded range, whereas here, each of the variables $(R_i \cdot p)^2$ lie in an unbounded range. Still, we do not expect these random variables to deviate too much from their mean because $(R_i \cdot p)$ has a normal distribution and we know that the normal distribution decays very rapidly as we go away from the mean by a distance more than its variance. One option would be to carry out the same steps as in the proof of the Chernoff–Hoeffding bound, and show that they work in the case of the sum of independent random variables with normal distribution.

Theorem 13.1 *Let $X_1, \ldots, X_k$ be i.i.d. $N(0, \sigma^2)$ random variables. Then, for any constant $\varepsilon < 1/2$,*

$$\Pr[(X_1^2 + \ldots + X_k^2)/k \geq (1+\varepsilon)\sigma^2] \leq e^{-\varepsilon^2 k/4}$$

and

$$\Pr[(X_1^2 + \ldots + X_k^2)/k \leq (1-\varepsilon)\sigma^2] \leq e^{-\varepsilon^2 k/4}.$$

It follows that if we pick k to be $12 \log n/\varepsilon^2$, then the probability that $||f(p) - f(q)||$ differs from $||p - q||$ by more than $(1 \pm \varepsilon)$ factor is at most $1/n^3$. Since we are only concerned with at most n^2 such pairs, the embedding has distortion at most $1 + \varepsilon$ with probability at least $1 - 1/n$. We now prove this theorem.

Proof: We prove the first inequality; the second one is similar. Let Y denote $(X_1^2 + \cdots + X_k^2)/k$. Then, $E[Y] = \sigma^2$. Therefore, as in the proof of Chernoff bounds,

$$\Pr[Y > (1+\varepsilon)\sigma^2] = \Pr[e^{sY} > e^{s(1+\varepsilon)\sigma^2}] \leq \frac{\mathbb{E}[e^{sY}]}{e^{s(1+\varepsilon)\sigma^2}} \tag{13.1.1}$$

where $s > 0$ is a suitable parameter; we have used Markov's inequality in the last inequality. Now, the independence of the variables $X_1, \ldots, X_k$ implies that

$$\mathbb{E}[e^{sY}] = \mathbb{E}[e^{\sum_{i=1}^{k} sX_i^2/k}] = \mathbb{E}[\prod_{i=1}^{k} e^{sX_i^2/k}] = \prod_{i=1}^{k} \mathbb{E}[e^{sX_i^2/k}].$$

For a parameter α and a $N(0, \sigma^2)$ normal random variable X,

$$E[e^{\alpha X^2}] = \frac{1}{\sqrt{2\pi}\sigma} \int_{-\infty}^{+\infty} e^{\alpha x^2} \cdot e^{-x^2/2\sigma^2} dx = (1 - 2\alpha\sigma^2)^{-1/2}$$

To evaluate the integral, we can use the result that $\int_{-\infty}^{+\infty} e^{-x^2/2} = \sqrt{2\pi}$. Therefore, we can express the right-hand side in Eq. (13.1.1) as

$$\frac{(1 - 2s\sigma^2/k)^{-k/2}}{e^{s(1+\varepsilon)\sigma^2}}$$

Now, we would like to find the parameter s such that this expression is minimized. By differentiating the expression with respect to s and setting it to 0, we see that the right value of s is $\frac{k\varepsilon}{2\sigma^2(1+\varepsilon)}$. Substituting this in the expression, we see that $\Pr[Y > (1+\varepsilon)\sigma^2]$ is at most $e^{k/2\ln(1+\varepsilon) - k\varepsilon/2}$. Using the fact that $\ln(1+\varepsilon) \leq \varepsilon - \varepsilon^2/2$ if $\varepsilon < 1/2$, we get the desired result. $\square$

13.2 Gaussian Elimination

Gaussian elimination is one of the most fundamental tools of solving a system of linear equations. As a by-product, it also tells us the rank of a matrix. This can be useful if we are given a set of points in a high-dimensional space but which belong to a low-dimensional subspace. We can find this low-dimensional subspace by computing Gaussian elimination of the matrix which has as its rows the coordinates of each of these points. We now explain the idea behind Gaussian elimination.

It is easier to understand the algorithm when A is an invertible $n \times n$ square matrix, and we need to solve the system of equations $Ax = b$, where b is a column vector of length n. If A was upper triangular, solving this system of equations is easy. The fact that A is invertible implies that all the diagonal entries of A would be non-zero. Therefore, we can first solve for x_n by $A_{n,n}x_n = b_n$. Having solved for x_n, we can solve for x_{n-1} by considering the equation $A_{n-1,n-1}x_{n-1} + A_{n-1,n}x_n = b_{n-1}$, and so on. Thus, we can solve this system of equations in $O(n^2)$ time. Of course, A may not be upper triangular in general. The idea behind the Gaussian elimination algorithm is to apply row operations to A such that it becomes upper triangular. A *row operation* on A would mean one of the following: (i) let A_i denote the ith row (vector) of A. Then, for a given row A_j, we would replace row A_i by $A_i - cA_j$, where c is a non-zero constant. It is easy to see that this operation is invertible

– if we add cA_j to the ith row of this new matrix, we will recover the original row vector A_i, or (ii) interchange two rows A_i and A_j of A. Again it is easy to see that this operation is invertible. The bottomline is that the row operations preserve the original solution. The algorithm is shown in Fig. 13.1.

At the beginning of iteration i, the matrix A satisfies the following properties: (i) rows 1 to $i-1$ of A look like an upper triangular matrix, that is, for any such row j, $j < i$, $A_{j,1}, \ldots, A_{j,j-1} = 0$ and $A_{j,j} \neq 0$, (ii) For all rows $j \geq i$, the first $i-1$ entries are 0. In iteration i, we would like to ensure that these properties hold for i as well. First assume that $A_{i,i} \neq 0$. In this case, we can subtract a suitable multiple of row i from each of the rows j, $j > i$, such that $A_{j,i}$ becomes 0. However, it may happen that $A_{i,i}$ is 0. In this case, we look for a row j, $j > i$, such that $A_{j,i} \neq 0$ – such a row must exist. Otherwise, $A_{j,i}$ would be 0 for all $j \geq 0$. But then the determinant of A would be 0 (see Exercise Problem 13.4). We started with A being invertible, and have applied operations to it which are invertible. So A should remain invertible, which would be a contradiction. Therefore, such a row A_j must exist. Hence, we first interchange A_i and A_j and perform the same operations as earlier. When the procedure terminates, A is reduced to an upper triangular matrix, where all diagonal entries are non-zero. Observe that the running time of this procedure is $O(n^3)$ since in iteration i, there are $O((n-i)^2)$ row operations.

---

1 **Input** Square $n \times n$ matrix A ;
2 **for** $i = 1, \ldots, n$ **do**
3   **if** $A_{i,i} = 0$ **then**
4   Let j be an index, $i < j \leq n$ such that $A_{j,i} \neq 0$.
5   Interchange rows A_i and A_j.
6
7   **for** $j = i+1, \ldots, n$ **do**
8   Replace A_j by $A_j - \frac{A_{j,i}}{A_{i,i}} A_i$.
9
10
11 Output A.

Figure 13.1 *Gaussian elimination algorithm*

To summarize, let $R_1, \ldots, R_k$ be the row operations applied to A. Each of these row operations can be represented by an invertible matrix (see Exercise Problem 13.5). Further, $R_k R_{k-1} \ldots R_1 \cdot A$ is upper triangular. Let us see how this can be used for solving a system of equations $Ax = b$. If we apply these row operations to both sides, we get $R_k R_{k-1} \ldots R_1 \cdot A x = R_k R_{k-1} \ldots R_1 \cdot b$. In other words, while we apply these row operations on A, we apply them simultaneously on b as well. As a result, we replace these systems of equations by an equivalent system of equations $Ux = b'$, where U is upper triangular and b' is obtained from b by these sequences of row operations. As mentioned earlier, this system of equations can be easily solved in $O(n^2)$ time.

While implementing this algorithm, we should worry about the following issue – it may happen that during iteration i, $A_{i,i} \neq 0$, but is very close to 0. In this case, computing $A_j - \frac{A_{j,i}}{A_{i,i}} A_i$ would lead to a large numerical error. Therefore, it is always a good idea to first find the index $j \geq i$ for which $|A_{j,i}|$ is largest, and then interchange rows A_i and A_j. We now consider the more general scenario when A may not be invertible (or a non-square matrix). It is useful to start with a notation: let $A[i:j,k:l]$ be the sub-matrix of A consisting of rows i till j and columns k till l. For example, in the Gaussian elimination algorithm described earlier, the sub-matrix $A[i:n,1:i-1]$ is 0 at the beginning of iteration i.

We can run the same algorithm as before for a general matrix A. The only problem would be that in iteration i, it is possible that $A_{i,i}, A_{i+1,i}, \ldots, A_{n,i}$ are all 0. However, notice that we could in principle bring any non-zero element in the sub-matrix $A[i:n,i:n]$ by performing row and column interchanges. For example, if $A_{k,l}$ is non-zero, where $i \leq k \leq n, i \leq l \leq n$, then we can interchange rows A_i and A_k, and similarly, columns A^i and A^l (here, A^j refers to column j of A). This will bring the non-zero element at location $A_{i,i}$ and the invariant for rows 1 up to $i-1$ will remain unchanged (i.e., they will continue to look like an upper triangular matrix). After this, we can continue the algorithm as before. Just as interchanging two rows corresponds to multiplying A by an invertible (permutation) matrix on the left, interchanging two columns of A corresponds to multiplying A by such a matrix on the right. Thus, we have shown the following result.

Theorem 13.2 *Given any $m \times n$ matrix A, we can find invertible row operation matrices $R_1, \ldots, R_k$, and invertible column interchange matrices $C_1, \ldots, C_l$ such that the $R_k \ldots R_1 \cdot A \cdot C_1 \ldots C_k$ has the following structure: if the rank of A is i, then the sub-matrix $A[1:i,1:n]$ is upper triangular with non-zero diagonal entries, and the sub-matrix $A[i+1:n,1:n]$ is 0.*

Since the row operation and the column interchange matrices are invertible, one can also get a basis for the sub-space spanned by the rows of A (see Exercise Problem 13.7).

13.3   Singular Value Decomposition and Applications

Singular value decomposition, often abbreviated as SVD, is a key tool in understanding data which is inherently low-dimensional. Consider an $n \times d$ matrix A, where each row of A can be thought of as a point in a d-dimensional Euclidean space. Suppose the points of A actually lie in a lower dimensional subspace (say, a plane through the origin). We can find the basis of this low-dimensional subspace by Gaussian elimination. However, in most real-life applications, the coordinates of the points may get perturbed. This could happen because of measurement errors, or even inherent limitations of the model being used. Thus, we would like to find a matrix $\tilde{A}$ which is low rank (i.e., rows span a low-dimensional subspace), and closely approximates A. One can think of $\tilde{A}$ as *de-noising* of A – we are eliminating errors or noise in the data to get its 'true' representation. As we will see in applications, there are many scenarios where we suspect that the data is low-dimensional. SVD turns out to be an important tool for finding such a low-rank matrix.

13.3.1 Some matrix algebra and the SVD theorem

Let A be an $n \times d$ matrix. It is often useful to think of A as a linear transformation from $\Re^d$ to $\Re^n$. Given a vector $\mathbf{x} \in \Re^d$, this linear transformation (denoted by T_A) maps it to the vector $A \cdot (x_1, \ldots, x_d)^T \in \Re^n$, where $(x_1, \ldots, x_d)$ are the coordinates of $\mathbf{x}$. Now, if we change the basis of any of the two Euclidean spaces, it changes the expression for A. Let us first review linear algebra to see how change of basis works.

Suppose the current basis of $\Re^d$ is given by linearly independent vectors $e_1, \ldots, e_d$. Thus, if a vector $\mathbf{x}$ has coordinates $(x_1, x_2, \ldots, x_d)$ with respect to this basis, then $\mathbf{x} = x_1 \cdot e_1 + \ldots + x_d e_d$. Now suppose we change the basis of $\Re^d$ to $e_1', \ldots, e_d'$. How does this change the representation of the linear transformation T_A? To understand this, let B be a $d \times d$ matrix whose ith column is the representation of e_i' in the basis $\{e_1, \ldots, e_d\}$, that is, for every $i = 1, \ldots, d$:

$$e_i' = B_{1i}e_1 + B_{2i}e_2 + \ldots + B_{di}e_d \tag{13.3.2}$$

It is easy to show that B must be an invertible matrix (see Exercise Problem 13.8). We would like to now answer the following question: what are the coordinates of the vector $\mathbf{x}$ in the new basis? If $(x_1', \ldots, x_d')$ are the coordinates of $\mathbf{x}$ in this new basis, we see that $\sum_{i=1}^d x_i' e_i' = \sum_{i=1}^d x_i e_i$. Now using the expression for e_i' obtained in Eq. (13.3.2) and the linear independence of vectors in a basis, we see that

$$x_i = \sum_{j=1}^d x_j' B_{ij}$$

which can be written more compactly as

$$(x_1, \ldots, x_d)^T = B \cdot (x_1', \ldots, x_d')^T \tag{13.3.3}$$

Now we can understand how the representation of T_A changes. Earlier, a vector $\mathbf{x}$ with coordinates $(x_1, \ldots, x_d)$ was getting mapped to a vector $A \cdot (x_1, \ldots, x_d)^T$. If A' is the new representation of T_A, then the same vector with new coordinates $(x_1', \ldots, x_d')$ as before will get mapped to $A' \cdot (x_1', \ldots, x_d')^T$. Since these two resulting vectors are same, we see that

$A' \cdot (x'_1, \ldots, x'_d)^T = A \cdot (x_1, \ldots, x_d)^T$. Using Eq. (13.3.3), this implies that $A' \cdot (x'_1, \ldots, x'_d)^T = A \cdot B \cdot (x'_1, \ldots, x'_d)^T$. Since we could have chosen $(x'_1, \ldots, x'_d)$ to be any vector, this equality can happen if and only if $A' = A \cdot B$. This expression shows how the matrix A changes when we change the basis of $\Re^d$. One can similarly show the following more general result.

Theorem 13.3 *Suppose we change the basis of domain $\Re^d$ and range $\Re^n$ with corresponding matrices B and C respectively. Then the matrix for the linear transformation T_A becomes $C^{-1}AB$.*

We will be interested in cases where the basis vectors are always orthonormal, that is, if $e_1, \ldots, e_d$ is a basis, then $< e_i, e_j > = 0$ if $i \neq j$, 1 if $i = j$. Here $< e_i, e_j >$ denotes the dot product of these two vectors. Orthonormal vectors are convenient to work with because if $\mathbf{x}$ is any vector with coordinates $(x_1, \ldots, x_d)$ with respect to such a basis, then $x_i = < \mathbf{x}, e_i >$. This follows from the fact that $\mathbf{x} = \sum_{j=1}^{d} x_j e_j$, and so, $< \mathbf{x}, e_i > = \sum_{j=1}^{d} x_j < e_j, e_i > = x_i$. Using this notation, let B be the $d \times d$ matrix corresponding to the orthonormal bases $\{e'_1, e'_2, \ldots, e'_d\}$ and $\{e_1, e_2, \ldots, e_d\}$. It immediately follows from Eq. (13.3.3) that the columns of B are orthonormal, that is, the dot product of a column of B with itself is 1, and with any other column is 0. To see this, observe that

$$< e'_i, e'_l > = < \sum_{j=1}^{d} B_{ji} e_j, \sum_{k=1}^{d} B_{kl} e_l > = \sum_{j=1}^{d} B_{ji} B_{jl},$$

where the last equality follows from the orthonormality of the basis vectors $e_1, \ldots, e_d$. The RHS in this equation is the dot product of columns i and l of B. If $i = l$, $< e'_i, e'_i > = 1$, and so, each column of B has length 1. If $i \neq l$, then the columns i and l of B are orthogonal to each other. Such a matrix is also called a unitary matrix. It follows that for a unitary matrix, B^T is the inverse of B (see Exercise Problem 13.10). The singular value decomposition theorem shows that given a linear transformation T_A from $\Re^d$ to $\Re^n$ with corresponding matrix A, one can find orthonormal bases in the range and the domain such that the matrix corresponding to this transformation becomes diagonal. More formally, we can state the following theorem.

Theorem 13.4 *Let A be any $n \times d$ matrix. Then there exist $d \times d$ and $n \times n$ unitary matrices V and U respectively such that $A = U\Sigma V^T$, where Σ is a $n \times d$ diagonal matrix. Further, if σ_i denotes the diagonal entry $\Sigma_{i,i}$, then $\sigma_1 \geq \sigma_2 \ldots \geq \sigma_{\min(d,n)}$, and the matrix Σ is uniquely defined by A.*

The theorem essentially says that any matrix can be thought of as a diagonal matrix once we suitably change bases. The decomposition of A into $U\Sigma V^T$ is called the singular value decomposition (SVD) of A, and the diagonal entries of Σ are also called the singular values of A. We give the proof of the SVD theorem in Section 13.3.5. We now give some interesting applications of this decomposition. Given the SVD decomposition of A, one can very easily read off several interesting properties of A. Since the rank of a matrix does not change if we multiply it (on the left or the right) by an invertible matrix, we get the following result by multiplying A with U^T on the left and with V on the right.

Corollary 13.1 *The rank of A is the number of non-zero singular values of A.*

Let $u_1, \ldots, u_n$ and $v_1, \ldots, v_d$ be the columns of U and V respectively. Notice that the SVD decomposition theorem implies that $AV = U\Sigma$ and so, $Av_i = \sigma_i u_i$ if $1 \leq i \leq \min(d, n)$, and $Av_i = 0$ if $i > n$. Therefore, if r denotes the number of non-zero singular values of A, it follows that $u_1, \ldots, u_r$ span the range of A. Indeed, if $x \in \Re^d$, then we can write x as a linear combination of $v_1, \ldots, v_d$. So assume that $x = \sum_{i=1}^{d} \alpha_i v_i$. Therefore, $Ax = \sum_{i=1}^{d} \alpha_i Av_i = \sum_{i=1}^{r} \alpha_i \cdot \sigma_i \cdot u_i$. Thus, $u_1, \ldots, u_r$ spans the range of A. One can show similarly that $v_{r+1}, \ldots, v_d$ spans the null-space[1] of A.

13.3.2 Low-rank approximations using SVD

As outlined in the beginning of this section, one of the principal motivations for studying SVD is to find a low-rank approximation to a matrix A, that is, given an $n \times d$ matrix A, find a matrix $\tilde{A}$ of rank k, where $k << d, n$, such that $\tilde{A}$ is close to A. We need to formally define when a matrix is close to another matrix (of the same dimension). Recall that a similar notion for vectors is easy to define: two vectors v and v' are close if $v - v'$ has small length (or norm). Similarly, we will say that $\tilde{A}$ is close to A if the difference matrix $A - \tilde{A}$ has small norm. It remains to define the meaning of 'norm' of a matrix. There are many ways of defining this (just as there are many ways of defining length of a vector, for example, ℓ_p norms for various values of p). One natural way to define this notion is by thinking of a matrix A as a linear transformation. Given a vector x, A maps it to a vector Ax. Intuitively, we would like to say that A has large norm if A *magnifies* the length of x by a large factor, that is, $||Ax||/||x||$ is large, where $|| \cdot ||$ refers to the usual 2-norm (or the Euclidean norm) of a vector. Thus, we define the norm of a matrix A, denoted by $||A||$ (sometimes also called the *spectral norm* of A) as

$$\max_{x:x \neq \mathbf{0}} \frac{||Ax||}{||x||}$$

that is, the maximum ratio by which A magnifies the length of a non-zero vector. Although the notation $|| \cdot ||$ is used for both the 2-norm for vectors and spectral norm for the matrices, the reader should be able to interpret it unambiguously in a given context. We can now define the low-rank approximation problem formally. Given an $n \times d$ matrix, and a non-negative parameter $k \leq d, n$, we would like to find a rank k matrix $\tilde{A}$ such that $||A - \tilde{A}||$ is minimized.

Before we go into the details of this construction, we observe that SVD of a matrix A immediately gives its norm as well. To prove this, we make some simple observations.

---

[1] All vectors v such that $Av = 0$.

Lemma 13.1 *Let A be an $n \times d$ matrix and B be a unitary $n \times n$ matrix. Then, $||A|| = ||BA||$. Similarly, if C is a unitary $d \times d$ matrix, then $||A|| = ||AC||$.*

Proof: The main observation is that a unitary matrix preserves the length of a vector, that is, if U is unitary, then $||Ux|| = ||x||$ for any vector x (of appropriate dimension). Indeed, $||Ux||^2 = (Ux)^T \cdot Ux = x^T U^T U x = ||x||^2$, because $U^T U = I$. Therefore, if x is any vector of dimension d, then, by considering the vector Ax, we get $||Ax|| = ||BAx||$. So, $\max_{x:x \neq 0} \frac{||Ax||}{||x||} = \max_{x:x \neq 0} \frac{||BAx||}{||x||}$.

For the second part, observe that if C is a unitary matrix, $||Cx|| = ||x||$. Therefore, by substituting $x = Cy$, we get

$$\max_{x:x \neq 0} \frac{||Ax||}{||x||} = \max_{y:y \neq 0} \frac{||ACy||}{||y||}$$

This implies that $||A|| = ||AC||$. $\square$

In particular, this result implies that $||A|| = ||\Sigma||$. But it is easy to see that $||\Sigma|| = \sigma_1$ (see Exercise Problem 13.11). Consider the matrix $\tilde{A} = U \Sigma_k V^T$, where Σ_k is obtained from Σ by zeroing out all diagonal entries after (and excluding) σ_k, that is, the non-zero entries in Σ_k are $\sigma_1, \ldots, \sigma_k$. As argued before, the rank of $\tilde{A}$ is $\leq k$ (it could be less than k if some of the singular values among $\sigma_1, \ldots \sigma_k$ are 0). Now, observe that $\Sigma - \Sigma_k$ has norm σ_{k+1}. Therefore, $||A - \tilde{A}|| = \sigma_{k+1}$. We claim that $\tilde{A}$ is the best rank k approximation to A, that is, for any other rank k matrix B, $||A - B|| \geq \sigma_{k+1}$. Thus, SVD gives an easy way of finding the best rank k approximation to A.

In order to prove this, we shall need some elementary linear algebra facts:

- Let V be a vector space of dimension n and W_1 and W_2 be two subspaces of V. If dimension of W_1 plus that of W_2 is strictly greater than n, then there must be a non-zero vector in $W_1 \cap W_2$.

- Let A be an $n \times d$ matrix. Recall that the nullspace of A is the set of vectors v such that $Av = 0$. Then, the rank of A plus the rank of the null space of A is d.

Armed with these two facts, we now show that $\tilde{A}$ is the best rank k approximation. Indeed, let B be a matrix of rank at most k. Let $v_1, \ldots, v_d$ be the columns of V. Let V_{k+1} be the sub-space spanned by $v_1, \ldots, v_{k+1}$. The second observation mentioned here shows that the dimension of $N(B)$, the nullspace of B, is at least $d - k$, and so, using the first observation, there is a non-zero vector x in $V_{k+1} \cap N(B)$. Observe that $(A - B)x = Ax$ because $Bx = 0$. Therefore, $||A - B|| \geq \frac{||Ax||}{||x||}$. Finally, observe that x is a linear combination of $v_1, \ldots, v_{k+1}$, each of which gets magnified (in length) by a factor of at least σ_{k+1} by A. Therefore, $||Ax||/||x|| \geq \sigma_{k+1}$ as well (see Exercise Problem 13.12). This shows that $||A - B|| \geq \sigma_{k+1}$.

SVD also yields optimal low rank approximation with respect to another popular matrix norm, called the Frobenius norm. The Frobenius norm of an $m \times n$ matrix A is defined as $(\sum_{i=1}^{m} \sum_{j=1}^{n} |A_{ij}|^2)^{1/2}$. This is the usual Euclidean norm of the vector obtained from A by linearly arranging the entries in a vector of length mn. We state the following theorem without proof.

Theorem 13.5 *Let $\tilde{A}$ denote the rank k matrix $U\Sigma_k V^T$. For any rank matrix B of rank k or less, $||A - \tilde{A}||_F \leq ||A - B||_F$.*

Let us understand this result in a more detailed manner. Consider the matrix $m \times n$ matrix A as representing m points, $a_1, \ldots, a_m$, where a_i is the ith row of A. Each of these points lie in the n-dimensional Euclidean space. The rows of a rank k matrix B span a k-dimensional subspace, call it S. If we think of the ith row b_i as an approximation of a_i, then $||a_i - b_i||$ represents the distance between a_i and its approximation b_i. Thus, the problem of minimizing $||A_B||_F^2$ for a rank k matrix B can be stated as follows: we want to find a rank k subspace and points b_i in this subspace such that $\sum_{i=1}^{n} ||a_i - b_i||^2$ is minimized. Clearly, b_i should be the orthogonal projection of a_i on this subspace. Theorem 13.5 states that the optimal subspace is given by the span of the rows of $\tilde{A}$, and the projection of a_i on this subspace is given by the ith row of $\tilde{A}$. Can we give an orthonormal basis of this rank k subspace?

We claim that this subspace is spanned by the first k columns of V (which also form an orthonormal basis for this subspace). To see this, let V_k and U_k be the submatrix of V and U respectively consisiting of their first k columns. Let Σ_k denote the $k \times k$ diagonal matrix with diagonal entries $\sigma_1, \ldots, \sigma_k$. It is easy to check that $\tilde{A} = U_k \Sigma_k V_k^T$. From this it follows that each row of $\tilde{A}$ is a linear combination of the columns of V_k.

13.3.3 Applications of low-rank approximations

As indicated earlier, SVD is often used to remove noise from data. We now illustrate this concretely with an application in text processing, also denoted as 'latent sematic indexing'. We are given a set of documents, and would like to perform several tasks – (i) given two documents, figure out how close they are, and (ii) given a query term, output all documents which are relevant to this query term. A popular way to represent a document, also called the 'bag of words' model, is to think of it as a multi-set of words appearing in it. In other words, we store the documents in a term-document matrix T, where the columns of T correspond to documents, and rows of T corresponds to words (or terms) that could occur in these documents.[2] Thus, the entry T_{ij}, for a document j and term i, stores the frequency of term i in document j.[3] Suppose there are n documents and m terms. Then we can think of each document as a vector in $\mathfrak{R}^m$ corresponding to the

---

[2] Typically, we only use the *root* word, so that all forms of this word based on tense, etc. are unified.

[3] There are more nuanced measures than frequency of a term, which increase the weights of 'rare' words, and decrease the weights of very frequent words, like 'the', but we ignore such issues for the sake of clarity.

column representing it in this matrix. Similarly, we can think of a term as a vector of size n. Now, we can compare two documents by the cosine-similarity measure. Given two documents i and j, this measure considers the angles between the vectors corresponding to i and j, that is,

$$\cos^{-1} \frac{<T_i, T_j>}{||T_i|| ||T_j||}$$

This gives a measure in the range $[-1, 1]$ of similarity between two documents. Now suppose we want to output all documents which are relevant for a set of terms $(t_1, t_2, \ldots, t_r)$. We first think of this set of terms as a document containing just these terms, and so, it can be thought of as a vector of length m as well (so, it is a bit vector, where we have 1 for coordinates corresponding to these terms). Now, we can again use the cosine-similarity measure to output all relevant documents.

This approach is very appealing in practice because one can ignore issues involving grammar and a plethora of experimental data suggests that it works very well in practice. However, there are several issues. The first issue is computational – the matrix T is huge because the number of possible words can easily go up to several tens of thousands. The second issue is more semantic in nature. It is possible that two documents use the same term frequently but could be about completely different topics. For example, the term 'jaguar' could mean either the animal jaguar, or the car brand Jaguar. Similarly, it is possible that two different terms (e.g., 'car' and 'auto') could mean the same entity, and so, two different documents involving these terms respectively should be considered similar. Both of these problems suggest that perhaps the vectors corresponding to text belong to a different 'semantic' space where there are a smaller number of 'semantic' terms instead of actual words. The SVD approach tries to find such a low-dimensional representation of these vectors.

Another way to think about this problem is as follows. Suppose there are only k different topics that a document can be referring to (e.g., automobile, cooking, science, fiction, etc.). Each of the documents is inherently a vector $(w_1, \ldots, w_k)$, where w_i denotes the relevance of topic i for this document (if the weights are normalized so that they add up to 1, then the weights can be thought of as probabilities). Similarly, each of the terms can be thought of as a vector of length k. Now the entry T_{ij} of the term-document matrix corresponding to document j and term i can be thought of as the dot product of the corresponding vectors (that is, for the document and the term) of length k (if we think of weights as probabilities, then this dot product is the probability of seeing term j in document i). It is also easy to see that the rank of T will be at most k in this case (see Exercise Problem 13.13). Of course, the actual matrix T may not be obtained in this manner, but we still hope to see a rank k matrix which represents most of the content in T. In this sense, it is very appealing to replace T by a low rank (that is, rank k)

representation. Also, observe that if we can replace T by such a low-rank representation, it may be possible to get representations of documents as vectors of low dimension.

We use SVD to get a low-rank representation of T, that is, if $T = U\Sigma V^T$, then define $\tilde{T} = U\Sigma_k V^T$, where Σ_k is obtained from Σ by zeroing out all diagonal entries after σ_k. Let U_k be the $m \times k$ matrix obtained by selecting the first k columns of U (define V_k similarly). If $\tilde{\Sigma}_k$ denotes the square $k \times k$ matrix with diagonal entries $\sigma_1, \ldots, \sigma_k$, then we can rewrite the expression for T as $\tilde{T} = U_k \tilde{\Sigma}_k V_k^T$. We can now give the following low-dimensional representation of each term and document: for a term i, let t_i be the ith row vector of $U_k \tilde{\Sigma}_k$. Similarly, for a document j, let d_j be the jth row of V_k. It is easy to check that the (i, j) entry of T is equal to the dot product of t_i and d_j. Thus, we can think of these vectors as low-dimensional representation of terms and documents. Observe that for a document j, which was originally represented by column vector T_j of T (of length m), the representation d_j is obtained by the operation $\tilde{\Sigma}_k^{-1} U_k^T T_j$. Therefore, given a query q (which is a vector of length m), we perform the same operation (that is, $\tilde{\Sigma}_k^{-1} U_k^T q$) to get a vector $\tilde{q}$ of length k. Now we can find the most relevant documents for q by computing the angles (cosine) between $\tilde{q}$ and the vectors d_j for all documents j.

By storing low-dimensional representation of these vectors, we save on the space and time needed to answer a query. Moreover, experimental data suggests that replacing T by a low-dimensional representation gives better results.

13.3.4 Clustering problems

A very common scenario that a business company often encounters can be stated as follows – Given a certain population distribution in a city, where would they open k outlets such that the business benefits the most. Translated into a concrete objective function, a company wants to open outlets in locations $L = \{L_1, L_2 \cdots L_k\}$ such that $\sum_{i=1}^{n} \text{dist}(a_i, L)$ is minimized where a_i denotes the location of customer i and $\text{dist}(a_i, L) = \min_{L_j \in L} \|a_i - L_j\|$ i.e., the Euclidean distance to the closest outlet from customer i. Often, we prefer the square of the Euclidean distance, viz., $\sum_{i=1}^{n} \text{dist}^2(a_i, L)$. This belongs to a class of optimization problems known as *facility location* where the solution depends on the choice of the distance metric. In the aforementioned problem, we chose the sum of squared distance which is known as k-means problem. For the sum of (absolute) distance, it is known as k-median. The chosen locations are referred to as facilities and the customers that are associated with the same facility (usually the closest) define *clusters*. If the clusters are known (or fixed), then it is known that

Claim 13.1 *The k-means objective function is optimized by setting up a facility in the geometric centroid for each of the k clusters.*

This is true for any Euclidean space and the proof is left as an exercise to the reader.

Finding optimal locations for potential customers may be thought of as converse of the previously described nearest neighbor problem where the locations are known. Not surprisingly, solving this problem even in low dimensions is known to be intractable, so it makes sense to explore efficient algorithms that yield close to the optimal answers.

The k-means problem suffers from the so-called *curse of dimensionality* – many known heuristics run in time exponential in the number of dimensions, and so, reducing the dimensionality of the underlying points is an important preprocessing step. For example, we know by the Johnson–Lindenstrauss lemma that we can project the points to an $O(\log n/\varepsilon^2)$ dimensional space while losing only $(1+\varepsilon)$-factor in the approximation ratio. In typical applications k is a small number. Can we reduce the dimension to k? We show that this is indeed possible if we are willing to lose a factor 2 in the approximation.

The trick is again to use the SVD. Construct the $n \times d$ matrix A representing the n points as the rows of this matrix. Let a_i denote the ith row (and so, the ith point) of A. Recall from Theorem 13.5 that the first k rows of V represent a basis for the subspace S for which $\sum_{i=1}^{n} \text{dist}(a_i, S)^2$ is minimized, where $\text{dist}(a_i, S)$ denotes the distance between a_i and its projection on S. Further, the projection $\tilde{a}_i$ of a_i on S is given by the ith row of the matrix $\tilde{A} = U_k \Sigma_k V_k^T$.

We claim that the optimal set of locations for the points $\tilde{a}_1, \ldots, \tilde{a}_n$ will also give a good solution for the k means problem for the points $a_1, \ldots, a_n$. To see this, let $\tilde{L} = \{\tilde{L}_1, \tilde{L}_2 \ldots \tilde{L}_k\}$ be the optimal locations in the subspace S for $\{\tilde{a}_1, \tilde{a}_2 \ldots \tilde{a}_n\}$. The following statement shows that these locations are good for the original set of points as well.

Lemma 13.2 *Let $L = \{L_1, \ldots, L_k\}$ be the optimal set of locations for $a_1, \ldots, a_n$. Then*

$$\sum_{i=1}^{n} dist(a_i, \tilde{L})^2 \leq 2 \cdot \sum_{i=1}^{n} dist(a_i, L)^2.$$

Since $\tilde{L}$ is subset of S, and $\tilde{a}_i$ is the orthogonal projection of a_i on this subspace, Pythagoras theorem shows that

$$\text{dist}(a_i, \tilde{L})^2 = ||a_i - \tilde{a}_i||^2 + \text{dist}(\tilde{a}_i, \tilde{L})^2. \tag{13.3.4}$$

We prove the lemma by bounding each of these terms separately. To bound the first term, consider the subspace S' spanned by the points in L. Since L has size k, S' has dimension at most k. By optimality of S (Theorem 13.5),

$$\sum_{i=1}^{n} ||a_i - \tilde{a}_i||^2 = \sum_{i=1}^{n} \text{dist}(a_i, S)^2 \leq \sum_{i=1}^{n} \text{dist}(a_i, S')^2 \leq \sum_{i=1}^{n} \text{dist}(a_i, L)^2,$$

where the last inequality follows from the fact that L is a subset of S'. This bounds the first term in Eq. (13.3.4). Let us bound the second term now.

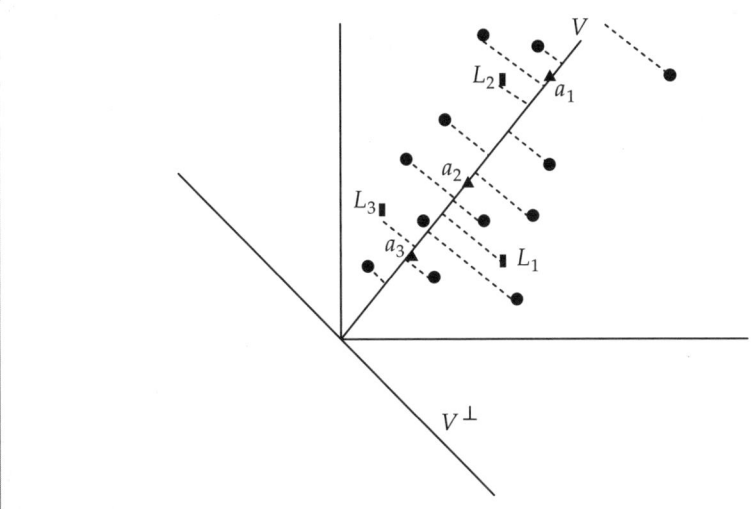

Figure 13.2   *A two-dimensional illustration of the SVD subspace V_1 of points represented by circular dots, where V_1 denotes the subspace spanned by the first column of V. The optimal locations are L_1, L_2, L_3. The optimal locations restricted to the subspace V_1 are a_1, a_2, a_3.*

Recall that $\tilde{L}$ is the optimal solution to the k-means instance given by the points $\tilde{a}_1, \ldots, \tilde{a}_n$. Let $\bar{L}$ denote the projection of L on the subspace S. Clearly, $\text{dist}(\tilde{a}_i, L) \geq \text{dist}(\tilde{a}_i, \bar{L})$. But the optimality of $\tilde{L}$ shows that

$$\sum_{i=1}^{n} \text{dist}(\tilde{a}_i, \tilde{L})^2 \leq \sum_{i=1}^{n} \text{dist}(\tilde{a}_i, \bar{L})^2 \leq \sum_{i=1}^{n} \text{dist}(a_i, L)^2,$$

where the last inequality follows from the fact that both $\tilde{a}_i$ and the points $\bar{L}_j$ are obtained by projecting a_i and L_j to the subspace S (see Exercise Probem 13.16). This bounds the second term in Eq. (13.3.4) and proves the lemma.

13.3.5   Proof of the SVD theorem

In this section, we give the main ideas behind a proof of the SVD theorem. The details are left as an exercise. Recall that the SVD theorem states that any $m \times n$ matrix A can be written as $U\Sigma V^T$, where U and V are unitary square matrices (dimensions $m \times m$ and $n \times n$ respectively) and Σ has non-zero entries in its diagonal only. Further, if σ_i denotes $\Sigma_{i,i}$, then $\sigma_1 \geq \sigma_2 \geq \ldots$. If $v_1, \ldots, v_n$ denote the columns of V, then the fact that V is unitary implies that $v_1, \ldots, v_n$ form an orthonormal basis of $\mathfrak{R}^n$. Similarly, $u_1, \ldots, u_m$, the columns of U, form an orthonormal basis of $\mathfrak{R}^m$. Finally, $Av_i = \sigma u_i$ for $1 \leq i \leq \min(m, n)$, and $Av_i = 0$ if $i > \min(m, n)$.

Let us see how to determine σ_1 first. Observe that $||Av_1|| = ||\sigma_1 u_1|| = \sigma_1$. Further, if x is any unit vector, then we claim that $||Ax|| \leq \sigma_1$. To see this, we write x as a linear combination of $v_1, \ldots, v_n$, that is, $\sum_{i=1}^{n} \alpha_i v_i$. Since $||x||^2 = \sum_{i=1}^{n} \alpha_i^2$, we know that $\sum_{i=1}^{n} \alpha_i^2 = 1$. Now, $Ax = \sum_{i=1}^{n} \alpha_i \cdot Av_i = \sum_{i=1}^{\min(m,n)} \alpha_i \sigma_i u_i$, where we have used the properties of the SVD theorem. Since the vectors u_i are also orthonormal, it follows that $||Ax||^2 = \sum_{i=1}^{\min(m,n)} \alpha_i^2 \sigma_i^2 \leq \sigma_1^2$. Thus, we can conclude that if the SVD theorem is true, then σ_1 must be the maximum value of $||Ax||$ over all unit vectors x.

With this intuition, we proceed with the proof. Define σ_1 as the highest value of $||Ax||$ over all unit vectors $x \in \mathfrak{R}^n$. Let v_1 be the unit vector which achieves this maximum.[4] So, $||Av_1|| = \sigma_1$. Hence, we can write Av_1 as $\sigma_1 u_1$, where u_1 is a unit vector. As the notation suggests, the vectors u_1 and v_1 should form the first column of U and V respectively. To complete the proof, we would argue that if V_1 denotes the subspace $\mathfrak{R}^n$ of vectors orthogonal to v_1 and U_1 denotes the corresponding subspace of vectors orthogonal to u_1, then A maps any vector in V_1 to a vector in U_1. Since the dimension of V_1 is going to be one less than the dimension of $\mathfrak{R}^n$, the proof will be completed by induction (details left as exercises).

Let x be any vector in V_1, and suppose, for the sake of contradiction that Ax does not belong to U_1, that is, Ax can be written as $\alpha u + \alpha_1 u_1$, where $u \in U_1$ and $\alpha_1 \neq 0$. By scaling, we can assume that x is a unit vector; similarly, by choosing α suitably, we can assume that u is a unit vector as well. Now consider the vector $v' = v_1 + \varepsilon x$, where ε is a small enough parameter which we will decide later. Since v_1 and x are orthogonal, $||v'||^2 = 1 + \varepsilon^2$. Further, $Av' = (\sigma_1 + \varepsilon \alpha_1) u_1 + \varepsilon \alpha u$. Since u and u_1 are orthonormal, we get $||Av'||^2 = (\sigma_1 + \varepsilon \alpha_1)^2 + \varepsilon^2 \alpha^2 \geq \sigma_1^2 + 2\varepsilon \alpha_1 \sigma_1$. Since $\alpha_1 \neq 0$, we choose ε with small enough absolute value such that $2\varepsilon \alpha_1 \sigma_1 > \varepsilon^2 \sigma^2$. Therefore, $||Av'||^2 > \sigma_1^2 (1 + \varepsilon^2) = \sigma_1^2 ||v'||^2$. We have found a vector v' such that the ratio $||Av'||/||v'||$ is strictly larger than σ_1, which is a contradiction. Therefore, any vector in V_1 must get mapped to U_1 by A. We can now finish the proof by induction.

Note that this proof is non-constructive. The proof requires us to find the unit vector which maximizes $||Ax||$. There are constructive methods for computing SVD; typically, these would take cubic time (in the dimension of the matrix). There are also connections between SVD and eigenvalues. To see this, first assume that $m \geq n$. If $A = U\Sigma V^T$, then $A^T = V\Sigma^T U^T$. Therefore, $A^T A = V\Sigma^T \Sigma V^T$. Now $\Sigma^T \Sigma$ is an $n \times n$ diagonal matrix, and its entries are $\sigma_1^2, \ldots, \sigma_n^2$. But note that if $A^T A = VDV^T$ for some diagonal matrix D and unitary matrix V, then the diagonal entries of D must be the eigenvalues of $A^T A$ (see Exercise Problems). Therefore, one way of computing SVD is via eigenvalue computation – form the matrix $A^T A$ and compute its eigenvalues. The (positive) square roots of the eigenvalues would give the singular values of A.

---

[4] One needs to use compactness arguments to show that such a vector exists, but we ignore this issue.

Further Reading

One of the first non-trivial applications of the Johnson–Lindenstrauss [71] result on random projection was to the problem of high-dimensional near neighbor searching by Indyk and Motwani [66]. The proof in this chapter follows the presentation by Dasgupta and Gupta [39]. Later, it was shown that the random projection can be approximated by an uniform choice over $\{+1, -1\}$ – see Achlioptas' paper [1]. The classical algorithms for SVD are iterative and take time $O(\min\{m \cdot n^2, n \cdot m^2\})$. Using sophisticated random projection techniques, the best known algorithms for *approximate* SVD take a time polynomial in ε and k, where ε is a measure of the approximation in a suitable norm and k corresponds to the best rank k approximation – see Frieze et al. [55].

Exercise Problems

13.1 Consider the 3 vertices of an equilateral triangle in the plane, each at distance 1 from the other two vertices. Show that it is not possible to map these 3 points to a line such that all pair-wise distances are 1.

13.2 If X is an $N(0,1)$ random variable, and a is a real number, prove that aX is distributed as $N(0, a^2)$. Using this, prove that if X, Y are two independent $N(0,1)$ random variables, and a and b are two real numbers, then $aX + bY$ has distribution $N(0, a^2 + b^2)$. Finally, use induction to prove that if $X_1, \ldots, X_d$ are d i.i.d. $N(0,1)$ random variables, then $a_1 X_1 + \ldots + a_d X_d$ has distribution $N(0, ||a||^2)$, where a denotes the vector $(a_1, \ldots, a_d)$.

13.3 Let f be the random projection on a line as defined in Section 3.1. Assuming ε is a small positive constant, prove that the probability that $||f(p) - f(q)||^2$ is within $(1 \pm \varepsilon)||p - q||^2$ is $\theta(\sqrt{\varepsilon})$.

13.4 Let A be a square matrix such that for the first $i - 1$ rows, the following property holds: for any j, $1 \le j \le i - 1$, the first $j - 1$ entries in row j are 0. Further, suppose $A_{j,i} = 0$ for all $j \ge i$. Prove that determinant of A is 0.

13.5 Let A be a square matrix. Find an invertible matrix P such that PA is the matrix obtained from A by interchanging rows A_i and A_j. Find an invertible matrix Q such that QA is obtained from A by replacing row A_j with $A_j - cA_i$, where c is a constant.

13.6 Let A be an $n \times d$ matrix whose rank is k ($k \le d$). Show how to use Gaussian elimination to find the low-dimensional representation of these points.

13.7 Show how Gaussian elimination can be used to obtain a basis for the rows of A.

13.8 Prove that the matrix B in Section 13.3.1 is invertible.

13.9 Suppose A and A' are two $n \times d$ matrices such that $Ax = A'x$ for all vectors x. Prove that $A = A'$.

13.10 Let B be a unitary matrix. Show that the inverse of B is same as the transpose of B.

13.11 Let D be an $n \times d$ diagonal matrix. Show that $||D|| = \max_{i=1}^{\min(d,n)} |D(i,i)|$.

13.12 Let $v_1, \ldots, v_s$ be a set of orthonormal vectors in $\Re^d$ and let A be an $n \times d$ matrix. Further, assume that $||Av_i|| \geq \sigma$ for $i = 1, \ldots, s$. If x is any non-zero vector in the span of $v_1, \ldots, v_s$, prove that $||Ax||/||x|| \geq \sigma$.

13.13 Let $v_1, \ldots, v_n, w_1, \ldots, w_m$ be a set of vectors in $\Re^k$. Construct a matrix T, where T_{ij} is equal to the dot product of v_i and w_j. Prove that the rank of T is at most k. Give an example to show that the rank of T can be strictly less than k.

13.14 Complete the proof of the SVD theorem by using induction on the subspaces U_1 and V_1.

13.15 Let B be an $n \times n$ square matrix. Suppose we can express B as $V^T D V$, where V is unitary and D is a diagonal matrix. Prove that the entries of D are eigenvalues of B.

13.16 Let S be a subspace in an n dimensional Euclidean space. Let a and b be two points in the Euclidean space, and let $\tilde{a}, \tilde{b}$ denote the orthogonal projection of a and b on S respectively. Prove that $||\tilde{a} - \tilde{b}|| \leq ||a - b||$.

13.17 Prove Claim 13.1.

14

CHAPTER

Parallel Algorithms

14.1 Models of Parallel Computation

There is a perpetual need for faster computation which is unlikely to be ever satisfied. With device technologies hitting physical limits, alternate computational models are being explored. The *Big Data* phenomenon precedes the coinage of this term by many decades. One of the earliest and natural direction to speed-up computation was to deploy multiple processors instead of a single processor for running the same program. The ideal objective is to speed-up a program p-fold by using p processors simultaneously. A common caveat is that an egg cannot be boiled faster by employing multiple cooks! Analogously, a program cannot be executed faster indefinitely by using more and more processors. This is not just because of physical limitations but dependencies between various fragments of the code, imposed by precedence constraints.

At a lower level, namely, in digital hardware design, parallelism is inherent – any circuit can be viewed as a parallel computational model. Signals travel across different paths and components and combine to yield the desired result. In contrast, a program is coded in a very sequential manner and the data flows are often dependent on each other – just think about a loop that executes in a sequence. Second, for a given problem, one may have to re-design a sequential algorithm to extract more parallelism. In this chapter, we focus on designing fast parallel algorithms for fundamental problems.

A very important facet of parallel algorithm design is the underlying architecture of the computer, viz., how do the processors communicate with each other and access data concurrently. Moreover, is there a common clock across which we can measure the actual running time? *Synchronization* is an important property that makes parallel algorithm design somewhat more tractable. In more generalized asynchronous models, there are additional issues like deadlock and even convergence, which are very challenging to analyze.

In this chapter, we will consider synchronous parallel models (sometimes called SIMD) and look at two important models – *parallel random access machine* (PRAM) and the *interconnection network* model. The PRAM model is the parallel counterpart of the popular sequential RAM model where p processors can simultaneously access a common memory called *shared* memory. Clearly, enormous hardware support is required to enable processors to access the shared memory concurrently which will scale with increasing number of processors and memory size. Nevertheless, we adopt a uniform access time assumption for reads and writes. The weakest model is called EREW PRAM or *exclusive read exclusive write* PRAM where all the processors can access memory simultaneously provided that there is no conflict in the accessed locations. Exclusiveness must be guaranteed by the algorithm designer. There are other varations as well, called CREW[1] and CRCW PRAMs that allow read conflicts and write conflicts. Although these are abstract models that are difficult to build, they provide conceptual simplicity for designing algorithms which can subsequently be mapped into the (weaker) realistic models that could lead to some slowdown.

Interconnection networks are based on some regular graph topology where the nodes are processors and the edges represent a physical link. The processors communicate with each other via messages passing through wired links where each link is assumed to take some fixed time. The time to send a message between two processors is proportional to the number of links (edges) in the route between the two processors. This could encourage us to add more links, but there is a tradeoff between the number of edges and the cost and area of the circuit, which is usually built as a VLSI circuit. Getting the right data to the right processor is the key to faster execution of the algorithm. This problem is commonly referred to as *routing*. Toward the end of this chapter we will discuss routing algorithms that provides a bridge between PRAM algorithms and the interconnection networks.

14.2   Sorting and Comparison Problems

14.2.1   Finding the maximum

This is considered to be a trivial problem in the sequential context and there are several ways of computing the maximum using $n - 1$ comparisons. A simple scan suffices where one maintains the maximum of the elements seen so far.

---

[1] **C** denotes concurrent and E denotes exclusive.

Claim 14.1 *Finding the maximum of n elements requires at least n − 1 comparisons.*

The proof is left as an exercise. We want to do many comparisons in parallel so that we can eliminate many elements from further consideration – every comparison eliminates the smaller element. We assume that in each round, each of the available processors compares a pair of elements. If we want to minimize the number of rounds, we can use $\binom{n}{2}$ processors to do all the pairwise comparisons and output the element that *wins* across all comparisons. The second phase of locating the element that has not *lost* requires more details in the parallel context and may require several rounds. But is the first phase itself efficient? We need roughly $\Omega(n^2)$ processors and so the total number of operations far exceeds the sequential bound of $O(n)$ comparisons and does not seem to be cost effective. The number of operations is often compared to the processor-time product which is a measure of the total computational bandwidth. A parallel algorithm is considered to be *efficient* if the number of operations is close to the processor-time product.[2]

Can we reduce the number of processors to $O(n)$. That seems unlikely as we can do at most $\frac{n}{2}$ comparisons in one round by pairing up elements and there will be at least $n/2$ potential maximums at the end of the first round. We can continue doing the same – pair up the *winners* and compare them in the second round and keep repeating this till we find the maximum. This is similar to a *knockout* tournament where after i rounds, there are at most $\frac{n}{2^i}$ potential winners. So after $\log n$ rounds, we can pick the maximum.

How many processors do we need?
If we do it in a straight forward manner by assigning one processor to each of the comparisons in any given round, we need $\frac{n}{2}$ processors (which is the maximum across all rounds). So the processor time product is $\Omega(n \log n)$; however, the total number of comparisons is $\frac{n}{2} + \frac{n}{4} + \ldots \leq n$, which is optimum. Hence, we must explore the reason for the inefficient use of processors.

One possibility is to reduce the number of processors to $p \ll n$ and slow down each round. For example, the $\frac{n}{2}$ first round comparisons can be done using p processors in roughly $\lceil \frac{n}{2p} \rceil$ rounds. This amounts to slowing down round i by a factor $\frac{n}{2^i \cdot p}$[3] so that the total number of rounds is

$$\frac{n}{p} \cdot \left(\frac{1}{2} + \frac{1}{2^2} + \ldots + \frac{1}{2^i}\right) \leq \frac{n}{p}$$

By definition, this is optimal work as the processor time product is linear. There is a caveat – we are ignoring any cost associated with assigning the available processors to the prescribed comparisons in each round. This is a crucial component for implementing parallel algorithms called *load balancing* which itself is a non-trivial parallel procedure

---

[2] Another popular measure is called FLOPS (floating point operations per second)
[3] We ignore the ceilings for simplicity.

requiring attention at the system level. We will sketch some possible approaches to this in the section on parallel prefix computation. For now, we ignore this component and therefore, we have a parallel algorithm for finding the maximum of n elements that require $O(\frac{n}{p})$ parallel time. But this tells us that we can find the maximum in $O(1)$ time using $p = \Omega(n)$! Clearly, we cannot do this in less than $O(\log n)$ rounds by the previous algorithm.

So there is a catch – when the number of comparisons falls below p, the time is at least 1, a fact that we ignored in the previous summation. So let us split the summation into two components – one when the number of camparisons is $\geq p$ and the subsequent ones when they are less than p. When they are less than p, we can run the first version in $O(\log p)$ rounds which is now an additive term in the expression for parallel time that is given by $O(\frac{n}{p} + \log p)$. It is now clear that for $p = \Omega(n)$, the running time is $\Omega(\log n)$; the more interesting observation is that it is minimized for $p = \frac{n}{\log n}$. This leads to a processor-time product $O(n)$ with parallel running time $O(\log n)$.

A simpler way to attain this bound will be to first let the $p = \frac{n}{\log n}$ processors sequentially find the maximum of (disjoint subsets of) $\log n$ elements in $\log n$ comparisons and then run the first version of $\frac{n}{\log n}$ elements using p processors in $\log(\frac{n}{\log n}) \leq \log n$ parallel steps. This has the added advantage that practically no *load balancing* is necessary as all the comparisons can be carried out by the suitably indexed processor. If a processor has index i, $1 \leq i \leq p$, we must pre-assign the comparisons to each of the processors. This is left as an Exercise Problem.

Can we reduce the number of rounds without sacrificing efficiency?
Let us re-visit the one-round algorithm and try to improve it. Suppose we have $n^{3/2}$ processors which is substantially less than n^2 processors. We can divide the elements into $\sqrt{n}$ disjoint subsets, and compute their maximum using n processors in a single round. After this round, we are still left with $\sqrt{n}$ elements which are candidates for the maximum. However, we can compute their maximum in another round using the one-round algorithm.[4] Taking this idea forward, we can express the algorithm in a recursive manner as follows.

The recurrence for parallel time can be written in terms of $T^{\|}(x, y)$ which represents the parallel time for computing the maximum of x elements using y processors. Then, we can write

$$T^{\|}(n, n) \leq T^{\|}(\sqrt{n}, \sqrt{n}) + T^{\|}(\sqrt{n}, n)$$

The second term yields $O(1)$ and with appropriate terminating conditions, we can show that $T^{\|}(n, n)$ is $O(\log \log n)$. This is indeed better than $O(\log n)$ and the processor time

---

[4] We are ignoring the cost of load balancing.

product can be improved further using the previous technqiues. The number of processors can be further reduced to $\frac{n}{\log\log n}$ and still retain $T^{\parallel}(n, n/\log\log n) = O(\log\log n)$. This is left as an Exercise Problem.

Can we improve the parallel running time further?

This is a very interesting question that requires a different line of argument. It turns out that with n processors, any algorithm requires $\Omega(\log\log n)$ rounds. We will provide a sketch of the proof. Consider a graph $G = (V, E)$, where $|V| = n$ and $|E| = p$. Every vertex corresponds to an element and an edge denotes a comparison between a pair of elements. We can think about the edges as the set of comparisons done in a single round of the algorithm. Consider an independent subset $W \subset V$. We can assign the largest $|W|$ values to the elements associated with W. Therefore, at the end of the round, there are still $|W|$ elements that are candidates for the maximum. In the next round, we consider the (reduced) graph G_1 on W and the sequence of comparisons in round i corresponds to the edges in G_{i-1}. Again, we can choose an independent set in this graph, and let them be the 'winners' of comparisons done in this round. The number of edges is bound by p. The following result on the size of the independent set of a graph, known as Turan's theorem will be useful in our context.

Lemma 14.1 *In an undirected graph with n vertices and m edges, there exists an independent subset of size at least $\frac{n^2}{m+n}$.*

Proof: We will outline a proof that is based on probabilistic reasoning. Randomly number the vertices V in the range 1 to n, where $n = |V|$ and scan them in an increasing order. A vertex i is added to the independent set I if all its neighbors are numbered higher than i. Convince yourself that I is an independent set. We now find a bound for $\mathbb{E}[|I|]$. A vertex $v \in I$ iff all the $d(v)$ neighbors ($d(v)$ is the degree of vertex v) are numbered higher than v and the probability of this event is $\frac{1}{d(v)+1}$. Let us define an indicator random variable $I_v = 1$ if v is chosen in I and 0 otherwise. Then,

$$\mathbb{E}[|I|] = \mathbb{E}[\sum_{v \in V} I_v] = \sum_{v \in V} \frac{1}{d(v) + 1}$$

Note that $\sum_v d(v) = 2m$ and that the aforementioned expression is minimized when all the $d(v)$ are equal, that is, $d(v) = \frac{2m}{n}$. Hence, $\mathbb{E}[|I|] \geq \frac{n}{2m/n+1} = \frac{n^2}{2m+n}$. Since the expected value of $|I|$ is at least $\frac{n^2}{2m+n}$, it implies that for at least one permutation, I attains this value and therefore, the lemma follows.   □

Let n_i, $i = 0, 1, 2 \ldots$ denote $|G_i|$ in the sequence $G = G_0, G_1, G_2, \ldots, G_i$ as defined by the independent sets in the algorithm. Then, from the previous claim, for $m = n$, one can show using induction that

$$n_i \geq \frac{n}{3^{2^i-1}}$$

Claim 14.2 *For $p = n$, show that after $j = \dfrac{\log\log n}{2}$ rounds, $n_j > 1$.*

A detailed proof is left as an Exercise Problem.

14.2.2 Sorting

Let us discuss sorting on the interconnection network model where each processor initially holds an element and after sorting, the processor indexed i, $1 \leq i \leq n$, should contain the rank i element. The simplest interconnection network is a linear array of n processing elements. Since the diameter is $n - 1$, we cannot sort faster than $\Omega(n)$ parallel steps since exchanging elements located at the two ends will have to move $n - 1$ steps.

An intuitive approach to sorting is to compare and exchange neighboring elements with the smaller element going to the smaller index. This can be done simultaneously for all (disjoint) pairs. To make this more concrete, we will define rounds with each round containing two phases – odd–even and even–odd. In the odd–even phase (Fig. 14.1), each odd numbered processor compares its element with the larger even number element and in the odd–even phase, each even numbered processor compares its element with that of the higher odd numbered processor.

Procedure Odd–even transposition sort for processor(i)

---

1 **for** $j = 1$ *to* $\lceil n/2 \rceil$ **do**
2   **for** $p = 1, 2$ **do**
3   **if** *If i is odd* **then**
4 | Compare and exchange with processor $i + 1$;
5   **else**
6   Compare and exchange with processor $i - 1$;
7   **if** *If i is even* **then**
8 | Compare and exchange with processor $i + 1$;
9   **else**
10   Compare and exchange with processor $i - 1$;

Figure 14.1   *Parallel odd–even transposition sort*

We repeat this over many rounds till the elements are sorted. To argue that it will indeed be sorted, consider the smallest element. In every comparison, it will start moving toward the processor numbered 1 which is its final destination. Once it reaches this processor, it will continue to be there. Subsequently, we can consider the next element which will finally reside in the processor numbered 2 and so on. Note that once elements reach their final destination and all the smaller elements have also reached their correct location, we can ignore them for future comparisons. Therefore, the array will be sorted after no more than n^2 rounds as it takes at most n rounds for any element to reach its final destination. This analysis is not encouraging from the perspective of speed-up as it only matches bubble sort. To improve our analysis, we must *track* the movements of elements simultaneously rather than 1 at a time. To simplify our analysis, we invoke the following result.

Lemma 14.2 (0–1 principle) *If any sorting algorithm sorts all possible inputs of 0s and 1s correctly, then it sorts all possible inputs correctly.*

We omit the proof here but we note that there are only 2^n possible 0–1 inputs of length n, whereas there are $n!$ permutations. The converse clearly holds.

So, let us analyze the algorithm mentioned earlier, called the **odd–even transposition** sort for inputs restricted to $\{0,1\}^n$. Such an input is considered sorted if all the 0s are to the left of all 1s. Let us track the movement of the leftmost 0 over successive rounds of comparisons. It is clear that the leftmost 0 will keep moving till it reaches processor 1.[5] If this 0 is in position k in the beginning, it will reach its final destination within at most $\lceil k/2 \rceil$ rounds. If we consider the next 0 (leftmost 0 among the remaining elements) to be denoted by 0_2, the only element that can block its leftward progress is the leftmost 0 and this can happen at most once. Indeed, after the leftmost 0 is no longer the immediate left neighbor of 0_2, this elements will keep moving left till it reaches its final destination. If we denote the sequence of 0s using 0_i for the ith zero from left, we can prove the following by induction.

Claim 14.3 *The element 0_i may not move for at most i phases ($\lceil i/2 \rceil$ rounds) in the beginning; subsequently, it moves in every phase until it reaches its final destination.*

A detailed proof of this claim is left as an exercise to the reader.

Since the final destination of 0_i is i, and it can be at most $n - i$ phase away from the final destination, the total number of phases for it to reach processor i is $i + n - i = n$. Note that this argument holds simultaenously all the elements and so all the 0s (and therefore, all 1s) are in their final positions within n phases or $\lceil n/2 \rceil$ rounds.

Next, we consider the two-dimensional mesh which is a widely used parallel architecture. For sorting, we can choose from some of the standard indexing schemes like row-major – every row i contains elements smaller than the next row, and the elements in a row are sorted from left to right. Column-major has the same property across columns, and snake-like row major has alternate rows that are sorted from left to right and the others from right to left.

---

[5] We are assuming that there is at least one 0 in the input; otherwise, there is nothing to prove.

Suppose we sort rows and columns in successive phases. Does this converge to a sorted array? No, you can construct an input where each row is sorted and every column is sorted (top to bottom) but not all elements are in their final position. A small change fixes this problem – sort rows according to snake-like row major (i.e., sort first row in increasing order from left to right, but the second row in increasing order from right to left, and so on) and the columns from top to bottom. The more interesting question is how many rounds of row/column sorts are required?

Each row/column can be sorted using the odd–even transposition sort. So if we need t iterations; then, the total parallel steps will be $O(t\sqrt{n})$ for a $\sqrt{n} \times \sqrt{n}$ array. To simplify our analysis, we will again invoke the 0–1 principle. First consider only two rows of 0s and 1s. Let us sort the first row from left to right and the second row from right to left. Then we do the column sort.

Lemma 14.3 *Either the top row will contain only 0s or the bottom row will contain only 1s – at least one of the conditions will hold.*

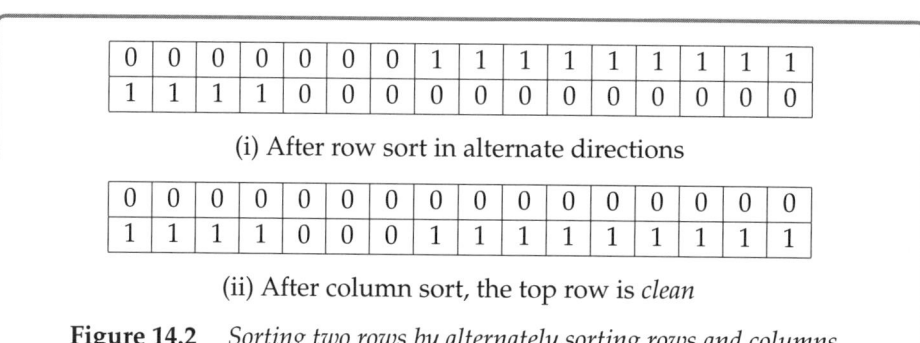

(i) After row sort in alternate directions

(ii) After column sort, the top row is *clean*

Figure 14.2 *Sorting two rows by alternately sorting rows and columns*

We define a row *clean* if it consists of only 0s or only 1s and *dirty otherwise*. According to this observation (prove it rigorously), after the row and column sort, at least one of the rows is *clean* so that in the next iteration (sorting rows), the array is sorted (Fig. 14.2). Now extend the analysis to an $m \times n$ array. After one round of row sort and column sort, at least half the rows are clean. Indeed each consecutive pair of rows produces at least one clean row and they continue to remain clean thereafter. In each iteration, the number of dirty rows reduce by at least a factor of 2, leading to $\log m$ iterations for all (but one) row to be clean. One more row sorting completes the ordering.

Lemma 14.4 *For an $m \times n$ array, alternately sorting rows and columns results in a snake-like sorted array after at most $\log m + 1$ iterations.*

This rather simple algorithm, called **Shearsort** (Fig. 14.3) is close to being optimal, within a factor of $O(\log n)$. Therefore, a $\sqrt{n} \times \sqrt{n}$ array can be sorted in $O(\sqrt{n}\log n)$ parallel steps. In the Exercise Problems, you will be led through an $O(\sqrt{n})$ algorithm based on a recursive variation of Shearsort.

Procedure Shearsort(m, n)

1 **for** $j = 1$ *to* $\lceil \log m \rceil$ **do**
2 $\quad$ Sort rows in alternating directions ;
3 $\quad$ Sort columns from top to bottom ;

Figure 14.3 *Shearsort algorithm for a rectangular mesh*

It is not difficult to extend Shearsort to higher dimensional meshes but it does not lead to an $O(\log n)$ time sorting algorithm in the hypercubic network. Obtaining an ideal speed-up sorting algorithm on an n-processor interconnection network is very challenging and requires many non-trivial ideas both in terms of algorithms and the network topology.

In the shared memory model like PRAM, one can obtain an $O(\log n)$ time algorithm by generalizing the idea of quicksort. This algorithm is called *partition* sort, and is shown in Fig. 14.4.

The analysis requires use of probabilistic inequalities like Chernoff bounds that enable us to obtain good control of the subproblem sizes for the recursive calls. Roughly speaking, if we can induce $O(\sqrt{n})$ bound on the size of the recursive calls when we partition n elements into $\sqrt{n}$ intervals (Fig. 14.4), the number of levels is bounded by $O(\log \log n)$.[6] Moreover, each level can be done in time $O(\log n_i)$, where n_i is the maximum subproblem size in level i. Then, the total parallel running time is bounded by $\sum_i \log(n^{1/2^i}) = O(\log n)$ (by ignoring some messy details).

Procedure Parallel partition sort

1 *Input* $X = \{x_1, x_2, \ldots, x_n\}$;
2 **if** $n \leq C$ **then**
3 $\quad$ Sort using any sequential algorithm
4 **else**
5 $\quad$ Choose a uniform random sample $\mathcal{R}$ of size $\sqrt{n}$;
6 $\quad$ Sort $\mathcal{R}$ - let $r_1, r_2 \ldots$ denote the sorted set ;
7 $\quad$ Let $X_i = \{x \in X | r_{i-1} \leq x \leq r_i\}$ be the ith subproblem ;
8 $\quad$ In **parallel** do ;
9 $\quad$ Recursively partition sort X_i for all i in parallel ;

Figure 14.4 *Partition sort in parallel*

---

[6] This is true for size of subproblems bounded by n^c for any $c < 1$.

In the following, we outline the proof for bounding the size of subprobems using a uniformly sampled subset $\mathcal{R} \subset S$, where $|S| = n$ and $\mathcal{R}$ has size about r.

Lemma 14.5 *Suppose every element of S is sampled uniformly and independently with probability $\frac{r}{n}$. Then, the number of unsampled elements of S in any interval induced by $\mathcal{R}$ is bounded by $O(\frac{n \log r}{r})$ with probability at least $1 - \frac{1}{r^{\Omega(1)}}$.*

Proof: Let $x_1, x_2, \ldots$ be the sorted sequence of S and let random variables

$$X_i = 1 \text{ if } x_i \text{ is sampled and } 0 \text{ otherwise } 1 \leq i \leq n$$

So the expected number of sampled elements $= \sum_{i=1}^{n} \Pr[X_i = 1] = n \cdot \frac{r}{n} = r$. Using Chernoff bounds, Eqs (2.2.7 , 2.2.8), we can bound the number of sampled elements by $r \pm O(\sqrt{r \log r})$ with probability $1 - \frac{1}{r^2}$. The actual number may vary but let us assume that $|R| = r$ that can be ensured by fusing some pairs of consecutive intervals. Suppose, the number of unsampled elements between two consecutive sampled elements $[r_i, r_{i+1}]$ is denoted by Y_i (Y_0 is the number of elements before the first sampled element). Since the elements are sampled independently, $\Pr[|Y_i| = k] = \frac{r}{n} \cdot (1 - \frac{r}{n})^k$ because there are k consecutive unsampled elements before a sampled element. It follows that

$$\Pr[|Y_i| \geq k] = \sum_{i=k}^{r} \frac{r}{n} \cdot \left(1 - \frac{r}{n}\right)^i \leq \left(1 - \frac{r}{n}\right)^k$$

For $k = \frac{cn \log r}{r}$, this is less than $e^{c \log r} \leq \frac{1}{r^c}$.

If any of the intervals has more than $k = c \frac{n \log r}{r}$ unsampled elements, then some pair of consecutive sampled elements $[r_i, r_{i+1}]$ has more than k unsampled elements between them; we have computed the probability of this event. So among the $\binom{r}{2}$ pairs of elements, the r consecutive pairs are the ones that are relevant events for us. In other words, the previous calculations showed that for a pair (r', r''),

$$\Pr[|(r', r'') \cap S| \geq k | r', r'' \text{ are consecutive}] \leq \frac{1}{r^c}$$

Since $\Pr[A|B] \geq \Pr[A \cap B]$, we obtain

$$\Pr[|(r', r'') \cap S| \geq k \text{ and } r', r'' \text{ are consecutive}] \leq \frac{1}{r^c}$$

So, for all the pairs, by the union bound, the probability that there is any consecutive sampled pair with more than k unsampled elements is $O(\frac{r^2}{r^c})$. For $c \geq 3$, this is less than $1/r$.

The reason that our sampling fails to ensure gaps less than $cn \log r / r$ is due to one of the following events

(i) Sample size exceeds $2r$ (ii) Given that sample size is less than $2r$, the gap exceeds k.

This works out as $O(\frac{1}{r})$ as the union of the probabilities. Note that we can increase c to further decrease failure probability and keep the union bound less than $1/r$ for r^2 samples.   □

The lemma shows that for an input of size $n = |X|$, the inputs for the next recursive call have sizes $O(\sqrt{n} \cdot \log n)$. This slightly changes our calculation for the number of phases (for which we had assumed that the recursive subproblems has size at most $\sqrt{n}$). Still it is easy to check that the number of recursive phases remains $O(\log \log n)$.

14.3 Parallel Prefix

Given elements $x_1, x_2, \ldots, x_n$ and an associative binary operator $\odot$, we want to compute

$$y_i = x_1 \odot x_2 \ldots x_i, \quad i = 1, 2, \ldots, n$$

Think about $\odot$ as addition or multiplication and while this may seem trivial in the sequential context, the prefix computation is one of the most fundamental problems in parallel computation and have extensive applications.

Note that $y_n = x_1 \odot x_2 \ldots x_n$ can be computed as a binary tree computation structure in $O(\log n)$ parallel steps. We need the other terms as well. Let $y_{i,j} = x_i \odot x_{i+1} \ldots x_j$. Then, we can express a recursive computation procedure as given in Fig. 14.5. Let $T^{\|}(x, y)$ represent

---

Procedure Prefix(a, b)

1 **if** *If* $b - a \geq 1$ **then**
2 $\quad$ $c = \lfloor \frac{a+b}{2} \rfloor$;
3 $\quad$ In **parallel** do
4 $\quad$ prefix (a,c) , prefix (c+1,b) ;
5 $\quad$ **end** parallel ;
6 $\quad$ Return (prefix (a,c) , $y_{a,c} \odot$ prefix (c+1 , b) (* $y_{a,c}$ is available in
$\quad\quad$ prefix (a,c) and is composed with each output of prefix$(c+1, b)$
$\quad\quad$ *)
7 **else**
8 $\quad$ Return x_a ;

Figure 14.5 *Parallel prefix computation: this procedure computes prefix of $x_a, x_{a+1}, \ldots, x_b$.*

the parallel time taken to compute the prefix of x inputs using y processors. For the algorithm mentioned here, we obtain the following recurrence

$$T^{\parallel}(n,n) = T^{\parallel}(n/2, n/2) + O(1)$$

The first term represents the time for two (parallel) recursive calls of half the size and the second set of outputs will be composed with the term $y_{1,n/2}$ that is captured by the additive constant. For example, given x_1, x_2, x_3, x_4, we compute in parallel prefix $(x_1, x_2) = x_1, x_1 \odot x_2$ and prefix$(x_3, x_4) = x_3, x_3 \odot x_4$. Then in a single parallel step, we compute $x_1 \odot x_2 \odot$ prefix$(x_3, x_4) = x_1 \odot x_2 \odot x_3, x_1 \odot x_2 \odot x_3 \odot x_4$ and return this result in line 6 along with prefix (x_1, x_2).[7] The solution is $T^{\parallel}(n,n) = O(\log n)$. Note that this is not optimal since the prefix can be computed sequentially in n operations, whereas the processor time product of our algorithm is $O(n \log n)$.

Let us try an alternate approach where we form n/k blocks of k inputs for a suitably chosen k and compute the values $x_i' = x_{(i-1)k+1} \odot x_{(i-1)k+2} \odot \ldots x_{ik}$ for $1 \leq i \leq \lfloor n/k \rfloor$. Now, we compute the prefix of x_i' using n/k processors which yields the prefix values y_i of the original inputs x_i for $i = k, 2k \ldots$. For the remaining elements within a block j, we can compute the prefixes sequentially by computing

$$y_{(j-1)k+\ell} = y_{(j-1)k} \odot x_{(j-1)k+1} \odot x_{(j-1)k+2} \ldots \odot x_{(j-1)k+\ell} \text{ for } 1 \leq \ell \leq k-1$$

This takes an additional k steps for each block and can be done simultaneously for all the n/k blocks using n/k processors.

For example, let us consider elements $x_1, x_2 \ldots x_{100}$ since $x_1 \odot x_2 \odot \ldots x_{20} = x_1' \odot x_2' = y_{20}$ and $k = 10$. Then,

$$y_{27} = x_1 \odot x_2 \ldots x_{27} = (x_1 \odot x_2 \ldots \odot x_{10}) \odot (x_{11} \odot x_{12} \ldots \odot x_{20}) \odot (x_{21} \odot \ldots x_{27})$$

$$= y_{20} \odot (x_{21} \odot x_{22} \ldots \odot x_{27})$$

The last term (in paranthesis) is computed within the block as prefix of 10 elements $x_{21}, x_{22} \ldots x_{30}$ and the rest is computed as prefix on the x_i's.

This approach is described formally in Fig. 14.6. The terminating condition takes $O(\log P)$ time using P processors for an input of size $\leq P$. An interesting variation would be to compute the prefixes of Z_is recursively (line 8 of the algorithm in Figure 14.6). The reader should complete the analysis of this algorithm by writing an appropriate recurrence and also choose an appropriate value of k for the best utilization of the P processors – see Exercise problems.

---

[7]The reader may note that although the procedure is defined using the parameters a, b, we are referring to it in the text as prefix (x_a, x_b).

Procedure Blocked prefix computation prefix(n, k, P)

1 **Input** $[x_1, x_2 \ldots x_n]$ $P =$ number of processors ;
2 **Output** Prefix $y_i = x_1 \odot x_2 \ldots x_i$ for $i = 1, \ldots, n.$;
3 **if** $n \geq P$ **then**
4   Divide $x_1, \ldots, x_n$ into blocks of size k ;
5   Compute prefix of each k-block independently in parallel ;
6   Let $Y_{ik+\ell} = x_{ik} \odot x_{ik+1} \ldots x_{ik+\ell}$, $1 \leq \ell \leq k-1$;
7   Let Z_i denote the last term in block i, i.e., $Z_i = x_{ik} \odot x_{ik+1} \ldots x_{ik+k-1}$;
8   Compute $y_{ik} = Z_1 \odot Z_2 \ldots Z_i$'s in parallel for all $1 \leq i \leq \lfloor n/k \rfloor$;
9   Compute $y_{i \cdot k+\ell} = Z_1 \odot Z_2 \ldots Z_i \odot Y_{ik+\ell}$ for all $i \leq n/k, \ell \leq k-1$;
10 **else**
11   Compute parallel prefix using the algorithm in Figure 14.5 ;

Figure 14.6 *Parallel prefix computation using blocking*

For the value $k = 2$, this scheme actually defines a circuit using gates to compute the operation $\odot$ (see Fig. 14.7).

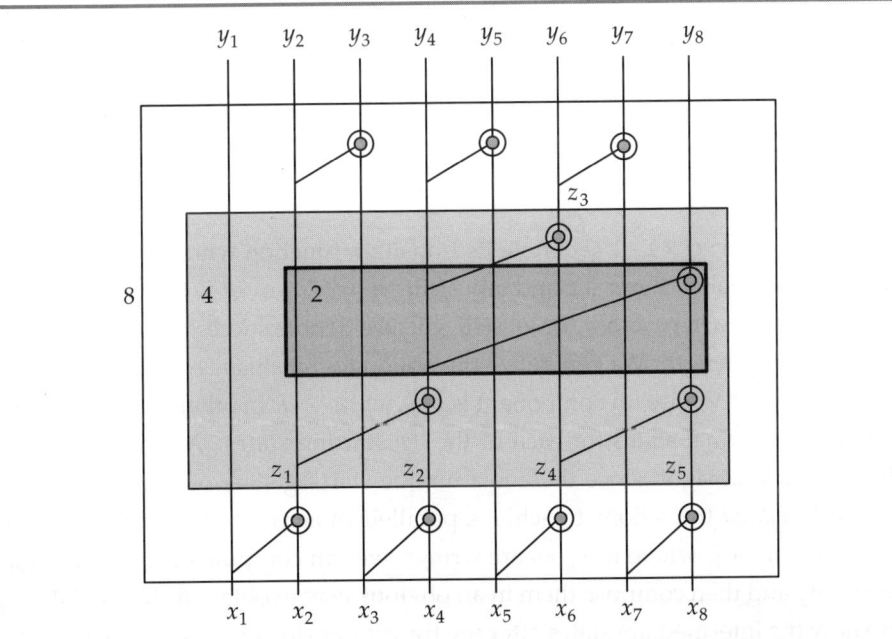

Figure 14.7 *Recursive unfolding of the prefix circuit with 8 inputs in terms of 4-input and 2-input circuits. These are indicated by the inner rectangles. The shaded circles correspond to the operation $\odot$.*

Parallel compaction

A very common scenario in parallel computation is periodic compaction of active processes so that the available processors can be utilized effectively. Even in a structured computation graph like a binary tree, at every level, half of the elements do not participate in future computation. The available processors should be equitably distributed to the active elements, so that the parallel time is minimized.

One way to achieve this is to think of the elements[8] in an array where we tag them as **0** or **1** to denote if they are dead or active. If we can compress the 1s to one side of the array, then we can find out how many elements are active, say m. If we have p processors, then we can distribute them equally such that every processor is allocated roughly $\frac{m}{p}$ active elements.

This is easily achieved by running parallel prefix on the elements with the operation $\odot$ defined as addition. This is known as *prefix sum*. The ith 1 from the left gets a label i and it can be moved to the ith location without any conflicts. Consider the array

$$1,0,01,0,1,0,1,0,0,0,1,1$$

After computing prefix sum, we obtain the labels y_is as

$$1,1,1,2,2,3,3,4,4,4,4,5,6$$

Then we can move the 1s to their appropriate locations and this can be done in parallel time n/p for $p \leq n/\log n$.

Simulating a DFA

Given a DFA M, let $\delta : Q \times \Sigma \to Q$ denote its transition function where Q is the set of states and Σ is a finite alphabet. For $a \in \Sigma$ and any string $w = a \cdot w_1$ over Σ, the transition function gets extended as $\delta(q, a \cdot w) = \delta(\delta(q,a), w)$, so δ will also denote successive transitions of M on a string of arbitrary length. We generalize the notion to *transition vector* $\delta_M(\bar{q}, w)$ where $\bar{q}$ is a vector of length $|Q|$ whose ith component is $\delta(q_i, w)$, $q_i \in Q$. In other words, the transition vector gives us the final states for each of the $|Q|$ starting states. Although this seems to be redundant at first sight, since there is a unique starting state of any DFA, it will help us in doing *lookahead* transitions to achieve parallelism across successive transitions. For example, if $w = w_1 \cdot w_2$, where w, w_1, w_2 are strings, we can compute $\delta_M(\bar{q}, w_1)$ and $\delta_M(\bar{q}, w_2)$ independently and then compose them in an obvious way to obtain $\delta_M(\bar{q}, w)$. Although we did not know the intermediate states after the transitions due to w_1, since we precomputed for all possible intermediate states, we can easily compose the transitions due to w_1 and w_2. For example, let $Q = \{q_0, q_1\}$ and the transition function be given by table

---

[8] They can also be thought of as labels of processes

	0	1
q_0	q_0	q_0
q_1	q_0	q_1

For $w = 1011$, $\delta_M(10) = \begin{bmatrix} q_0 \\ q_0 \end{bmatrix}$ and $\delta_M(11) = \begin{bmatrix} q_0 \\ q_1 \end{bmatrix}$. We have dropped $\bar{q}$ from the notation since it is implicit.

This yields $\delta_M(1011) = \delta_M(10) \odot \delta_M(11) = \begin{bmatrix} q_0 \\ q_0 \end{bmatrix}$, where $\odot$ denotes the *composition* of the transition functions. Alternately, we can express $\delta_M(a)$ as a $|Q| \times |Q|$ matrix A where $A_{i,j}^a = 1$ if $\delta(q_i, a) = q_j$ and 0 otherwise. Let $w = w_1 w_2 \ldots w_k$ where $w_i \in \Sigma$ and we will use the notation $w_{i,j}$ to denote the substring $w_i \cdot w_{i+1} \ldots w_j$. It can be easily verified that

$$\delta_M(w_{i,j}) = A^{w_i} \otimes A^{w_{i+1}} \otimes \ldots A^{w_j}$$

where $\otimes$ corresponds to matrix multiplication. Since the number of states is fixed, we can bound the cost of multiplication by $O(1)$. We will need the following property to reduce this problem to prefix computation.

Claim 14.4

$$(A^{w_1} \otimes A^{w_2}) \otimes A^{w_3} = A^{w_1} \otimes (A^{w_2} \otimes A^{w_3})$$

which follows from the associativity of matrix multiplication.

This shows that the composition of the generalized transition function $\delta_M()$ is associative, so we can use the prefix computation to compute all the intermediate states in $O(\log n)$ time using $n/\log n$ processors. This gives us the intermediate states for all possible starting states from which we choose the one corresponding to the actual starting state.

The addition of two binary numbers can be easily represented as state transition of a finite state machine. For example, if the numbers are 1011 and 1101 respectively, then one can design a DFA for an adder that takes an input stream (11, 10, 01, 11), which are the pairs of bits starting from the LSB. The successive transitions are made according to the previous carry, so there are two states corresponding to carry 0 and carry 1. Once the carry bits are known, then the sum bits can be generated in constant time. The reader is encouraged to work out the remaining details for the design of the *carry save adder*.

14.4 Basic Graph Algorithms

Many efficient graph algorithms are based on DFS (depth first search) numbering. A natural approach to designing parallel graph algorithms will be to design an efficient parallel algorithm for DFS. This turns out to be very challenging and no simple solutions are known; there is evidence to suggest that it may not be possible. This has led to interesting alternate techniques for designing parallel graph algorithms. We will consider the problem of constructing connected components in an undirected graph.

14.4.1   List ranking

A basic parallel subroutine involves finding the distance of every node of a given linked list to the end of the list. For concreteness and simplicity, we represent the list using an array $A[1 \ldots n]$, where $A[i] = j$ if node x_j is a successor of node x_i in the list. The first element of the list is referred to as the *tail* that has no predecessor and all other elements have exactly one successor. The head of the list is identified as k such that $A[k]$ contains k, that is, it points to itself. The purpose of list ranking is to find the distance of every element from the head of the list where the distance of the head to itself is considered as 0.

A sequential algorithm can easily identify the tail of the list (the integer in $\{1, \ldots, n\}$ which does not appear in the array) and simply traverse the list in n steps. For the parallel algorithm, let us initially assume that we have a processor for every element and each processor executes the algorithm in Fig. 14.8. To analyze the algorithm, let us re-number the list elements such that x_0 is the head of the list and x_i is at distance i from x_0. The crux of the algorithm is a *doubling* strategy. After $j \geq 1$ steps, the processor responsible for x_i, say p_i points to an element k such that k is 2^{j-1} steps away from x_i. So, in the next step, the distance doubles to 2^j. Of course, it cannot be further than the head of the list, so we have to account for that. Notice that when a processor points to the head, all the smaller numbered processors must also have reached the head. Moreover, they will also have the correct distances.

Procedure Parallel list ranking(p_i)

1   *Initialize* If $A[i] \neq i$ then $d[i] = 1$ else d[i] = 0 ;
2   **while** $A[i] > 0$ **do**
3   $\quad$ $A[i] \leftarrow A[A[i]]$;
4   $\quad$ $d[i] \leftarrow d[i] + d[A[i]]$;
5   Return $d[i]$;

Figure 14.8   *Parallel list ranking*

Table 14.1 *Consecutive snapshots of the list ranking algorithm on 15 elements. The $A(i)$s and $d(i)$s represent the value of the pointers and distances from x_i after i iterations. The x_is are shown to be in consecutive locations for convenience. Actually they are arbitrarily permuted but the progress of the algorithm remains as depicted in the table.*

i	x_0	x_1	x_2	x_3	x_4	x_5	x_6	x_7	x_8	x_9	x_{10}	x_{11}	x_{12}	x_{13}	x_{14}
$A(0)$	0	0	1	2	3	4	5	6	7	8	9	10	11	12	13
$d(0)$	0	1	1	1	1	1	1	1	1	1	1	1	1	1	1
$A(1)$	0	0	0	1	2	3	4	5	6	7	8	9	10	11	12
$d(1)$	0	1	2	2	2	2	2	2	2	2	2	2	2	2	2
$A(2)$	0	0	0	0	0	1	2	3	4	5	6	7	8	9	10
$d(2)$	0	1	2	3	4	4	4	4	4	4	4	4	4	4	4
$A(3)$	0	0	0	0	0	0	0	0	0	1	2	3	4	5	6
$d(3)$	0	1	2	3	4	5	6	7	8	8	8	8	8	8	8
$A(4)$	0	0	0	0	0	0	0	0	0	0	0	0	0	0	0
$d(4)$	0	1	2	3	4	5	6	7	8	9	10	11	12	13	14

Lemma 14.6 *After j iterations, $A[i] = \max\{i - 2^j, 0\}$. Equivalently, the distance function from i, $d(i)$ is given by $\min\{2^j, i\}$.*

Proof: Initially element i points to element $i - 1$. Note that for any i, once $A[i] = 0$, it remains 0 in future iterations. The same is also true for $d(i) = i$. Let $l(i)$ be defined as $2^{l(i)-1} < i \leq 2^{l(i)}$, for example, $l(8) = 3$ and $l(9) = 4$. We shall prove the following by induction on the number of iterations j :

(i) all elements x_i with $l(i) \leq j$ have $A(i) = 0$ and $d(i) = i$ and (ii) for $j < l(i)$, $A[i] = i - 2^j$, $d(i) = 2^j$.

Moreover all elements i with $l(i) = j$ satisfy property (i) for the first time in iteration j. For the base case $i = 0$, it is clearly true since x_0 keeps pointing to itself and $d[0]$ never changes.

Suppose the induction hypothesis is true for all iterations $< j$ where $j \geq 1$. For all elements k for which $l(k) \geq j$, from induction hypothesis, in iteration $j - 1$, all such elements k will have $A[k] = k - 2^{j-1}$. In particular, for $l(k) = j$, at the end of iteration $j - 1$, $A[k] = k'$ where $l(k') \leq j - 1$. Indeed the largest value of k, with $l(k) = j$ is $2^j - 1$ and $2^j - 1 - 2^{j-1} = 2^{j-1} - 1$ so $l(2^{j-1} - 1) = j - 1$. Since all elements i with $l(i) \leq j - 1$ point to x_0 by iteration $j - 1$, in iteration j, all elements k with $l(k) = j$ will have $A[k] = 0$ and $d[k] = k$ after the updates in iteration j. So the induction hypothesis holds for iteration j, thereby completing the proof.

If $2^{l(i)} < i$, by an analogous argument, $A[i]$ will increase by 2^j after j iterations. During the last iteration, the length would not double but increase additively by $i - 2^{l(i)}$. □

Since $l(n) \leq \log n$, the overall algorithm terminates in $O(\log n)$ steps using n processors. It is a challenging exercise to reduce the number of processors to $\frac{n}{\log n}$ so that the efficiency becomes comparable to the sequential algorithm.

Claim 14.5 *The number of processors in the list ranking algorithm can be reduced to $\frac{n}{\log n}$ without increasing the asymptotic time bound.*

For this, one can make use of a very useful randomized technique for symmetry-breaking. Suppose one could splice out every alternate element from the list, and solve the list ranking on the shortened list, then, one could also re-insert the deleted elements and compute their ranks easily from the adjacent elements. There are two main difficulties.

(i) Identifying alternate elements: Since the list elements are not in consecutive memory locations, we cannot use simple methods like *odd/even* locations. This is a classic example of *symmetry-breaking*.

(ii) We do not have a processor for every element – so we have to do *load balancing* to extract the ideal speed-up for every round of splicing.

To tackle the first problem, we label elements as *male/female* independently with equal probability. Subsequently, we remove any male node that does not have adjacent male nodes – probability of this is $\frac{1}{8}$ for any node, so the expected number of such nodes is $\frac{n}{8}$. These nodes can be spliced easily as they cannot be adjacent – while it is not half of the nodes, it is a constant fraction and serves our purpose. This technique is sometimes referred to as *random mate* in the literature.

For *load balancing*, we can use parallel prefix at the end of every round. Over $O(\log \log n)$ rounds, we can reduce the number of active elements to $\frac{n}{\log n}$ following which we can apply the previous algorithm. The details are left as an Exercise Problem.

In a PRAM model, each iteration can be implemented in $O(1)$ parallel steps, but it may require considerably more time in the interconnection network since the array elements may be far away and so the pointer updates cannot happen in $O(1)$ steps.

The aforementioned algorithm can be generalized to a tree where each node contains a pointer to its (unique) parent. The root points to itself. A list is a special case of a degenerate tree.

14.4.2 Connected components

Given an undirected graph $G = (V, E)$, we are interested to know if there is a path from u to w in G, where $u, w \in V$. The natural method to solve this problem is to compute maximal subsets of vertices that are connected to each other.[9]

---

[9] This is an equivalence relation on vertices.

Since it is difficult to compute DFS and BFS numbering in graphs in parallel, the known approaches adopt a strategy similar to computing minimum spanning trees. This approach is somewhat similar to Boruvka's algorithm described earlier in Figure 4.9. Vertices start out as singleton components and interconnect among each other using incident edges. A vertex u *hooks* to another vertex w using the edge (u, w) and intermediate connected components are defined by the edges used in the hooking step. The connected components are then merged into a single *meta-vertex* and this step is repeated until the meta-vertices do not have any edges going out. These meta-vertices define the connected components. There are several challenges in order to convert this high level procedure into an efficient parallel algorithm.

C1 What is an appropriate data structure that can maintain the meta-vertices?

C2 What is the hooking strategy so that the intermediate structures can be contracted into a meta-vertex? This will require choosing among multiple options to hook.

C3 How can we reduce the number of parallel phases?

Let us address these issues one by one. For C1, we pick a representative vertex from each component, called the *root* and let other vertices in the same component point to the root. This structure is called a *star* and it can be thought of as a (directed) tree of depth 1. The root points to itself. This is a very simple structure and it is easy to verify if a tree is a star. Each vertex can check if it is connected to the root (which is unique because it points to itself). With sufficient number of processors, it can be done in a single parallel phase.

We will enable only the root vertices to perform the hooking step, so that the intermediate structures have a unique root (directed trees) that can be contracted into a star. Note that a root vertex could hook to a non-root vertex according to this policy. We still have to deal with the following complications.

How do we prevent two (root) vertices hooking on to each other? This is a typical problem of *symmetry-breaking* in a parallel algorithm where we want to select one among the many (symmetric) possibilities to succeed using some discriminating properties. In this case, we can follow a convention that the smaller numbered vertex can hook on to a larger numbered vertex. We are assuming that all vertices have a unique id between $1 \ldots n$. Moreover, among the eligible vertices that it can hook onto, it will choose one arbitrarily.[10] This still leaves open the possibility of several vertices hooking to the same vertex but that does not affect the successful progress of the algorithm.

Let us characterize the structure of the subgraph formed by hooking. The largest numbered vertex in a component cannot hook to any vertex. Each vertex has at most one directed edge going out and there cannot be a cycle in this structure. If we perform

---

[10]This itself requires symmetry-breaking in disguise but we will appeal to the model supporting concurrent writes.

shortcut operations for every vertex similar to list ranking, then the directed tree will be transformed into a star.

For the hooking step, all the edges going out of a tree are involved, as the star can be considered as a meta-vertex. If all the directed trees are stars and we can ensure that a star combines with another one in every parallel hooking phase, then the number of phases is at most $\log n$. The number of stars will decrease by a factor of two (except those that are already maximal connected components). This would require that we modify the hooking strategy so that every star gets a chance to combine with another.

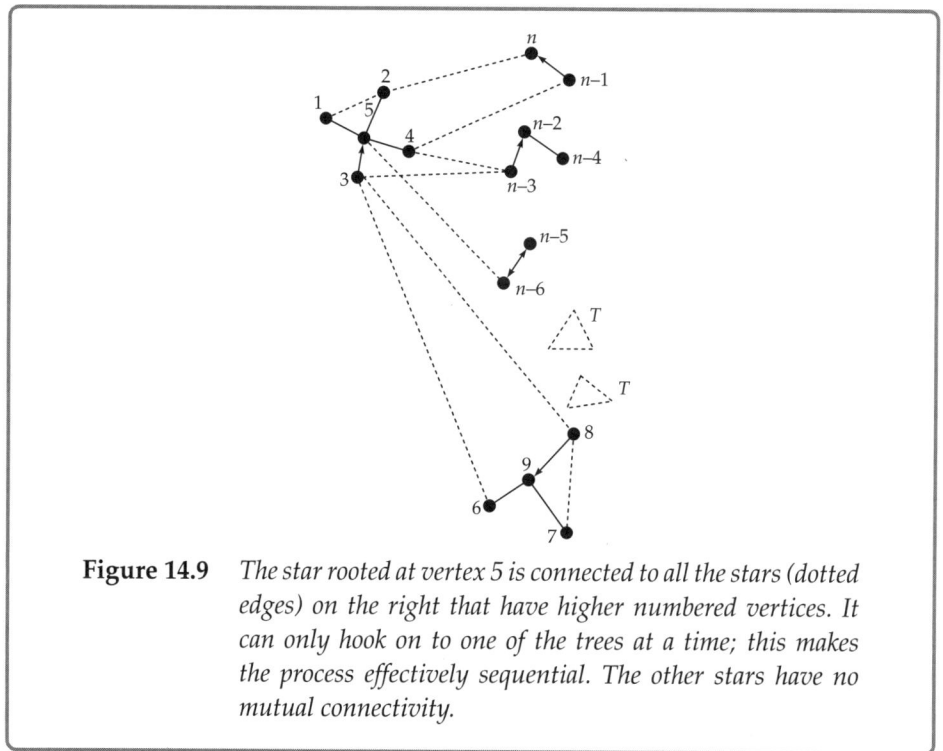

Figure 14.9   *The star rooted at vertex 5 is connected to all the stars (dotted edges) on the right that have higher numbered vertices. It can only hook on to one of the trees at a time; this makes the process effectively sequential. The other stars have no mutual connectivity.*

Figure 14.9 gives an example where only one star gets hooked in every step because of the symmetry-breaking rule. Therefore, we can add another step where a star that could not combine since the root had a larger number (but it lost out to other large numbered roots) can hook to a smaller numbered root. Since the smaller numbered root must have hooked to some *other* tree (since the present tree continues to be a star), this cannot create any cycles and is therefore safe.

The algorithm is described formally in Fig. 14.10.

For ease of understanding, the reader can assume that there is a processor assigned to each vertex and to each edge. In reality, with fewer processors, we will need to repeatedly use load balancing between iterations.

Procedure Parallel graph connectivity(G)

1 *Initialize* For all $v \in V$, we set $p(v) = v$, *allstar* = FALSE ;
2 **while** *NOT allstar* **do**
3   **for** $(u,v) \in E$ *in parallel* **do**
4   **if** *Isroot*($p(u)$) *and* ($p(u) < p(v)$) **then**
5             $p(p(u)) \leftarrow p(v)$ (hook root of u to $p(v)$);
6   **if** *IsStar* (v) **then**
7             $p(p(v)) \leftarrow p(u)$ (since v belongs to star, hook the root v to
                $p(u)$) ;
8   **for** $v \in V$ **do**
9         $p(v) \leftarrow p(p(v))$ (pointer jumping to reduce height);
10   *allstar* = TRUE (Check if all components are stars) ;
11   **for** *all vertices v in parallel* **do**
12   **if** *NOT IsStar*(v) **then**
13   *allstar* = FALSE

Function IsStar(w)

1 **if** $p(w) \neq p(p(w))$ **then**
2   (the tree containing w is not a star) ;
3   Set the tag of w to FALSE.

Function IsRoot(v)

1 **if** $p(v) = v$ **then**
2   true
3 **else**
4   false

Figure 14.10   *Parallel connectivity: We assume that there is a processor assigned to every vertex $v \in V$ and to every edge $(u,w) \in E$. The global variable allstar is TRUE only when all components are stars.*

The algorithm maintains a parent pointer $p(v)$ for each node v – this is meant to maintain the star data structure. For any root node r, $p(r)$ is set to r itself – the function IsRoot() in Fig. 14.10 checks if a node is a root node or not. Similarly, the function IsStar(w) checks if the parent of w is root. As the algorithm progresses, it may happen that in a particular component, all the vertices do not have their parent pointers to the root. Steps 8–9 fix this aspect by moving each of the $p(v)$ pointers closer to the corresponding root. In Steps 4–5, we ensure that if add the edge (u, v), then the parent of u is the root of this component. Indeed, we want to assign $p(p(u))$ to $p(v)$, and so, it better be the case that $p(p(u))$ points to $p(u)$ itself. A similar check is done in Steps 6–7.

The singleton vertices require special handling since these do not have any children. The function IsStar() could encounter a problem since it is unable to distinguish between stars with depth 1 and the depth 0 nodes. The reader can verify that it could lead to creation of cycles as singleton vertices could hook onto each other. To avoid this we initialize all the singleton nodes v by creating an artificial child node v' that makes it a star. Then we have stars hooking to another star creating a depth two tree that avoids the previous complication of cycles. The extra nodes v' do not otherwise affect the algorithm.

The nodes maintain the tree structure using the 'parent' pointers $p()$. The root of any component will point to itself. The procedure IsStar() is executed at all nodes in parallel, and tags those nodes v for which $p(p(v))$ is not equal to $p(v)$ – this will happen only when the pointer $p(v)$ is not pointing to the corresponding root vertex.

The analysis is based on a potential function that captures the progress of the algorithm in terms of the heights of the trees. Once all the connected components are hooked together, the algorithm can take at most $\log n$ iterations to transform them into stars, based on our analysis of pointer jumping.

We define a potential function $\Phi_i = \sum_{T \in \mathcal{F}} d_i(T)$, where $d_i(T)$ is the depth of a tree T (star has depth 1) in iteration i. Here $\mathcal{F}$ denotes the forest of trees. Note that a tree contains vertices from a single connected component. We can consider each of the components separately and calculate the number of iterations that it takes to form a single star from the component starting with singleton vertices. The initial value of Φ is $|C|$, where $C \subset V$ is a maximal component and finally, we want it to be 1, that is, a single star.

If T_1 and T_2 are two trees that combine in a single tree T after hooking, it is easily seen that $\Phi(T) \leq \Phi(T_1) + \Phi(T_2)$. For any tree (excluding a star), the height must reduce by a factor of almost $1/2$. Actually, a tree of depth 3, reduces to 2, which is the worst case. Hence, $\Phi(C) = \sum_{T \in C} d(T)$ must reduce by a factor $2/3$ in every iteration, resulting in overall $O(\log n)$ iterations. The total number of operations in each iteration is proportional to $O(|V| + |E|)$.

The overall running time would vary by a $\log n$ factor depending on the model of concurrency used. The present description assumes CRCW model – in particular, the functions IsStar and Isroot involves several processors trying to write and read from the same location. A CRCW model can be mapped to an EREW model at the cost of $O(\log n)$ overhead per step. The details are beyond the scope of this discussion.

The reader is encouraged to analyze a variation of this algorithm, where we perform repeated pointer jumping in step 3, so as to convert a tree into a star before we proceed to the next iteration. It is left as an Exercise Problem to compare the two variants of the connectivity algorithm.

14.5   Basic Geometric Algorithms

The Quickhull algorithm described in Section 7.5 is a good candidate for a parallel algorithm as most of the operations can be done simultaneously. These are $O(1)$ time left-turn tests involving the sign of a 3×3 determinant, based on which some of the points are eliminated from further consideration. The subproblems are no more than $\frac{3}{4}$ of the original problem implying that there are $O(\log n)$ levels of recursion. The number of operations in each level is proportional to $O(n)$ and so if each level of recursion can be done in $t(n)$ parallel time, the total time will be $O(t(n) \cdot \log n)$. If $t(n)$ is $O(1)$, it would lead to an $O(\log n)$ time parallel algorithm which is often regarded as the best possible algorithm because of a number of related lower bounds. Although the left-turn tests can be done in $O(1)$ steps, the partitioning of the point sets into contiguous locations in an array is difficult to achieve in $O(1)$ time. Without this, we will not be able to apply the algorithm recursively or work with points in contiguous locations. We know that compaction can be done in $O(\log n)$ time using prefix computation, so we will settle for an $O(\log^2 n)$ time parallel algorithm.

The number of processors is $O(n/\log n)$; this will enable us to do $O(n)$ left-turn tests in $O(\log n)$ time. Unlike the (sequential) Quickhull algorithm, the analysis is not sensitive to the output size. For this, we will relate the parallel running time with the sequential bounds to obtain an improvement of the following kind.

Theorem 14.1 *There is a parallel algorithm to construct a planar convex hull in $O(\log^2 n \cdot \log h)$ parallel time and total work $O(n \log h)$ where n and h are the input and output sizes respectively.*

We will describe a very general technique for load distribution in a parallel algorithm. Suppose there are T parallel phases in an algorithm where there is no dependence between operations carried out within a phase. If there are p processors available, then by sharing the m_i tasks in phase i tasks equally among them, $1 \le i \le T$, tasks in phase i can be completed in time $O(\lceil \frac{m_i}{p} \rceil)$. Hence, the total parallel time is given by $\sum_i^T O(\lceil \frac{m_i}{p} \rceil)$ $= O(T) + O(\frac{\sum_i m_i}{p})$.

To this, we also need to add the time for load balancing based on prefix computation, namely, $O(\frac{m_i}{p})$ for phase i as long as $m_i \ge p \log p$. So, this implies that each of the $O(\log n)$ phases requires $\Omega(\log p)$ steps since $m_i/p \ge \log p$. We can state the result as follows.

Lemma 14.7 (Load balancing) *In any parallel algorithm that has T parallel phases with m_i operations in phase i, the algorithm can be executed in $O(T \log p + \frac{\sum_i m_i}{p})$ parallel steps using p processors.*

Let us apply the previous result in the context of the Quickhull algorithm. There are $\log n$ parallel phases and in each phase, there are at most n operations as the points belonging to the different subproblems are disjoint. From the analysis of the sequential algorithm, we know that $\sum_i m_i = O(n \log h)$, where h is the number of output points. Then an application of the aforementioned load balancing technique using p processors will result in a running time of $O(\log n \cdot \log p) + O(\frac{n \log h}{p})$. Using $p \leq \frac{n}{\log^2 n}$ processors yields the required bound of Theorem 14.1.

Note that using $p = \frac{n \log h}{\log^2 n}$ would yield a superior time bound of $O(\log^2 n)$; however since h is an unknown parameter, we cannot use it in the algorithm description.

14.6 Relation between Parallel Models

The PRAM model is clearly stronger than the *interconnection network* since all processors can access any data in $O(1)$ steps from the shared memory. More formally, any single step of an interconnection network can be simulated by the PRAM in one step. The converse is not true since data redistribution in the network could take time proportional to its diameter.

The simplest problem related to redistribution of data is called 1–1 *permutation routing*. Here every processor is a source and a destination of exactly one data item. The ideal goal is to achieve this routing in time proportional to $\mathcal{D}$ which is the diameter. There are algorithms for routing in different architectures that achieve this bound.

One of the simplest algorithms is the greedy algorithm where the data item is sent along the shortest route to its destination. A processor can send and receive one data item to/from each of its neighbors in one step.

Claim 14.6 *In a linear array of n processors, permutation routing can be done in n steps.*

One can look at the direction of the movement of the packets – either leftward or rightward and one can argue about the number of steps taken being proportional to the distance between the source and the destination. The details are left as an Exercise Problem.

If a processor has multiple data items to be sent to any specific neighbor, then only one data item is transmitted at any time while the rest must wait in a queue. In any routing strategy, the maximum queue length must have an upper bound for scalability. In the case of linear array, the queue length can be bounded by a constant.

To simulate a PRAM algorithm on interconnection network, one needs to go beyond permutation routing. More specifically, one must be able to simulate concurrent read and concurrent write. There is a rich body of literature that describes emulation of PRAM algorithms on low diameter networks like hypercubes and butterfly networks that take $O(\log n)$ time using constant size queues. This implies that PRAM algorithms can run on interconnection networks incurring no more than a logarithmic slowdown.

14.6.1 Routing on a mesh

Consider an $n \times n$ mesh of n^2 processors whose diameter is $2n$. Let a processor be identified by (i, j), where i is the row number and j is the column number. Let the destination of a data packet starting from (i, j) be denoted by (i', j').

A routing strategy is defined by the following.

(i) Path selection.

(ii) Priority scheme between packets that contend for the same link.

(iii) Maximum queue size in any node.

In the case of a linear array, path selection is unique and priority is redundant since there were never two packets trying to move along a link in the same direction (we assume that links are bidirectional allowing two packets to move simultaneously in opposite directions on the same link). There is no queue build up during the routing process.

However, if we change the initial and final conditions by allowing more than one packet to start from the same processor, then a priority order has to be defined between contending packets. Suppose we have n_i packets in the ith processor, $1 \le i \le n$, and $\sum_i n_i \le cn$ for some constant c. A natural priority ordering is defined as *furthest destination first*. Let n'_i denote the number of packets that have destinations in the ith node. Clearly, $\sum_i n_i = \sum_i n'_i$ and let $m = \max_i\{n_i\}$ and $m' = \max_i\{n'_i\}$. The greedy routing achieves a routing time of cn steps using a queue size $\max\{m, m'\}$. We outline an argument for $c = 1$ that can be extended to the more general case.

Here is an analysis using the *furthest destination first* priority scheme. For a packet starting from the ith processor, with destination j, $j > i$, it can be delayed only by packets with destination in $[j+1, n]$. If $n'_i = 1$ for all i, then there are exactly $n - j - 1$ such packets so that the packet will reach its destination within $n - j - 1 + (j - i) = n - i - 1$ steps. This can be easily argued starting from the rightmost moving packet and the next packet which can be delayed at most once and so on. When n'_i exceeds 1, then a packet can get delayed by $\sum_{i=j}^{i=n} n'_i$ steps. The queue sizes do not increase while packets are being received and sent by processors but will require additional storage when a packet reaches its destination.

We can extend the previous strategy to routing on a mesh. Let us use a path such that a packet reaches the correct column and then goes to the destination row. If we allow unbounded queue size, then it can be easily done using two phases of one-dimensional routing, requiring a maximum of $2n$ steps. However, the queue sizes could become as large as n. For example, all packets in (r, i) may have destinations (i, r) for a fixed r and $1 \le i \le n$. To avoid this situation, let us *distribute* the packets within the same column such that the packets that have to reach a specific column are distributed across different rows.

A simple way to achieve this is for every packet to choose a random intermediate destination within the same column. From our previous observations, this routing would take at most $n + m$ steps, where m is the maximum number of packets in any (intermediate) destination. Subsequently, the time to route to the correct column will depend on the maximum number of packets that end up in any row. The third phase of routing will take no more than n steps since every processor is a destination of exactly one packet. Figure 14.11 illustrates the path taken by a packet in a three phase routing.

To analyze phases 1 and 2, we will get a bound on the expected number of packets that choose a given destination and the number of packets in a row that are destined for any specific column. This will enable us to bound the sizes of the queues required and the maximum number of packets that have to be routed in phase 2. Since the destinations are chosen uniformly at random, the probability that a processor in row r is chosen in phase 1 is $\frac{1}{n}$. Let X_i be a 0–1 random variable which is 1 if row r is chosen by the data packet i and 0 otherwise. Then the number of packets that will end up in the processor in row r is a random variable $X = \sum_{i=1}^{i=n} X_i$. So,

$$\mathbb{E}[X] = \mathbb{E}[\sum_i X_i] = \sum_i \Pr[X_i = 1] = 1$$

The random destinations can be thought of as independent Bernoulli trials, so their sum is a binomial random variable with expectation = 1. From Chernoff bounds, Eq. 2.2.5,

$$\Pr[X \geq \Omega(\log n / \log \log n)] \leq \frac{1}{n^3}$$

that bounds the maximum number of packets that end up in any processor at the end of phase 1 using the union bound over n^2 processors.

Let Y_r represent the number of packets that end up in row r at the end of phase 1. Then, by extending the aforementioned argument, the expected number of packets $\mathbb{E}[Y_r] = n^2 \cdot \frac{1}{n} = n$. From Chernoff bounds, Eq. 2.2.7, it follows that

$$\Pr[Y_r \geq n + \Omega(\sqrt{n \log n})] \leq \frac{1}{n^2}$$

This bounds the routing time of phase 2 within row r by $n + O(\sqrt{n \log n}$ from our previous discussion on greedy routing using *furthest destination first* priority.

To bound the queue size for phase 2, we also need to bound the number of packets (in row r) that have destinations in column j for any $1 \leq j \leq n$. Let n_j denote the number of packets in column j initially that have destinations in column C where $\sum_j n_j = n$. How many of them choose row r as the random destination? If $X_{j,C}$ represents the number of such packets in column j, then by using the previous arguments $\mathbb{E}[X_{j,C}] = \frac{1}{n} \sum_j n_j = 1$ since each of the n_j packets independently, choose their (random) destinations as row r in phase 1. Again by Chernoff bounds,

$$\Pr[X_{j,C} \geq \Omega(\log n / \log \log n)] \leq \frac{1}{n^3}$$

which holds for all rows r and all columns C using union bound.

So, all the queue sizes can be bounded by $\log n/\log\log n$ in phase 1 and phase 2. Moreover the routing time in phase 2, can be bounded by $n + O(\sqrt{n\log n})$ in phase 2. Thus the total routing time over all the 3 phases can be bound by $3n + O(\sqrt{n\log n})$ using $O(\log n/\log\log n)$ sized queues.

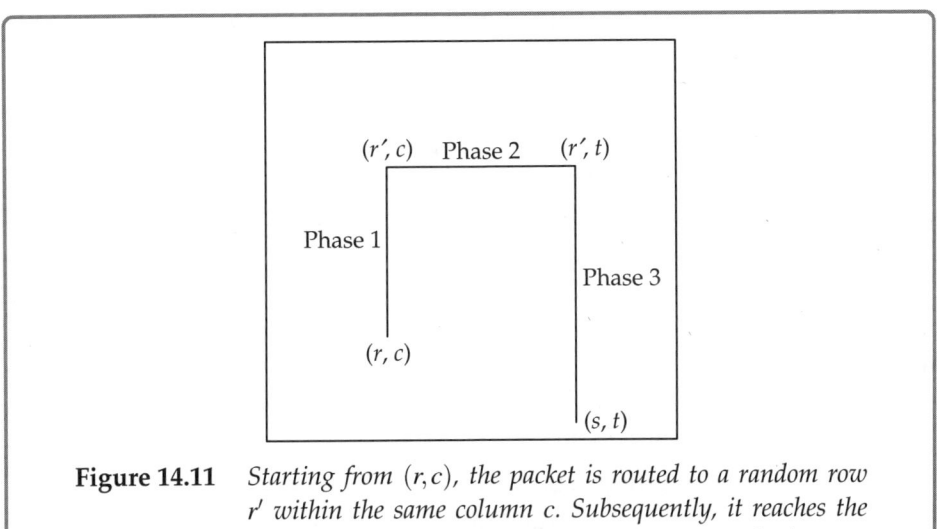

Figure 14.11 *Starting from (r,c), the packet is routed to a random row r' within the same column c. Subsequently, it reaches the destination column t and finally the destination (s,t).*

This bound can be improved to $2n + o(n)$ routing time and $O(1)$ queue size by using more sophisticated analysis and overlapping phases 2 and 3, that is, a packet begins its phase 3 as soon as it completes phase 3, rather than wait for all the other packets to complete phase 2.

Further Reading

Early work in the area of parallel algorithms was inspired by the discovery of some elegant sorting networks like *shuffle exchange* based on *bitonic sort* by Stone [137] and *odd–even merge sort* by Batcher [16]. Knuth [83] provides a detailed account of much of the early foundational work in parallel sorting network and the importance of 0–1 principle of sorting (Lemma [14.2]) – for further generalization, see Rajasekaran and Sen's work [120]. The quest for an n processor $O(\log n)$ time parallel algorithm led to some exciting developments starting from Reischuk's sort [124] to Flashsort [123] and the AKS sorting network [9] based on expander graphs. Cole [33] managed to come up with an elegant adaption of merge sort on the PRAM model.

These triggered almost two decades of hectic exploration of the power of parallel algorithms across all domains of problem areas like graphs, geometry, algebra, numerical, and so on. The early efforts focused on defining and resolving basic parallel algorithmic techniques like prefix [87] and list ranking [50]. The two textbooks that the reader would find extremely informative and rigorous are Leighton's [89] that discusses algorithms for interconnection networks and Ja'Ja' [68] that provides an account of PRAM-based algorithms. Further, the reader would also find an edited collection of articles by many leading researchers in Reif's book [122].

Limitations of parallel algorithms is a fascinating area of CS theory. After the early work of Valiant [145] who obtained the $\Omega(\log\log n)$ rounds lower bound for extremal selection in the parallel comparison tree problem, researchers [18, 156] showed that even by using polynomial number of processors, addition of n numbers cannot be done faster than $O(\frac{\log n}{\log\log n})$ time in CRCW PRAM. This was a remarkable fundamental result that established information computational limitations of parallel computation even without restrictions in network speed. It also led to the definition of interesting parallel complexity classes like $\mathcal{NC}$ and $\mathcal{RNC}$ that correspond to problems that admit polylogarithmic time using a polynomial number of processors (the latter correponds to randomized algorithms). After Reif [121] showed that lexicographic DFS is a P-complete problem, the interesting question in parallel complexity theory is if $P = \mathcal{NC}$?

Further variatons of the *Shearsort* algorithm can be found in Scherson and Sen [128]. Permutation routing and emulation of PRAM algorithms occupied considerable attention of researchers – see Leighton et al.'s paper [90] for a detailed summary. The parallel connectivity algorithm is based on the description by Shiloach and Vishkin [133].

While the *big data* phenomena has captured a lot of attention in contemporary applications, the effort taken for massively parallel algorithms and architectures was an early recognition of this aspect even though the applications had not caught up with it. Even to this day, communication between processors is considered to be a bottleneck in achieving the near ideal speed-up. This led researchers to experiment with a number of architectures and theoretical models ([38, 146]) to bridge the gap between predicted complexity and actual speed-ups. The recent popularity of *multicore* architectures and GPU is testimony to the fact that we are still in search of an acceptable model for building and designing parallel computers. Without efficient algorithms, these architectures will only be able to boast of high FLOPS but not have proven superiority over standard sequential algorithms that are optimized cleverly using many code optimization tools.

Exercise Problems

14.1 Analyze the parallel running time of the recursive version of the algorithm described in Fig. 14.6 by writing the appropriate recurrence.

What is the best value of the block size k for minimizing the parallel running time using an optimal number of processors?

14.2 Show that $n-1$ comparisons are necessary to find the maximum of n elements.

14.3 Prove the 0–1 principle of sorting.

Note that necessary direction is obvious, but the sufficiency of the condition needs to be proved rigorously. You may want to prove it by contradiction.

14.4 If a processor has index i, $1 \leq i \leq p$, find out a way of preassigning the comparisons to each of the processors for the problem of finding the maximum of n elements as described in Section 14.2.1.

Use the binary tree numbering to label the designated comparison.

14.5 Show how to reduce the number of processors further to $\frac{n}{\log\log n}$ and still retain $T^{\|}(n, n/\log\log n) = O(\log\log n)$ for finding maximum of n elements.

14.6 In the parallel comparison model for finding the maximum element, for $p = n$, show that after $j = \frac{\log\log n}{2}$ rounds, $n_j > 1$. This implies that any deterministic algorithm that correctly identifies the maximum using n processors will require $\Omega(\log\log n)$ parallel rounds.

14.7 Given two sorted sequences A and B of n elements, design an $O(\log\log n)$ time optimal speed-up merging algorithm in the CRCW PRAM model.

14.8 (i) Refine the idea of computing the minimum of n elements in $O(1)$ time using $n^{3/2}$ processors to $n^{1+\varepsilon}$ CRCW processors for any $0 < \varepsilon < 1$.

(ii) Show how to compute the minimum of n elements with n CRCW processors in $O(1)$ **expected** time using a randomized algorithm.

Note that this overcomes the deterministic lower bound.

14.9 Given a set S of n elements and $1 \leq k \leq n$, design an $O(\log n)$ time $\frac{n}{\log n}$ processors PRAM algorithm to find the kth ranked element in S. Clearly you cannot use sorting but feel free to use randomized techniques.

Hint: Selecting a sample of size m can be done by independent coin tossing in parallel.

14.10 Given an array A of 0–1 elements, design an $O(\frac{\log n}{\log\log n})$ time $O(n)$ operations CRCW algorithm that computes the sum of the bits. Note that this shows that $\Omega(\log n)$ is not the lower bound for such a problem.

Hint: Use a k-ary tree for an appropriate k and table look-up to add k bits. The precomputation phase should also be analysed carefully.

14.11 Recall the ANSV problem defined earlier in Exercise 3.21. Design a polylog time $O(n)$ processors CRCW PRAM algorithm for the ANSV problem.

14.12 Consider an array A of n integers and another set of integers $i_1, i_2, \ldots, i_k$, where $1 = i_1 < i_j < i_{j+1} < x_k = n+1$. Describe an optimal $O(\log n)$ time PRAM algorithm to compute the partial sums $S_j = \sum_{t=i_j}^{i_{j+1}-1} x_t$ for all $1 \leq j \leq k-1$.

For example, for inputs $4, 2, 8, 9, -3$ and indices $1, 2, 4, 6$, the answer is $4, 2 + 8 = 10$, $9 - 3 = 6$. The normal prefix sum can be done with $i_1 = 1, i_2 = n + 1$.

14.13 Consider the following algorithm to sort an $n \times n$ array of numbers. Assume for simplicity that $n = 2^k$.

- Sort the four $n/2 \times n/2$ subarrays recursively according to some indexing scheme.
- Rotate every alternate row of smaller subarrays by $n/2$ positions right/left.
- Run 3 iterations of shearsort.

Prove that this algorithm correctly sorts and analyze the parallel running time in some appropriate model.

14.14 Consider an $N \times N \times N$ three-dimensional mesh. Design an efficient sorting algorithm for N^3 numbers where each $O(1)$ memory processor holds one element initially and finally. You can choose any pre-defined indexing scheme for the sorted permutation. Justify why your algorithm is optimal or near-optimal.

14.15 Reduce the number of processors in the list ranking algorithm to $\frac{n}{\log n}$ without increasing the asymptotic time bound by using the technique of *random-mate* described in the end of the section 14.4.1 on list ranking. As an intermediate step, first design an optimal speed-up algorithm that runs in $O(\log n \cdot \log \log n)$ parallel time.

You may also need to use a faster (than logarithmic time) prefix computation to achieve the $O(\log n)$ bound. For this purpose, you can assume that the prefix computation of n elements can be done in $O(\frac{\log n}{\log \log n})$ time using an optimal number of n operations.

14.16 Generalize the list ranking algorithm to a tree and analyze the performance. Here we are interested in finding the distance of every node to the root node.

14.17 Given n integers in the range $[1, \log n]$

(i) Show how to sort in $O(\log n)$ time using $n/\log n$ processors in a PRAM model.

(ii) Show how to extend this to sorting numbers in the range $[1, n^2]$ in $O(\frac{\log^2 n}{\log \log n})$ time using $n/\log n$ processors.

Note that the processor time product is $O(n \log n)$.

14.18 Analyze the following variation of the parallel connectivity algorithm. Each directed tree is contracted to a star following the hooking step. Instead of the adjacency matrix, use a list or array data structure to implement the algorithm using $O(|E| + |V|)$ processors and polylog parallel time.

Compare the two variants of the connectivity algorithm.

14.19 Design a polylogarithmic time algorithm using a polynomial number of processors for the shortest path problem in graphs. Although this is not very attractive in terms of resources, it still establishes that the shortest path problem belongs to the class $\mathcal{NC}$.

Modify your algorithm to do topological sort within the same bounds.

Hint: You may want to explore matrix-based computations.

14.20 Given a tree (not necessarily binary) on n nodes, an **Euler tour** visits all the edges exactly twice – one in each direction.

(i) Show how to find an Euler tour in $O(1)$ time using n processors. Finding a tour implies defining the successor of every vertex where a vertex may be visited several times (proportional to its degree).

(ii) Given an unrooted tree, define a parent function $p(v)$ for every vertex for a designated root r. Note that $p(r) = r$. Design a PRAM algorithm that computes the parent function in $O(\log n)$ time using $n/\log n$ processors.

(iii) Find the postorder numbering of a rooted tree in $O(\log n)$ time using $n/\log n$ processors in the PRAM model.

14.21 Show that in a linear array of n processors, permutation routing can be done in n steps. Each processor is a source and destination of exactly one packet.

14.22 Consider a linear array of n processors p_i $1 \leq i \leq n$, where initially processor p_i holds n_i packets. Moreover, $\sum_i n_i = n$, such that each processor is a destination of exactly one packet. Analyze the greedy routing algorithm with *furthest destination first* queue discipline for this problem, giving rigorous and complete proofs of the sketch given in the chapter.

14.23 **Odd–Even Merge sort:** Recall the description of odd–even merge sort in Chapter 3.

Design an efficient parallel sorting algorithm based on this idea and analyze the running time as well as the total work complexity.

14.24 For a lower triangular $n \times n$ matrix A, design a fast parallel algorithm to solve the system of equations $A \cdot \bar{x} = \bar{b}$, where $\bar{x}, \bar{b}$ are n element vectors.

Note that by using straightforward back-substitution, your algorithm will take at least n phases. To obtain a polylogarithmic time algorithm, use the following identity. The lower triangular matrix A can be written as

$$A = \begin{bmatrix} A_1 & 0 \\ A_2 & A_3 \end{bmatrix}$$

where A_1, A_3 are lower triangular $n/2 \times n/2$ matrices. Then you can verify that

$$A^{-1} = \begin{bmatrix} A_1^{-1} & 0 \\ -A_3^{-1}A_2A_1^{-1} & A_3^{-1} \end{bmatrix}$$

15

Memory Hierarchy and Caching

15.1 Models of Memory Hierarchy

Designing memory architecture is an important component of computer organization that tries to achieve a balance between computational speed and memory speed, viz., the time to fetch operands from memory. Computational speeds are much faster since the processing happens within the chip; whereas, a memory access could involve off chip memory units. To bridge this disparity, the modern computer has several layers of memory, called cache memory that provides faster access to the operands. Because of technological and cost limitations, cache memories offer a range of speed–cost tradeoffs. For example, the L1 cache, the fastest cache level is usually also of the smallest size. The L2 cache is larger, say by a factor of ten but also considerably slower. The secondary memory which is the largest in terms of size, e.g., the disk could be 10,000 times slower than the L1 cache. For any large size application, most of the data resides on disk and is transferred to the faster levels of cache when required.[1]

This movement of data is usually beyond the control of the normal programmer and managed by the operating system and hardware. By using empirical principles called *temporal* and *spatial* locality of memory access, several replacement policies are used to maximize the chances of keeping the operands in the faster cache memory levels. However, it must be obvious that there will be occasions when the required operand is

---

[1] We are ignoring a predictive technique called pre-fetching here.

not present in L1; one has to reach out to L2 and beyond and pay the penalty of higher access cost. In other words, memory access cost is not uniform as discussed in the beginning of this book but for simplicity of the analysis, we had pretended that it remains same.

In this chapter, we will do away with this assumption; however, for simpler exposition, we will deal with only two levels of memory – *slow* and *fast* where the slower memory has infinite size while the faster one is limited, say, of size M and significantly faster. Consequently, we can pretend that the faster memory has zero (negligible) access cost and the slower memory has access cost 1. For any computation, the operands must reside inside the cache. If they are not present in the cache, they must be fetched from the slower memory, paying a unit cost (scaled appropriately). To offset this transfer cost, we are allowed to transfer a contiguous chunk of B memory locations. This applies to both reads and writes to the slower memory. The model is known as the *external memory model* with parameters M, B and will be denoted by $C(M, B)$. Note that $M \geq B$ and in most practical situations $M \geq B^2$.

We will assume that the algorithm designer can use the parameters M, B to design appropriate algorithms to achieve higher efficiency in $C(M, B)$. Later, we will discuss that even without the explicit use of M, B, one can design efficient algorithms, called *cache oblivious* algorithms. To focus better on the memory management issues, we will not account for the computational cost and only try to minimize the cost of memory transfers between cache and secondary memory. We will also assume that appropriate instructions are available to transfer a specific block from the secondary memory to the cache. If there is no room in the cache, then we have to replace an existing block in the cache and we can choose a cache block to be evicted.[2] A very simple situation is to add n elements stored as n/B memory blocks where initially they are all in the secondary memory. Clearly, we will encounter at least n/B memory transfers just to read all the elements.

We plan to study and develop techniques for designing efficient algorithms for some fundamental problems in this two-level memory model and focus on issues that are ignored in conventional algorithms. We would also like to remind the reader that memory management is a salient feature of any operating system where various cache replacement policies have been proposed to minimize cache misses for a given pattern of memory access, viz., *first in first out (FIFO)*, *least recently used (LRU)* etc. There is also an optimal replacement policy, OPT that achieves the minimum among all possible replacement policy for any given sequence of memory access. The OPT policy, also known as *clairvoyant* and discovered by Belady, evicts the variable that is not needed for the longest time in future. This makes it difficult for implementation as the access pattern

---

[2] This control is usually not available to programmers in user mode and is left to the operating system responsible for memory management.

may not be known in advance and it may be data-dependent. The goal in this chapter is to develop algorithmic techniques for specific problems so that we can minimize the worst case number of cache misses where the memory access pattern is dependent on the algorithm. In other words, there is no pre-determined memory access pattern for which we are trying to minimize cache misses. On the contrary, we would like to find the best access pattern for a given problem which is distinct from the system based optimization.

15.2 Transposing a Matrix

Consider a $p \times q$ matrix A that we want to transpose and store in another $q \times p$ matrix A'. Initially, this matrix is stored in the slower secondary memory and arranged in a *row-major* pattern. Since the memory is laid out in a linear array, a *row-major* format stores all the elements of the first row, followed by the second row elements, and so on. The *column major* layout stores the elements of column 1 followed by column 2, and so on. Therefore, computing $A' = A^T$ involves changing the layouts from row-major to column-major.

The straightforward algorithm for transpose involves moving an element $A_{i,j}$ to $A'_{j,i}$ for all i, j. In the $C(M, B)$ model, we would like to accomplish this for B elements simultaneously since we always transfer B elements at a time. Recall that matrices are laid out in row-major form. One idea would be to repeatedly take B elements from a row of A (assuming p, q are multiples of B) and then transfer them simultaneously to $A' = A^T$. But these B elements would lie in different blocks of A' and so each such transfer will require B memory transfers. This is clearly an inefficient scheme, but it is not difficult to improve it with a little thought. Partition the matrix into $B \times B$ submatrices (see Fig. 15.1) and denote these by $A_{t(a,b)}$ $1 \le a \le p/B$ $1 \le b \le q/B$ for matrix A. These submatrices define a *tiling* of the matrix A and the respective tiles for A' are denoted by $A'_{t(a,b)}$ $1 \le a \le q/B$ $1 \le b \le q/B$.

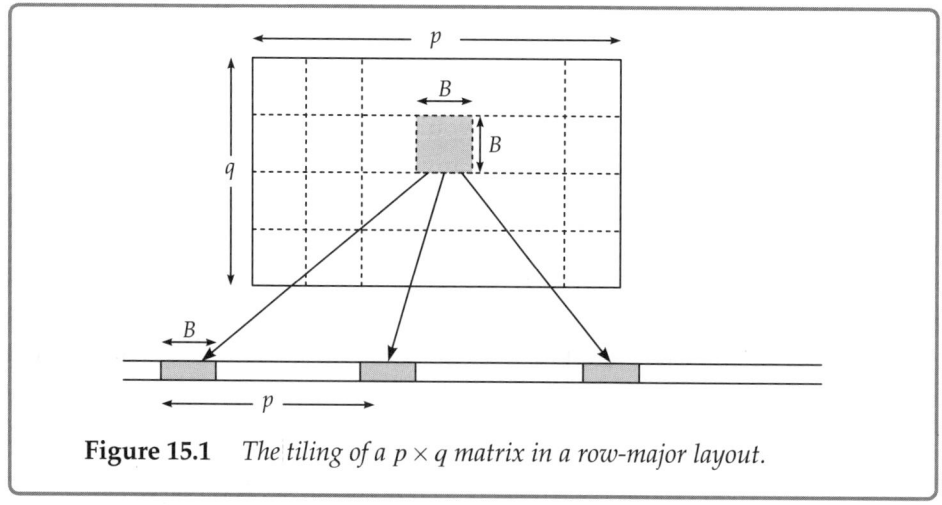

Figure 15.1   *The tiling of a $p \times q$ matrix in a row-major layout.*

The idea now is to first read each such submatrix in the cache (assume that $M > B^2$). Any such submatrix can be read using B memory transfers because each row of such a submatrix is stored in consecutive locations. Then we compute its transpose in the cache, and write back the submatrix to A', again using B memory transfers (in each memory transfer, we write one row of the corresponding submatrix of A'). See Fig. 15.1 for an illustration. The details of this algorithm are given in Fig. 15.2. It is easy to check that the the algorithm requires $O(pq/B)$ memory transfers, which is also optimal.

Procedure Computing transpose efficiently in for matrix $A(p,q)$

---

1 *Input A* is a $p \times q$ matrix in row-major layout in external memory ;
2 **for** $i = 1$ *to* p/B **do**
3   **for** $j = 1$ *to* q/B **do**
4   move $A_{t(i,j)}$ to the cache memory C using the Transfer function ;
5   Compute the transpose $A_{t(i,j)}^T$ within C in a conventional element-wise manner ;
6   move $A_{t(i,j)}^T$ to $A'_{t(i,j)}$ in the main memory using Transfer function .
7 A' contains the transpose of A in the external memory ;

Function Transfer$(D_{t(k,l)}, r, s)$

---

1 *Input* transfer a $B \times B$ submatrix located at $k \cdot B - 1, l \cdot B - 1$ of an $r \times s$ matrix to cache memory ;
2 **for** $i = 1$ *to* B **do**
3   move block starting at $(k \cdot B + i) \cdot s + B \cdot l$ into the ith block in C ;

4 *Comment* A similar procedure is used to transfer from C to the external memory ;

Figure 15.2   *Transposing a matrix using minimal transfers*

15.2.1 Matrix multiplication

Given matrices X, Y having n rows and n columns, we can first transpose Y since that changes the matrix layout to a column ordering. Subsequently, when we compute all the n^2 row–column dot products, the contiguous elements are fetched as blocks.

Let us analyze the straightforward approach of multiplying rows of X with columns of Y. For simplicity, assume that B divides M. We can fetch roughly $M/2$ elements of a row of X and the same number of elements from a column of Y using $\frac{M}{2B}$ I-Os. Multiply corresponding elements and sum them, that is, compute $Z_{i,j} = \sum_k X_{i,k} \cdot Y_{k,j}$ by repeating the aforementioned computation for the sub-rows and sub-columns of size $M/2$. A quick calculation for this simple approach shows that $O(n^2 \cdot \frac{M}{2B} \cdot \frac{n}{M}) = O(n^3/B)$ I-Os are incurred. This may look reasonable at first glance since $O(n^3)$ operations are needed to multiply the matrices X and Y. However, this is the number of I-Os and there is no direct dependence on M which is the size of the internal memory ! Suppose, $M \geq 3n^2$, then clearly, we can can read all the elements of X, Y in the internal memory using $O(n^2/B)$ I-Os, generate the product matrix Z internally and write it back to the external memory using the same number of I-Os, thus totaling $O(n^2/B)$ I-Os. This is significantly superior and we are making good use of the large internal memory. This should motivate us to look beyond the *simple matrix multiplication (SMM)* procedure. Consider the algorithm given in Fig. 15.3.

Procedure Tiled matrix multiplication TMM(X, Y, Z, s)

1 *Input* X, Y is a $n \times n$ matrix in row-major layout in external memory
2 Let D^s denote a tiling of matrix D of size $s \times s$ where $D^s_{\alpha,\beta}$ denotes
    the elements $\{D_{i,j} | \alpha s \leq i \leq (\alpha+1)s - 1, \beta s \leq j \leq (\beta+1)s - 1\}$;
3 $Y \leftarrow Y^T$;
4 **for** $\alpha = 1$ *to* n/s **do**
5   **for** $\beta = 1$ *to* n/s **do**
6   **for** $k = 1$ *to* n/s **do**
7   Transfer $X^s_{\alpha,k}, Y^s_{k,\beta}, Z^s_{\alpha,\beta}$ to the cache memory ;
8              $Z^s_{\alpha,\beta} \leftarrow Z^s_{\alpha,\beta} + SMM(X^s_{\alpha,k}, Y^s_{k,\beta})$;
9   Transfer $Z^s_{\alpha,\beta}$ to external memory ;

Figure 15.3 *Computing the product $Z = X \cdot Y$ using tiles of size s*

The reader should recognize that this variation of the matrix multiplication expressed in terms of blocks of size $s \times s$ is indeed correct. Let us analyze the number of I-Os required. By using the previous algorithm for transpose, we can use Step 2 in $O(n^2/B)$ I-Os. In the main loop, we are performing a matrix multiplication of size $s \times s$ using standard methods and if we choose s such that all the matrices X^s, Y^s, Z^s can fit into the cache, then there are no further I-Os. The inside nested loop is executed n^3/s^3 times where each execution involves transferring three sub-matrices that requires $O(s^2/B)$ I-Os. Therefore, the total number of I-Os is bounded by $O(\frac{n^3}{Bs})$ I-Os. The largest s that can be chosen is about $\sqrt{M}$ so that three submatrices can fit with the cache memory. This leads to overall $O(\frac{n^3}{B\sqrt{M}})$ I-Os. Note that for $M = n^2$, we get an optimal number $O(n^2/B)$ I-Os.

This method can be generalized to non-square matrices $X^{m \times n}$ and $Y^{n \times k}$, so that the number of I-Os required is $O(\frac{mnk}{B\sqrt{M}})$.

15.3 Sorting in External Memory

In this section, we consider the problem of sorting a set of n elements in the $C(M,B)$ model. We would like to remind the reader that unlike the traditional sorting algorithms, the number of comparisons is not relevant in this model. We will adapt Merge sort to this model by choosing a larger degree of merge. See Exercise Problem 15.2 at the end of this chapter regarding how the degree of merge affects the complexity.

Let us briefly recall how the traditional merge sort can be thought of as sorting n numbers using $\log n$ passes over it. Suppose the numbers are stored in an array A of size n. In the first pass, we consider pairs of consecutive numbers $A[2i], A[2i+1]$ and arrange them in the sorted order. After j passes, we have the invariant that sub-arrays of type $A[2^j \cdot l + 1], A[2^j \cdot l + 2], \ldots, A[2^j \cdot l + 2^j]$, where l is a non-negative integer, are sorted. In the next pass, we consider pairs of such consecutive length 2^k sub-sequences and merge them into a sorted sub-sequence of size 2^{j+1}. Thus, after $\log n$ passes, we would have sorted the entire data. The idea now is to consider not only pairs of consecutive sub-sequences, but choose a suitable parameter k and merge k consecutive sorted sub-sequences. Note that now we would need $\log_k n$ passes. Let us see how we can merge k such sorted subsequences. Recall the merging algorithm – we need to keep the leading (smallest) blocks of each sequence in the main memory, and choose the smallest element among them for the next output. To economize memory transfer, we want to read and write contiguous chunks of B elements, so we write only after B elements are output. Note that the smallest B elements must occur among the leading blocks (smallest B elements) of the sorted sequence. Since all the $k+1$ sequences, including the k input and 1 output sequence, must be within the cache, the largest value of k is $O(M/B)$. We need some extra space to store the data structure for merging (a k-ary min-heap) but we will not discuss any details of this implementation since it can be done using any conventional approach within the cache memory. So we can assume that $k = \frac{M}{cB}$ for some appropriate constant $c > 1$.

We shall first analyze the number of memory block transfers it takes to merge k sorted sequences of lengths ℓ each. As previously discussed, we maintain the leading block of each sequence in the cache memory and fetch the next block after this is exhausted. So we need $\ell/B = \ell'$ block transfers for each sequence which may be thought of as the number of blocks in the sequence (if it is not a multiple of B, then we count the partial block as an extra block). Likewise, the output is written out as blocks and this must be the sum of all input sequences, which is $k \cdot \ell'$. In other words, the number of block transfers for merging is proportional to the sum of the sequences being merged. This implies that for each pass over the data, the total merging cost is proportional to n/B.

For $k = \Omega(M/B)$, there are $\log_{M/B}(n/B)$ levels of recursion as the smallest size of a sequence is at least B. So, the total number of block transfers is $O(\frac{n}{B} \log_{M/B}(n/B))$ for sorting n elements in $C(M,B)$.

Recall that this is only the number of memory block transfers – the number of comparisons remain $O(n \log n)$ like conventional Merge sort. For $M > B^2$, note that $\log_{M/B}(n/B) = O(\log_M(n/B))$.

15.3.1   Can we improve the algorithm?*

In this section, we give a lower bound for sorting n elements in the $C(M,B)$ model. We relate sorting to permutation. Given a permutation π of n elements, we want the algorithm to rearrange the elements in this permutation. Any lower bound on permutation is also applicable to sorting since permutation can be done by sorting on the destination index of the elements. If $\pi(i) = j$, then one can sort on js, where $\pi()$ is the permutation function.

We will make some assumptions to simplify the arguments for the lower bound. These assumptions can be removed with some loss of constant factors in the final bound. There will be exactly one copy of any element, viz., when the element is fetched from slower memory then there is no copy left in the slower memory. Likewise, when an element is stored in the slower memory, then there is no copy in the cache. With a little thought, the reader can convince herself that maintaining multiple copies in a permutation algorithm is of no use since the final output has only one copy that can be traced backward as the relevant copy.

The proof is based on a simple counting argument on how many orderings are possible after t block transfers. For a worst-case bound, the number of possible orderings must be at least $n!$ for n elements. The argument does not apply to generating any specific permutation but the total number of orderings that can be generated by bounding the number of I-Os. We do not insist that the elements must be in contiguous locations. If $\pi(i) > \pi(j)$, then $R_i > R_j$, where R_i is the final location of the ith element for all pairs i, j.

A typical algorithm has the following behavior.

1. Fetch a block from the slow memory into the cache.

2. Perform computation within the cache to facilitate the permutation.

3. Write out a block from the cache to the slower memory.

Note that Step 2 does not require block transfers and is *free* since we are not counting operations within the cache. So we would like to count the additional orderings generated by Steps 1 and 3.

Once a block of B elements is read into the cache, it can induce additional orderings with respect to the $M - B$ elements present in the cache. This number is $\frac{M!}{B! \cdot (M-B)!} = \binom{M}{B}$, which is the relative orderings between $M - B$ and B elements. Further, if these B elements were not written out before, that is, these were never present in cache before, then there are $B!$ ordering possible among them. (If the block was written out in a previous step, then they were in cache together and these orderings would have been already accounted for.) So this can happen at most n/B times, viz., only for the initial input blocks.

In Step 3, during the tth output, there are at most $n/B + t$ places relative to the existing blocks. There were n/B blocks to begin with and $t - 1$ previously written blocks, so the tth block can be written out in $n/B + t$ intervals relative to the other blocks. Note that there may be arbitrary gaps between blocks as long as the relative ordering is achieved.

From the previous arguments, we can bound the number of attainable orderings after t memory transfers by

$$(B!)^{n/B} \cdot \prod_{i=0}^{i=t-1} (n/B + i) \cdot \binom{M}{B}$$

If T is the worst-case bound on the number of block transfers, then

$$(B!)^{n/B} \cdot \prod_{i=1}^{i=T} (n/B + i) \cdot \binom{M}{B} \quad \leq \quad (B!)^{n/B} \cdot (n/B + T)! \cdot \binom{M}{B}^T$$

$$\leq \quad B^n \cdot (n/B + T)^{n/B+T} \cdot e^{-n} \cdot (M/B)^{BT}$$

using Stirling's approximation $n! \sim (n/e)^n$ and $\binom{n}{k} \leq (en/k)^k$.

From the last inequality, it follows that

$$e^{-n} \cdot B^n \cdot (n/B + T)^{n/B+T} \cdot (M/B)^{BT} \geq n! \geq (n/e)^n$$

Taking logarithm on both sides and re-arranging, we obtain

$$BT \log(M/B) + (T + n/B) \cdot \log(n/B + T) \geq n \log n - n \log B = n \log(n/B) \qquad (15.3.1)$$

Since any algorithm must read all the numbers, we know that $n/B \leq T$. Therefore, $(T + n/B) \log(n/B + T) \leq 4T \log(n/B)$; we can re-write this inequality as

$$T (B \log(M/B) + 4 \log(n/B)) \geq n \log(n/B)$$

For $4 \log(n/B) \leq B \log(M/B)$, we obtain $T = \Omega(\frac{n}{B} \log_{M/B}(n/B))$. For $\log(n/B) > 4 \cdot B \log(M/B)$, we obtain $T = \Omega(n \frac{\log(n/B)}{\log(n/B)}) = \Omega(n)$.

Theorem 15.1 *Any algorithm that permutes n elements in $C(M.B)$ uses $\Omega(\frac{n}{B} \cdot \log_{M/B}(n/B))$ block transfers in the worst case.*

As a consequence of Theorem 15.1, the lower bound for sorting matches the bound for the Merge sort algorithm and hence, the algorithm cannot be improved in asymptotic complexity. Using some elegant connections between permutation networks with FFT graphs, the aforementioned result also implies a similar bound for FFT computation in external memory.

15.4 Cache Oblivious Design

Consider the problem of searching a large static dictionary in an external memory of n elements. If we use a B-tree type data structure, then we can easily search using $O(\log_B n)$ memory transfers. This can be explained by effectively doing a B-way search. Each node of the B-tree contains B records that we can fetch using a single block transfer.

Building a B-tree requires the explicit knowledge of B. Since we are dealing with a static dictionary, we can consider doing a straightforward B-ary search in a sorted data set. Still, it requires the knowledge of B. What if the programmer is not allowed to use the parameter B? Consider the following alternative of doing a $\sqrt{n}$-ary search presented in Fig. 15.4.

Procedure Search(x, S)

1 *Input* A sorted set $S = \{x_1, x_2, \ldots, x_n\}$;
2 **if** $|S| = 1$ **then**
3 $\quad$ return Yes or No according to whether $x \in S$
4 **else**
5 $\quad$ Let $S' = \{x_{i\sqrt{n}}\}$ be a subsequence consisting of every $\sqrt{n}$th
$\quad\quad$ element of S ;
6 $\quad$ Search (x, S') ;
7 $\quad$ Let $p, q \in S'$ where $p \leq x < q$;
8 $\quad$ Return Search $(x, S \cap [p, q])$ – search the relevant interval of S' ;

Figure 15.4   *Searching a dictionary in external memory*

The analysis of this algorithm in $C(M, B)$ depends crucially on the elements being in contiguous locations. Although S is initially contiguous, S' is not, so the indexing of S has to be done carefully in a recursive fashion. The elements of S' must be indexed before the elements of $S - S'$ and the indexing of each of the $\sqrt{n}$ subsets of $S - S'$ will also be indexed recursively. Figure 15.5 shows the numbering of a set of 16 elements.

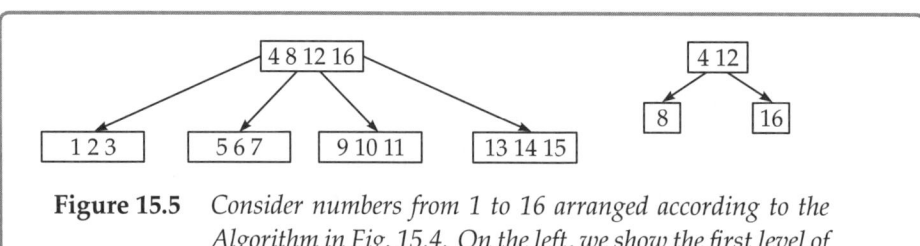

Figure 15.5   *Consider numbers from 1 to 16 arranged according to the Algorithm in Fig. 15.4. On the left, we show the first level of recursion. On the right we show the actual order in which $4, 8, 12, 16$ are stored.*

The number of memory transfers $T(n)$ for searching satisfies the following recurrence

$$T(n) = T(\sqrt{n}) + T(\sqrt{n}) \quad T(k) = O(1) \quad \text{for } k \leq B$$

since there are two calls to subproblems of size $\sqrt{n}$. This yields $T(n) = O(\log_B n)$. Note that although the algorithm did not rely on the knowledge of B, the recurrence made effective use of B, since searching within contiguous B elements requires one memory block transfer (and at most two transfers if the memory transfers are not aligned with block boundaries). After the block resides within the cache, no further memory transfers are required; although, the recursive calls continue till the terminating condition is satisfied.

15.4.1 Oblivious matrix transpose

We consider the problem of transposing a matrix. Recall that a common assumption[3] is that M is $\Omega(B^2)$. Given any $m \times n$ matrix A, we use a recursive approach for transposing it into an $n \times m$ matrix $B = A^T$. Two cases arise, whether A has more rows than columns or vice versa.

$$\left[\begin{array}{cc} A_1^{m \times n/2} A_2^{m \times n/2} \end{array} \right] \quad \Rightarrow \quad \left[\begin{array}{c} B_1^{n/2 \times m} \\ B_2^{n/2 \times m} \end{array} \right] \qquad \text{where } n \geq m \text{ and } B_i = A_i^T$$

$$\left[\begin{array}{c} A_1'^{m/2 \times n} \\ A_2'^{m/2 \times n} \end{array} \right] \quad \Rightarrow \quad \left[\begin{array}{cc} B_1'^{n \times m/2} & B_2'^{n \times m/2} \end{array} \right] \qquad \text{where } m \geq n \text{ and } B_i' = A_i'^T$$

The formal algorithm based on the previous recurrence is described in Fig. 15.6.

Procedure Transpose(A, B)

1 *Input A is an $m \times n$ matrix ;*
2 **if** $\max\{m, n\} \leq c$ **then**
3 | perform transpose by swapping elements
4 **if** $n \geq m$ **then**
5 | Transpose (A_1, B_1) ; Transpose (A_2, B_2)
6 **else**
7 | Transpose (A_1', B_1'); Transpose (A_2', B_2')

Figure 15.6 *Algorithm for matrix transpose*

---

[3]also called *tall cache*

When $m, n \leq B/4$, then there are no more cache misses, since each row (column) can occupy at most two cache lines (Fig. 15.7). The algorithm actually starts moving elements from external memory when the recursion terminates at size $c \ll B$. Starting from that stage, until $m, n \leq B/4$, there are no more cache misses since there is enough space for submatrices of size $B/4 \times B/4$. The other cases of the recurrence addresses the recursive cases corresponding to splitting across columns or rows – whichever is larger. Therefore, the number of memory block transfers $Q(m, n)$ for an $m \times n$ matrix satisfies the following recurrence.

$$Q(m, n) \leq \begin{cases} 4m & n \leq m \leq B/4 \text{ in cache} \\ 4n & m \leq n \leq B/4 \text{ in cache} \\ 2Q(m, \lceil n/2 \rceil) & m \leq n \\ 2Q(\lceil m/2 \rceil, n) & n \leq m \end{cases}$$

The reader is encouraged to find the solution of the recurrence (Exercise Problem 15.11). When the matrix has less than B^2 elements ($m \leq n \leq B$ or $n \leq m \leq B$), the recursive algorithm brings all the required blocks – a maximum of B, transposes them within the cache and writes them out. All this happens without the explicit knowledge of the parameters M, B but requires support from the memory management policy. For example, consider the base case. When we are reading m rows (from different blocks), the algorithm should not evict an earlier fetched row while we read the next row. In particular, the recurrence is valid for the *least recently used* (LRU) policy. Since the algorithm is parameter oblivious, there is no explicit control on the blocks to be replaced and hence, its inherent dependence on the replacement policy. The good news is that the LRU policy is known to be competitive with respect to the ideal *optimal* replacement policy

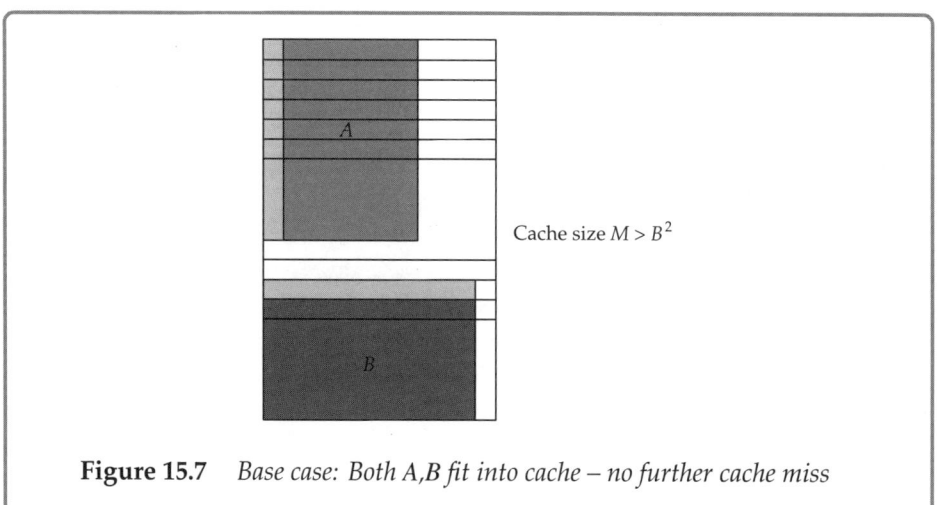

Cache size $M > B^2$

Figure 15.7   *Base case: Both A,B fit into cache – no further cache miss*

Theorem 15.2 *Suppose OPT is the number of cache misses incurred by an optimal algorithm on an arbitrary sequence of length n with cache size p. Then, the number of misses incurred by the LRU policy on the same sequence with cache size $k \geq p$ can be bounded by $\frac{k}{k-p} \cdot OPT$.*[4]

Observe that the algorithm OPT *knows* the entire sequence of length n beforehand, whereas the LRU policy sees this sequence in an online manner, i.e., when it receives the request for a page, it needs to decide which page to evict (assuming that cache is full) without knowing which pages will be requested. It follows from the aforementioned theorem that for $k = 2p$, the number of cache misses incurred by the LRU policy is within a factor two of the optimal replacement.

We can pretend that the available memory is $M/2$, which preserves all the previous asymptotic calculations. The number of cache misses by the LRU policy will be within a factor two of this bound. Theorem 15.2 is a well-known result in the area of *competitive algorithms*, which is somewhat out of the scope of the discussion here but we present a proof of the theorem.

Consider a sequence of n requests $\sigma_i \in \{1, 2, \ldots, N\}$, which can be thought of as the set of cache lines. We further divide this sequence into subsequences $s_1, s_2, s_3, \ldots$, such that every subsequence has $k + 1$ distinct requests from $\{1, 2, \ldots, N\}$ and the subsequence is of minimal length, viz., it ends the first time when we encounter the $k + 1$st distinct request without including this request. For example, suppose $k = 3$ and suppose there are 5 pages. Consider the request sequence $1, 2, 1, 2, 4, 4, 1, 2, 3, 3, 4, 5, 4, 3, 3, 1, 2$, where the integer i refers to the fact that page i is being requested. Here we will define the subsequences as $1, 2, 1, 2, 4, 4; 1, 2, 3, 3; 4, 5, 4, 3, 3; 1, 2$. The LRU policy will incur at most k misses in each subsequence (Fig. 15.8). Now consider any policy (including the optimal policy) that has cache size p, where $k > p$. In each phase, it will incur at least $k - p$ misses since it has to evict at least that many items to handle k distinct requests. Here we are assuming that out of the k distinct requests, there are p cache lines from the previous phase and it cannot be any better. In the first phase, both policies will incur the same number of misses (starting from an empty cache).

$$\sigma_{i_1} \sigma_{i_1+1} \cdots \sigma_{i_1+r_1} \mid \sigma_{i_2} \sigma_{i_2+1} \cdots \sigma_{i_2+r_2} \mid \sigma_{i_3} \sigma_{i_3+1} \cdots \sigma_{i_3+r_3} \mid \cdots \mid \sigma_{i_t} \sigma_{i_t+1} \cdots$$

Figure 15.8 *The subsequence $\sigma_{i_1} \sigma_{i_1+1} \cdots \sigma_{i_1+r_1} \sigma_{i_2}$ have $k+1$ distinct elements, whereas the subsequence $\sigma_{i_1} \sigma_{i_1+1} \cdots \sigma_{i_1+r_1}$ have k distinct elements.*

---

[4] A more precise ratio is $k/(k - p + 1)$.

Let f^i_{LRU} denote the number of cache misses incurred by LRU policy in subsequence i and f^i_{OPT} denote the number of cache misses by the optimal policy. Then, $\sum_{i=1}^t f^i_{LRU} \le (t-1) \cdot k$ and $\sum_{i=1}^t f^i_{OPT} \ge (p-k) \cdot (t-1) + k$. Their ratio is bounded by

$$\frac{\sum_{i=1}^t f^i_{LRU}}{\sum_{i=1}^t f^i_{OPT}} \le \frac{(t-1) \cdot k + k}{(t-1) \cdot (p-k) + k} \le \frac{(t-1) \cdot k}{(t-1) \cdot (k-p)} = \frac{k}{k-p}$$

Further Reading

The *external memory* model was formally introduced by Aggarwal and Vitter [3], who presented a version of Merge sort that uses a maximum $O(\frac{N}{B} \log_{M/B} N/B)$ I/Os. Further, they showed that this is the best possible algorithm by proving a tight lower bound. Our description of the algorithm and the lower bound is based on their presentation. Prior to this model, there had been very interesting work on *IO* complexity, which did not have the notion of memory blocks. The area of external sorting on tapes and disks has been historically significant and one of the first lower bounds was given by Floyd [48] on matrix transpose. Hong and Kung [70] introduced the notion of pebbling games that led to many non-trivial lower bounds for the *IO* complexity. The model formulated by Aggarwal and Vitter [3] was further refined to multiple levels of cache – for example, see that by Aggarwal et al. [4].

The cache oblivious model was formally introduced by Frigo et al. [56] who presented a number of techniques that matched the performance of the cache-aware counterparts. One of the non-trivial algorithms was a cache-oblivious sorting algorithm called *Funnel sort* using a $\sqrt{n}$-way recursive Merge sort algorithm. The tall cache assumption is quite crucial for the optimality of the bounds. Sen et al. [131] present a general technique for efficient emulation of the external memory algorithms on *limited set-associative* cache models that have fixed mapping of memory to cache, and restricts efficient use of cache. Vitter [150] provides a comprehensive survey of the algorithms and data structures for external memory.

Subsequent work on memory hierarchy expanded the scope to multiprocessors that have their own cache memory as well as access to shared memory. The local accesses are much faster. Arge et al. [12] formalized the *parallel external memory* (PEM) model and presented a cache-aware Merge sort algorithm that runs in $O(\log n)$ time and has optimal cache misses. Blelloch et al. [30] presented a resource-oblivious distribution sort algorithm that incurs sub-optimal cache cost in the private-cache multicore model. A somewhat different model was given by Valiant [144]; the model was designed for a BSP-style version of a cache-aware, multi-level multicore which was difficult to compare directly with the previous results. Recently, Cole and Ramachandran [34] presented a new optimal Merge sort algorithm (SPMS) for resource-oblivious multicore models.

Exercise Problems

15.1 For $M = O(B)$, what is the IO complexity (number of block transfers) to transpose an $n \times n$ matrix?

15.2 In Merge sort, we partition the input into two (almost) equal halves and sort them recursively. In the external memory merge sort, we partition into $k \geq 2$ parts and do a k-ary merge, analyze the number of comparisons required even though it is not a metric for design.

15.3 Design an efficient version of partition sort (quicksort with multiple pivots) for the external memory model with parameters M and B. Show that it is comparable to Merge sort.

Hint: You may want to use the sampling lemma used for PRAM-based partition sort.

15.4 Show that the average case lower bound for permutation is asymptotically similar to the worst-case bound.

15.5 A k-transposition permutes $n = k \cdot \ell$ elements as follows

$x_1 x_2 x_3 x_\ell x_{\ell+1}, x_{\ell+2}, \ldots x_{\ell \cdot k}, x_{\ell \cdot k+1} \ldots x_{\ell \cdot k + \ell}$ are mapped to

$x_1, x_{\ell+1}, x_{2 \cdot \ell+1} \ldots x_{\ell k}, x_2, x_{\ell+2} \ldots x_{2\ell k} \ldots$

Show how to do this in an external memory model using $O(\frac{n}{B} \log_{M/B} k)$ IOs.

15.6 Describe a cache-efficient algorithm for computing the matrix product

$$C^{m \times n} = X^{m \times n} \cdot Y^{k \times n}$$

for parameters M, B.

15.7 Describe a cache-efficient impementation of Shearsort in the external memory model with parameters M, B (See Section 14.2.2 for a discussion on Shearsort).

15.8 Describe a cache-efficient algorithm for constructing planar convex hull of n points in the external memory model.

15.9 Describe a cache-efficient algorithm for finding the maximal elements of n points on the plane in the external memory model.

15.10 Describe a cache-efficient algorithm for computing *all nearest smaller value* problem (defined in Exercise Problem 3.21) in the IO model.

15.11 Consider the recurrence $Q(m,n)$ used in the analysis of Section 15.4.1. Show that $Q(m,n) \leq O(mn/B)$ from this recurrence. You may want to rewrite the base cases to simplify the calculations.

15.12 Design a cache-oblivious algorithm for computing matrix transpose for the case $M \geq B^{3/2}$. Recall that the method described in the chapter assumes that $M \geq B^2$.

15.13 Design a cache-oblivious algorithm for multiplying an $N \times N$ matrix by an N vector.

15.14 The FFT computation based on the butterfly network in Fig. 9.2 is a very important problem with numerous applications. Show how to accomplish this in $O(\frac{n}{B}\log_{M/B}(n/B))$ IOs in $C(M,B)$.

Hint: Partition the computation into FFT sub-networks of size M.

15.15 *****Parallel disk model (PDM):** Consider a realistic extension of the external memory model where there are D disks capable of simultaneous input and output. That is, during any read operation, blocks of size B can be read in parallel from each of the D disks and similarly, D blocks of size B can be written in parallel. This is more constrained compared to the ability of simultaneously accessing any set of D blocks from a disk – which yields a virtual block size DB.

  (i) Design an algorithm to transpose an $N \times N$ matrix that is faster than the single disk model by a factor D.

  (ii) Re-design the Merge sort algorithm to exploit this D fold increase in IO capability.

15.16 *Discuss methods to design an efficient external memory priority queue data structure. This must support the operations delete-min , insert and delete efficiently. Use this to implement an external memory heap sort. The reader will have devise fast amortized versions of the priority queue operations to make heapsort match the performance of the external memory merge sort since a straightforward application of n delete-min operations would result in an $O(n\log_B(n/B))$ I-Os.

15.17 Consider the paging problem where we have a cache of size p and are given a sequence of page requests at the beginning (i.e., we know the entire sequence of page in advance). Prove that the following algorithm (called Furthest In Future) is optimal – when we need to evict a page, we evict the one which is requested farthest in future.

16

CHAPTER

Streaming Data Model

16.1 Introduction

In this chapter, we consider a new model of computation where the data arrives as a very long sequence of elements of unknown length. Such a setting has become increasingly important in scenarios where we need to handle huge amounts of data and do not have space to store all of it, or do not have time to scan the data multiple times. As an example, consider the amount of traffic encountered by a network router – it sees millions of packets every second. We may want to compute some properties of the data seen by the router, for example, the most frequent (or the top ten) destinations. In such a setting, we cannot expect the router to store details about each of the packet – this would require terabytes of storage capacity, and even if we could store all this data, answering queries on them will take too much time. Similar problems arise in the case of analyzing web-traffic, data generated by large sensor networks, etc.

In the data streaming model, we assume that the data arrives as a long stream $x_1, x_2, \ldots, x_m$, where the algorithm receives the element x_i at step i (see Fig. 16.1). Further, we assume that the elements belong to a universe $U = \{e_1, \ldots, e_n\}$. Note that the stream can have the same element repeated multiple times.[1] Both the quantities m and n are assumed to be very large, and we would like our algorithms to take sub-linear space

---

[1] There are more general models which allow both insertion and deletion of an element. We will not discuss these models in this chapter, though some of the algorithms discussed in this chapter extend to this more general setting as well.

(sometimes, even logarithmic space). This implies that the classical approach where we store all the data and can access any element of the data (say, in the RAM model) is no longer valid here because we are not allowed to store all of the data. This also means that we may not be able to answer many of the queries exactly. Consider, for example, the following query – output the most frequent element in the stream. Now consider a scenario where each element arrives just once, but there is one exceptional element which arrives twice. Unless we store all the distinct elements in the stream, identifying this exceptional element seems impossible. Therefore, it is natural to make some more assumptions about the nature of output expected from an algorithm. For example, here we would expect the algorithm to work only if there is some element which occurs much more often than other elements. This is an assumption about the nature of the data seen by the algorithms. At other times, we would allow the algorithm to output approximate answers. For example, consider the problem of finding the number of distinct elements in a stream. In most practical settings, we would be happy with an answer which is a small constant factor away from the actual answer.

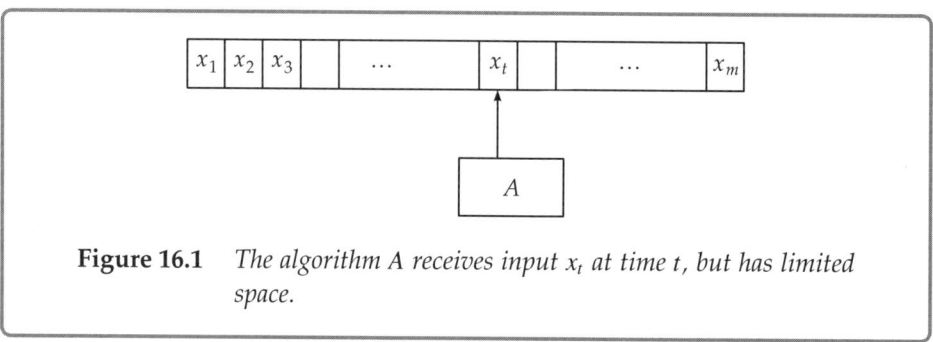

Figure 16.1 *The algorithm A receives input x_t at time t, but has limited space.*

In this chapter, we consider some of the most fundamental problems studied in the streaming data model. Many of these algorithms will be randomized. In other words, they will output the correct (or approximate) answer with high probability.

16.2 Finding Frequent Elements in a Stream

In this section, we consider the problem of finding *frequent* elements in a stream. As indicated earlier, this happens to be a very useful statistic for many applications. The notion of frequent elements can be defined in many ways:

- **Mode:** The element (or elements) with the highest frequency. In case there are multiple modes, any of them is an acceptable output.

- **Majority:** An element with more than 50% occurrence – note that there may not be any such element.

- **Threshold:** Find out all elements that occur more than f fraction of the length of the stream, for any $0 < f \leq 1$. Finding majority is a special case with $f = 1/2$.

Observe that the aforementioned problems are hardly interesting from a classical algorithmic design perspective because they can be easily reduced to sorting. Designing more efficient algorithms requires further thought (for example, finding the mode). Accomplishing the same task in a streaming environment with limited memory presents interesting design challenges. Let us first review a well-known algorithm for *majority* among n elements known as the *Boyer–Moore voting algorithm*. Recall that a majority element in a stream of length m is an element which occurs more than $m/2$ times in the stream. If no such element exists, the algorithm is allowed to output *any* element. This is always acceptable if we are allowed to scan the array once more because we can check if the element output by the algorithm is indeed the majority element. Therefore, we can safely assume that the array has a majority element.

The algorithm is described in Fig. 16.2. The procedure scans the array sequentially [2] and maintains one counter variable. It also maintains another variable maj which stores the (guess for) majority element. Whenever the algorithm sees an element which is identical to the one stored in maj, it increases the counter variable, otherwise it decreases it. If the counter reaches 0, it *resets* the variable maj to the next element. It is not obvious why it should return the majority element *if it exists*. If there is no such element, then it can return any arbitrary element.

Procedure Finding majority of m elements in array(a)

1   $count \leftarrow 0$;
2   **for** $i = 1$ *to* m **do**
3   **if** $count = 0$ **then**
4      $maj \leftarrow a[i]$ (* initialize maj *)
5   **if** $maj = a[i]$ **then**
6      $count \leftarrow count + 1$
7   **else**
8      $count \leftarrow count - 1$;

9   Return maj ;

Figure 16.2   *Boyer–Moore majority voting algorithm*

---

[2] Often, we will think of the stream as a long array which can be scanned only once. In fact, there are more general models which allow the algorithm to make a few passes over the array.

As mentioned earlier, we begin by assuming that there is a majority element, denoted by M. We need to show that when the algorithm stops, the variable maj is same as M. The algorithm tries to delete elements without affecting the majority. More formally, we will show that at the beginning of each step t (i.e., before arrival of x_t), the algorithm maintains the following invariant: let S_t denote the *multi-set* consisting of the elements $x_t, x_{t+1}, \ldots, x_m$ and count number of copies of the element maj, that is, $S_t = \{x_t, \ldots, x_m\} \cup \underbrace{\{\mathtt{maj}, \mathtt{maj} \ldots\}}_{\mathtt{count}}$. We

shall prove that for all times t, M will be the majority element of S_t. This statement suffices because at the end of the algorithm (when $t = m$), S_t will be a multi-set consisting of copies of the element maj only. The invariant shows that M will be the majority element of S_t, and so, it must be the same as the variable maj when the algorithm terminates.

Example 16.1 *Consider an input consisting of a b b c c a a b a a a. Then, the values of* maj *and* count *after each iteration are*

$$(a,1), (a,0), (b,1), (b,0), (c,1), (c,0), (a,1), (a,0), (a,1), (a,2), (a,3)$$

We will prove this invariant by induction over t. Initially, S_0 is same as the input sequence, and so, the statement follows trivially. Suppose this fact is true at the beginning of step t. A key observation is that if M is the majority of a set of elements, it will remain in majority if some other element $x \neq M$ is deleted along with an instance of the majority element – this is implicitly done by reducing count. Indeed, if M occurs m_1 times, $m_1 > m/2$, then $m_1 - 1 > (m-2)/2$. So, if $x_t \neq$ maj, we decrement count. Hence, S_{t+1} is obtained from S_t by removing x_t and **at most** one copy of M. Note that it is also possible that neither maj nor x_t equals M. From this observation, M continues to be the majority element of S_{t+1}.

The other case is when x_t happens to be the same as maj. Here, the set $S_{t+1} = S_t$ since we replace x_t by one more copy of maj in the variable count. So the invariant holds trivially. This shows that the invariant holds at all times, and eventually, the algorithm outputs the majority element, M. In case, there is no majority element, the algorithm can output anything and so, if we want to verify if the element output is actually the majority, we have to make another pass through the array.

This idea can be generalized to finding out elements whose frequency exceeds $\frac{m}{k}$ for any integer k, see Fig. 16.3. Note that there can be at most $k - 1$ such elements. So instead of one counter, we shall use $k - 1$ counters. When we scan the next element, we can either increment the count if there exists a counter for the element or start a new counter if the number of counters used is less than $k - 1$. Otherwise, we decrease the counts of all the existing counters. If any counter becomes zero, we discard that element and instead assign a counter for the new element. In the end, the counters return elements that have non-zero counts. As before, these are potentially the elements that have frequencies at least $\frac{m}{k}$ and we need a second pass to verify them.

The proof of correctness is along the same lines as the majority. Note that there can be at most $k-1$ elements that have frequencies exceeding $\frac{m}{k}$, that is, a fraction $\frac{1}{k}$. So, if we remove such an element along with $k-1$ distinct elements, it still continues to be at least $\frac{1}{k}$ fraction of the remaining elements: $n_1 > \frac{m}{k} \Rightarrow n_1 - 1 > \frac{m-k}{k}$.

Procedure Algorithm for threshold(m,k)

---

1   cur : current element of stream ;
2   S: current set of elements with non-zero counts, $|S| \le k$;
3   **if** $cur \in S$ **then**
4   increment counter for cur
5   **else**
6   **if** $|S| < k$ **then**
7   Start a new counter for cur, update S
8   **else**
9   decrement all counters ;
10   If a counter becomes 0 delete it from S
11 Return S ;

Figure 16.3   *Misra–Gries streaming algorithm for frequent elements*

The previous algorithms have the property that the data is scanned in the order it is presented and the amount of space is proportional to the number of counters where each counter has $\log m$ bits. Thus, the space requirement is logarithmic in the size of the input. This algorithm can be used for *approximate counting* – see Exercise Problems.

16.3 Distinct Elements in a Stream

The challenging aspect of this problem is to count the number of distinct elements d in the input stream with limited memory s, where $s \ll d$. If we were allowed space comparable to d, then we could simply hash the elements and count the number of non-zero buckets. Uniformly sampling a subset of elements from the stream could be misleading. Indeed, if some elements occur much more frequently than others, then multiple occurrence of such elements would be picked up by the uniform sample and it does not provide any significant information about the number of distinct elements.

Instead, we will hash the incoming elements uniformly over a range, $[1,p]$ such that if there are d distinct elements, then they will be roughly p/d apart where $p > n \ge d$. If g is the gap between two consecutive hashed elements, then we can estimate $d = p/g$. Think of throwing d balls in p bins randomly which will be scattered evenly on the average. Alternately, we can use the position of the first hashed position as estimate of g. This is the underlying idea behind the algorithm given in Fig. 16.4. The algorithm keeps track of the smallest value to which an element gets hashed (in the variable Z). Again, the idea is that if there are d distinct elements, then the elements get mapped to values in the array that are roughly p/d apart. So, p/Z should be a good approximation of d.

Procedure Distinct elements in a stream $S(m,n)$

---

1   *Input* A stream $S = \{x_1, x_2, \ldots, x_m\}$, where $x_i \in [1, n]$;
2   Suppose p is a prime in the range $[n, 2n]$. Choose $0 \le a \le p-1$ and
     $0 \le b \le p-1$ uniformly at random ;
3   $Z \leftarrow \infty$;
4   **for** $i = 1$ *to* m **do**
5      $Y = (a \cdot x_i + b) \mod p$;
6   **if** $Y < Z$ **then**
7         $Z \leftarrow Y$

8   Return $\lceil \frac{p}{Z} \rceil$;

Figure 16.4   *Counting number of distinct elements*

This procedure will be analyzed rigorously using the property of the universal hash family discussed earlier in Section 6.3. The parameter of interest will be the *expected* gap between consecutive hashed elements. Our strategy will be to prove that the Z lies between $k_1 p/d$ and $k_2 p/d$ with high probability, where k_1 and k_2 are two constants. It will then follow that the estimate p/Z is within a constant factor of d.

Let $Z_i = (a \cdot x_i + b) \mod p$ be the sequence of hashed values from the stream. Then, we can claim the following.

Claim 16.1 *The numbers Z_i, $1 \le i \le m$ are distributed uniformly at random in the range $[0, p-1]$ and are also pair-wise independent, viz., for $i \ne k$*

$$\Pr[Z_i = r, Z_k = s] = \Pr[Z_i = r] \cdot \Pr[Z_k = s] = \frac{1}{p^2}$$

Proof: For some fixed $i_0 \in [0, p-1]$ and $x \in [1, n]$, we want to find the probability that x is mapped to i_0. So,

$$
\begin{aligned}
i_0 &\equiv (ax+b) \mod p \\
i_0 - b &\equiv ax \mod p \\
x^{-1}(i_0 - b) &\equiv a \mod p
\end{aligned}
$$

where x^{-1} is the multiplicative inverse of x in the multiplicative prime field modulo p and it is unique since p is prime.[3] For any fixed b, there is a unique solution for a. As a is chosen uniformly at random, the probability of this happening is $\frac{1}{p}$ for any *fixed* choice of b. Therefore, this is also the unconditional probability that x is mapped to i_0.

For the second part, consider $i_0 \neq i_1$. We can consider $x \neq y$ such that x, y are mapped respectively to i_0 and i_1. We can write the simultaneous equations similar to the previous one.

$$\begin{bmatrix} x & 1 \\ y & 1 \end{bmatrix} \cdot \begin{bmatrix} a \\ b \end{bmatrix} \equiv_p \begin{bmatrix} i_0 \\ i_1 \end{bmatrix}$$

The 2×2 matrix is invertible for $x \neq y$ and therefore, there is a unique solution corresponding to a fixed choice of (i_0, i_1). The probability that a, b matches the solution is $\frac{1}{p^2}$ as they are chosen uniformly at random. $\square$

Recall that d denotes the number of distinct elements in the stream. We will show the following.

Claim 16.2 *For any constant $c \geq 2$,*

$$Z \in \left[\frac{p}{cd}, \frac{cp}{d}\right] \text{ with probability } \geq 1 - \frac{2}{c}$$

Proof: Note that if $Z = p/d$, then the algorithm returns d, which is the number of distinct elements in the stream. Since Z is a random variable, we will only be able to bound the probability that it is within the interval $\left[\frac{p}{cd}, \frac{cp}{d}\right]$ with significant probability implying that the algorithm will return an answer in the range $[p/c, pc]$ with significant probability. Of course, there is a risk that it falls outside this window and that is the inherent nature of a *Monte Carlo* randomized algorithm.

First, we will find the probability that $Z \leq s - 1$ for some arbitrary s. For the sake of notational simplicity, assume that the d distinct elements are $x_1, x_2, \ldots, x_d$. Let us define a family of indicator random variables in the following manner

$$X_i = \begin{cases} 1 & \text{if } (ax_i + b) \mod p \leq s - 1 \\ 0 & \text{otherwise} \end{cases}$$

So the total number of x_i that map to numbers in the range $[0, s-1]$ equals $\sum_{i=1}^{d} X_i$ (recall that we assumed that $x_1, \ldots, x_d$ are distinct). Let $X = \sum_{i=1}^{d} X_i$ and therefore, using linearity of expectation

$$\mathbb{E}[X] = \mathbb{E}[\sum_i X_i] = \sum_i \mathbb{E}[X_i] = \sum_i \Pr[X_i = 1] = d \cdot \Pr[X_i = 1] = \frac{sd}{p}$$

---

[3] By our choice of p, $x \not\equiv 0 \mod p$

The last equality follows from the previous result as there are s (viz., $0, 1, \ldots, s-1$) possibilities for x_i to be mapped and each has probability $\frac{1}{p}$.

If we choose $s = \frac{p}{cd}$ for some constant c, then $\mathbb{E}[X] = 1/c$. From Markov's inequality, $\Pr[X \geq 1] \leq \frac{1}{c}$, implying that with probability greater than $1 - 1/c$, no x_i will be mapped to numbers in the range $[0, \lceil \frac{p}{cd} \rceil]$. This establishes that $\Pr[Z \leq \frac{p}{cd}] \leq \frac{1}{c}$.

For the other direction, we will use Chebyshev's inequality (Eq. (2.2.4)), which requires computing the variance of X, which we shall denote by $\sigma^2(X)$. We know that

$$\sigma^2[X] = \mathbb{E}[(X - \mathbb{E}[X])^2] = \mathbb{E}[X^2] - \mathbb{E}^2[X]$$

Since $X = \sum_{i=1}^d X_i$, we can calculate (assume that all indices i and j vary from 1 to d)

$$\mathbb{E}[X^2] = \mathbb{E}\left[\left(\sum_{i=1}^d X_i\right)^2\right]$$

$$= \mathbb{E}\left[\sum_{i=1}^d X_i^2 + \sum_{i \neq j} X_i \cdot X_j\right]$$

$$= \mathbb{E}\left[\sum_{i=1}^d X_i^2\right] + \mathbb{E}\left[\sum_{i \neq j} X_i \cdot X_j\right]$$

$$= \sum_{i=1}^d \mathbb{E}[X_i^2] + \sum_{i \neq j} \mathbb{E}[X_i] \cdot \mathbb{E}[X_j]$$

which follows from linearity of expectation and pairwise independence of X_i and X_j.[4] It is easily seen that $\mathbb{E}[X_i^2] = \mathbb{E}[X_i]$ since X_i is 0-1 valued, so the expression simplifies to $d \cdot \frac{s}{p} + d(d-1) \cdot \frac{s^2}{p^2}$. This yields the expression for

$$\sigma^2(X) = \frac{sd}{p} + \frac{d(d-1)s^2}{p^2} - \frac{s^2 d^2}{p^2} = \frac{sd}{p} \cdot \left(1 - \frac{s}{p}\right) \leq \frac{sd}{p}$$

For $s = \frac{cp}{d}$, the variance is bounded by c. From Chebyshev's inequality, we know that for any random variable X,

$$\Pr[|X - \mathbb{E}[X]| \geq t] \leq \frac{\sigma^2(X)}{t^2}$$

Using $t = \mathbb{E}[X] = \frac{sd}{p} = c$, we obtain $\Pr[|X - \mathbb{E}[X]| \geq \mathbb{E}[X]] \leq \frac{c}{c^2} = \frac{1}{c}$. The event $|X - \mathbb{E}[X]| \geq \mathbb{E}[X]$ is the union of two disjoint events, namely

---

[4] This needs to be rigorously proved using Claim 16.1 on pair-wise independence of (x_i, x_j) being mapped to (i_0, i_1). We have to technically consider all pairs in the range $(0, s-1)$.

(i) $X \geq 2\mathbb{E}[X]$ and

(ii) $\mathbb{E}[X] - X \geq \mathbb{E}[X]$, or $X \leq 0$

Clearly, both events must have probability bounded by $\frac{1}{c}$ and specifically, the second event implies that the probability that none of the m elements is mapped to the interval $[0, \frac{cp}{d}]$ is less than $\frac{1}{c}$. Using the union bound $\Pr[Z \leq p/(cd) \cup Z] \geq (cp)/d \leq \frac{1}{c} + \frac{1}{c} = \frac{2}{c}$, we obtain the required result.   □

So the algorithm outputs a number that is within the range $[\frac{d}{c}, cd]$ with probability $\geq 1 - \frac{2}{c}$.

16.4 Frequency Moment Problem and Applications

Suppose the set of elements in a stream $S = \{x_1, \ldots, x_m\}$ belong to a universe $U = \{e_1, \ldots, e_n\}$. Define the frequency f_i of element e_i as the number of occurrences of e_i in the stream S. The kth frequency moment of the stream is defined as

$$F_k = \sum_{i=1}^{n} f_i^k.$$

Note that F_0 is exactly the number of distinct elements in the stream. F_1 counts the number of elements in the stream, and can be easily estimated by keeping a counter of size $O(\log m)$. The second frequency moment F_2 captures the non-uniformity in the data – if all n elements occur with equal frequency, that is, m/n (assume that m is a multiple of n for the sake of this example), then F_2 is equal to m^2/n; whereas, if the stream contains just one element (with frequency m), then F_2 is m^2. Thus, larger values of F_2 indicate non-uniformity in the stream. Higher frequency moments give similar statistics about the stream – as we increase k, we are putting more emphasis on higher frequency elements.

The idea behind estimating F_k is quite simple: suppose, we sample an element uniformly at random from the stream, call it X. Suppose X happens to be the element e_i. Conditioned on this fact, X is equally likely to be any of the f_i occurrences of e_i. Now, we observe how many times e_i occurs in the stream from now onward. Say it occurs r times. What can we say about the expected value of r^k? Since e_i occurs f_i times in the stream, the random variable r is equally likely to be one of $\{1, \ldots, f_i\}$. Therefore,

$$\mathbb{E}[r^k | X = e_i] = \frac{1}{f_i} \sum_{j=1}^{f_i} j^k$$

It follows from the aforementioned expression that $\mathbb{E}[r^k - (r-1)^k | X = e_i] = \frac{1}{f_i} \cdot f_i^k$. Now, we remove the conditioning on X, and we have

$$\mathbb{E}[r^k - (r-1)^k] = \sum_i \mathbb{E}[r^k - (r-1)^k | X = e_i] \Pr[X = e_i] = \frac{1}{f_i} \cdot f_i^k \cdot \sum_i \frac{f_i}{m} = \frac{1}{m} \cdot F_k$$

Therefore, the random variable $m(r^k - (r-1)^k)$ has expected value F_k.

The only catch is that we do not know how to sample a uniformly random element of the stream. Since X is a random element of the stream, we want

$$\Pr[X = x_j] = \frac{1}{m}$$

for all values of $j = 1, \ldots, m$. However, we do not know m in advance, and so cannot use this expression directly. Fortunately, there is a more clever sampling procedure, called *reservoir sampling*, described in Fig. 16.5 (also see Section 2.3.2). Note that in iteration i, the algorithm just tosses a coin with probability of heads equal to $1/i$. It is left as an exercise problem to show that at any step i, X is indeed a randomly chosen element from $\{x_1, \ldots, x_i\}$.

Procedure Reservoir sampling

1   $X \leftarrow x_1$;
2   **for** $i = 2$ *to* m **do**
3   Sample a binary random variable t_i, which is 1 with probability $1/i$ using Reservoir sampling ;
4   **if** $t_i = 1$ **then**
5         $X \leftarrow x_i$
6   Return X

Procedure Estimating F_k

1   $X \leftarrow x_1, r \leftarrow 1, i \leftarrow 1$;
2   **while** *not end-of-stream* **do**
3      $i \leftarrow i + 1$;
4   Sample a binary random variable t_i, which is 1 with probability $1/i$ using Reservoir sampling ;
5   **if** $t_i = 1$ **then**
6         $X \leftarrow x_i, r \leftarrow 1$
7   **else**
8   **if** $X = x_i$ **then**
9         $r \leftarrow r + 1$;
10   Return $m\left(r^k - (r-1)^k\right)$;

Figure 16.5   *Combining reservoir sampling with the estimator for F_k*

We now need to show that this algorithm gives a good approximation to F_k with high probability. So far, we have only shown that there is a random variable, namely $Y :=$

$m(r^k - (r-1)^k)$, which is equal to F_k in expectation. But now we want to compute the probability that Y lies within $(1 \pm \varepsilon)F_k$. In order to do this, we need to estimate the variance of Y. If the variance is not too high, we can hope to use Chebyshev's bound. We know that the variance of Y is at most $\mathbb{E}[Y^2]$. Therefore, it is enough to estimate the latter quantity. Since we are going to use Chebyshev's inequality, we would like to bound $\mathbb{E}[Y^2]$ in terms of $(\mathbb{E}[Y])^2$, which is the same as F_k^2. The first few steps for estimating $\mathbb{E}[Y^2]$ are identical to those for estimating $\mathbb{E}[Y]$:

$$
\begin{aligned}
\mathbb{E}[Y^2] &= \sum_{i=1}^{n} \mathbb{E}[Y^2 | X = e_i] \cdot \Pr[X = e_i] = \sum_{i=1}^{n} m^2 \cdot \mathbb{E}\left[\left(r^k - (r-1)^k \right)^2 | X = e_i \right] \cdot \frac{f_i}{m} \\
&= \sum_{i=1}^{n} m f_i \cdot \frac{1}{f_i} \sum_{j=1}^{f_i} (j^k - (j-1)^k)^2 = m \cdot \sum_{i=1}^{n} \sum_{j=1}^{f_i} (j^k - (j-1)^k)^2
\end{aligned}
\tag{16.4.1}
$$

We now show how to handle the expression $\sum_{j=1}^{f_i} (j^k - (j-1)^k)^2$. We first claim that

$$
j^k - (j-1)^k \leq k \cdot j^{k-1}
$$

This follows from applying the mean value theorem to the function $f(x) = x^k$. Given two points $x_1 < x_2$, the mean value theorem states that there exists a number $\theta \in [x_1, x_2]$ such that $f'(\theta) = \frac{f(x_2) - f(x_1)}{x_2 - x_1}$. We now substitute $j-1$ and j for x_1 and x_2 respectively, and observe that $f'(\theta) = k\theta^{k-1} \leq kx_2^{k-1}$ to get

$$
j^k - (j-1)^k \leq k \cdot j^{k-1}
$$

Therefore,

$$
\sum_{j=1}^{f_i} (j^k - (j-1)^k)^2 \leq \sum_{j=1}^{f_i} k \cdot j^{k-1} \cdot (j^k - (j-1)^k) \leq k \cdot f_i^{k-1} \sum_{j=1}^{f_i} (j^k - (j-1)^k) \leq k \cdot f_\star^{k-1} \cdot f_i^k
$$

where $f_\star$ denotes $\max_{i=1}^{n} f_i$. Substituting this in Eq. (16.4.1), we get

$$
\mathbb{E}[Y^2] \leq k \cdot m \cdot f_\star^{k-1} F_k
$$

Recall that we wanted to bound $\mathbb{E}[Y^2]$ in terms of F_k^2. So we need to bound $m \cdot f_\star^{k-1}$ in terms of F_k. Clearly,

$$
f_\star^{k-1} = (f_\star^k)^{\frac{k-1}{k}} \leq F_k^{\frac{k-1}{k}}
$$

In order to bound m, we apply Jensen's inequality[5] to the convex function x^k to get

$$
\left(\frac{\sum_{i=1}^{n} f_i}{n} \right)^k \leq \frac{\sum_{i=1}^{n} f_i^k}{n}
$$

---

[5] For any convex function f, $\mathbb{E}[f(X)] \geq f(\mathbb{E}[X])$ for a random variable X.

which implies that

$$m = \sum_{i=1}^{n} f_i \leq n^{1-1/k} \cdot F_k^{1/k}$$

Combining all of these inequalities, we see that

$$\mathbb{E}[Y^2] \leq k \cdot n^{1-1/k} \cdot F_k^2$$

If we now use Chebyshev's bound, we get

$$\Pr[|Y - F_k| \geq \varepsilon F_k] \leq \frac{\mathbb{E}[Y^2]}{\varepsilon^2 F_k^2} \leq k/\varepsilon^2 \cdot n^{1-1/k}$$

The expression on the right-hand side is (likely to be) larger than 1, and so this does not give us much information. The next idea is to further reduce the variance of Y by keeping several independent copies of it, and computing the average of all these copies. More formally, we maintain t i.i.d. random variables $Y_1, \ldots, Y_t$, each of which has the same distribution as that of Y. If we now define Z as the average of these random variables, linearity of expectation implies that $\mathbb{E}[Z]$ remains F_k. However, the variance of Z now becomes $1/t$ times that of Y (see Exercise Problems).

Therefore, if we now use Z to estimate F_k, we get

$$\Pr[|Z - F_k| \geq \varepsilon F_k] \leq \frac{k}{t \cdot \varepsilon^2} \cdot n^{1-1/k}$$

If we want to output an estimate within $(1 \pm \varepsilon) F_k$ with probability at least $1 - \delta$, we should pick t to be $\frac{1}{\delta \varepsilon^2} \cdot n^{1-1/k}$. It is easy to check that the space needed to update one copy of Y is $O(\log m + \log n)$. Thus, the total space requirement of our algorithm is $O(\frac{1}{\delta \varepsilon^2} \cdot n^{1-1/k} \cdot (\log m + \log n))$.

16.4.1 The median of means trick

We now show that it is possible to obtain the same guarantees about Z; but we need to keep only $O\left(\frac{1}{\varepsilon^2} \cdot \log\left(\frac{1}{\delta}\right) \cdot n^{1-1/k}\right)$ copies of the estimator for F_k. Note that we have replaced the factor $1/\delta$ by $\log(1/\delta)$. The idea is that if we use only $t = \frac{4}{\varepsilon^2} \cdot n^{1-1/k}$ copies of the variable Y in the earlier analysis, then we will get

$$\Pr[|Z - F_k| \geq \varepsilon F_k] \leq 1/4$$

Although this is not good enough for us, what if we keep several copies of Z (where each of these is average of several copies of Y)? In fact, if we keep $\log(1/\delta)$ copies of Z, then at least one of these will give the desired accuracy with probability at least δ – indeed, the probability that all of them are at least εF_k far from F_k will be at most $(1/2)^{\log(1/\delta)} \leq \delta$.

But we will not know *which* one of these copies is correct! Therefore, the plan is to keep slightly more copies of Z, say about $4\log(1/\delta)$. Using Chernoff bounds, we can show that with probability at least $1-\delta$, roughly a majority of these copies will give an estimate in the range $(1\pm\varepsilon)F_k$. Therefore, the *median* of all these copies will give the desired answer. This is called the 'median of means' trick.

We now give details of this idea. We keep an array of variables Y_{ij}, where i varies from 1 to $\ell := 4\log(1/\delta)$ and j varies from 0 to $t := \frac{2}{\varepsilon^2}\cdot n^{1-1/k}$. Each row of this array (i.e., elements Y_{ij}, where we fix i and vary j) will correspond to one copy of the estimate described here. So, we define $Z_i = \sum_{j=1}^{t}Y_{ij}/t$. Finally, we define Z as the median of Z_i, for $i=1,\dots,\ell$. We now show that Z lies in the range $(1\pm\varepsilon)F_k$ with probability at least $1-\delta$. Let E_i denote the event: $|Z_i - F_k| \geq \varepsilon F_k$. We already know that $\Pr[E_i] \leq 1/4$. Now, we want to show that the number of such events will be close to $\ell/4$. We can use Chernoff bounds to prove that the size of the set $\{i : E_i \text{ occurs}\}$ is at most $\ell/2$ with at least $(1-\delta)$ probability (see Exercise Problems).

Now assume this happens. If we look at the sequence $Z_i, i=1,\dots,\ell$, at least half of them will lie in the range $(1\pm\varepsilon)F_k$. The median of this sequence will also lie in the range $(1\pm\varepsilon)F_k$ for the following reason: if the median is (say) above $(1+\varepsilon)F_k$, then at least half of the events E_i will occur, which is a contradiction. Thus, we have shown the following result:

Theorem 16.1 *We can estimate the frequency moment F_k of a stream with $(1\pm\varepsilon)$ multiplicative error with probability at least $1-\delta$ using $O\left(\left(\frac{1}{\varepsilon^2}\cdot\log\left(\frac{1}{\delta}\right)\cdot n^{1-1/k}\right)\cdot(\log m + \log n)\right)$ space.*

16.4.2 The special case of second frequency moment

It turns out that we can estimate the second frequency moment F_2 using logarithmic space only (the aforementioned result shows that space requirement will be proportional to $\sqrt{n}$). The idea is again to have a random variable whose expected value is F_2, but now we will be able to control the variance in a much better way. We will use the idea of universal hash functions (refer to Section 6.3). We will require binary hash functions, that is, they will map the set $U = \{e_1,\dots,e_n\}$ to $\{-1,+1\}$. By generalizing the notion of pair-wise independent universal hash function, a set of functions H is said to be k-universal if for any set S of indices of size at most k, and values $a_1,\dots,a_k \in \{-1,+1\}$,

$$\Pr_{h\in H}\left[\wedge_{i\in S}x_i = a_i\right] = \frac{1}{2^{|S|}}$$

where h is a uniformly chosen hash function from H. We can construct such a set H which has $O(n^k)$ functions; a hash function $h \in H$ can be stored using only $O(k\log n)$ space (see Exercise Problems at the end of the chapter). We will need a set of 4-universal hash functions. Thus, we can store the hash function using $O(\log n)$ space only.

The algorithm for estimating F_2 is shown in Fig. 16.6. It maintains a running sum X – when the element x_t arrives, it first computes the hash value $h(x_t)$, and then adds $h(x_t)$ to X (so, we add either $+1$ or -1 to X). Finally, it outputs X^2. It is easy to check that the expected value of X^2 is indeed F_2. First observe that if f_i denotes the frequency of element e_i, then $X = \sum_{i=1}^{n} f_i \cdot h(e_i)$. Therefore, using linearity of expectation,

$$\mathbb{E}[X^2] = \sum_{i=1}^{n} \sum_{j=1}^{n} f_i f_j \mathbb{E}[h(e_i)h(e_j)]$$

Procedure Second frequency moment

---

1 $X \leftarrow 0, h \leftarrow$ uniformly chosen ± 1 hash function from a 4-universal family; **for** $i = 1$ *to* m **do**
2 $\quad \big\lfloor \; X \leftarrow X + h(x_i)$
3 Return X^2

Figure 16.6   *Estimating F_2*

This sum splits into two parts: if $i = j$, then $h(e_i)h(e_j) = h(e_i)^2 = 1$; and if $i \neq j$, then the fact that H is 4-universal implies that $h(e_i)$ and $h(e_j)$ are pair-wise independent random variables. Therefore, $\mathbb{E}[h(e_i)h(e_j)] = \mathbb{E}[h(e_i)] \cdot \mathbb{E}[h(e_j)] = 0$, because $h(e_i)$ is ± 1 with equal probability. So,

$$\mathbb{E}[X^2] = \sum_{i=1}^{n} f_i^2 = F_2$$

As before, we want to show that X^2 comes close to F_2 with high probability. We need to bound the variance of X^2, which is at most $\mathbb{E}[X^4]$. As previously, we expand the fourth power of the expression of X:

$$\mathbb{E}[X^2] = \sum_{i,j,k,l=1}^{n} f_i f_j f_k f_l \mathbb{E}[h(e_i)h(e_j)h(e_k)h(e_l)]$$

Each of the summands is a product of 4 terms – $h(e_i), h(e_j), h(e_k), h(e_l)$. Consider such a term. If an index is distinct from the remaining three indices, then we see that its expected value is 0. For example, if i is different from j, k, l, then $\mathbb{E}[h(e_i)h(e_j)h(e_k)h(e_l)] = \mathbb{E}[h(e_i)]\mathbb{E}[h(e_j)h(e_k)h(e_l)]$ (we are using 4-universal property here – any set of 4 distinct hash values are mutually independent). But $\mathbb{E}[h(e_i)] = 0$, and so the expected value of the whole term is 0. Thus, there are only two cases when the summand need not be 0: (i) all the four indices i, j, k, l are same – in this case, $\mathbb{E}[h(e_i)h(e_j)h(e_k)h(e_l)] = \mathbb{E}[h(e_i)^4] = 1$, because $h(e_i)^2 = 1$, or (ii) exactly two of i, j, k, l take one value and the other two indices take another value – for example, $i = j$, $k = l$, but $i \neq k$. In this case, we again get $\mathbb{E}[h(e_i)h(e_j)h(e_k)h(e_l)] = \mathbb{E}[h(e_i)^2 h(e_k)^2] = 1$. We can simplify

$$\mathbb{E}[X^4] = \sum_{i=1}^{n} f_i^4 + \sum_{i=1}^{n} \sum_{j \in \{1,\dots,n\} \setminus \{i\}} f_i^2 f_j^2 \leq 2F_2^2$$

Thus, we see that the variance of the estimator X^2 is at most $2\mathbb{E}[X^2]^2$. The remaining ideas for calculations are the same as in the previous section (see Exercise Problems).

16.5   Proving Lower Bounds for Streaming Model

The primary challenges in the streaming model are limited space and our inability to go back and forth on the input data, that is, the single pass restriction. The latter has been relaxed to allow multiple passes to understand the relative complexity of certain problems, in particular graph problems that are very difficult to solve in one or even constant number of passes. However, in reality, the single pass is the most realistic setting and one would like to derive lower bounds on this problem. This is further complicated by our willingness to allow approximation and use of randomness, so the lower bounds must address these aspects along with any tradeoffs.

The most successful approach has been through *communication complexity* that has many other applications for proving information theoretic lower bounds. There are two parties in this model, typically named *Alice* and *Bob*. Alice holds an input denoted by X and Bob has an input denoted by Y. The goal of the model is to compute a function $f : X \times Y \to \mathbb{Z}$ by exchanging as little information as possible. The inputs X and Y can be thought of as bit strings of length n. For example, we may want to compute $X + Y$ which is the sum of the inputs or even simpler functions like $(X + Y) \mod 2$ etc. If one of the parties sends her input to the other party then it is trivially computed, but it involves communication of n bits. While computing the sum may necessitate knowing the entire input, for computing the sum modulo 2, one of the parties can just send over the parity, i.e., one bit to the other, who can now easily compute the answer correctly as in the same line with the expression $(X + Y) \mod 2 = X \mod 2 + Y \mod 2$.

In the most general setting, the communication is expected to happen in $k \geq 1$ rounds where alternately Alice and Bob send some inputs to the other party – this can help in adaptive computation of the function f. The final answer is supposed to be with Bob and the *communication complexity* of f is the total number of bits exchanged in all the rounds. In our scenario, we will assume $k = 1$, so it is only Alice who will communicate some of her input to Bob who is entrusted to compute the answer.

To give the reader a flavor of this model, consider computation of the equality function

$$Equal : X \times Y \to \{0,1\} = \begin{cases} 1 & \text{if } X = Y \\ 0 & \text{otherwise} \end{cases}$$

Since the case of n bits is trivial, let us see even when Alice sends $n-1$ or fewer bits, if Bob can compute $Equal$ correctly. The total number of possible messages that Alice can send to Bob is $\sum_{i=1}^{n-1} 2^i = 2^n - 2$ since each bit can be 0 or 1; she must send at least one bit (otherwise the problem is trivial). Since Alice holds n bits that has 2^n possibilities, for at least two distinct inputs $x_1, x_2 \in \{0,1\}^n$, Alice sends the same message to Bob. So the answer computed by Bob for his inputs $Y = \{x_1\}$ and $Y = \{x_2\}$ must be identical since it depends only on the message sent by Alice, but $Equal(x_1, x_2) \neq Equal(x_2, x_2)$ so clearly the function will not be correctly computed for all inputs.

An interesting variation is to allow use of randomization. Alice will send her input modulo some random prime p where p is much smaller than n bits. Bob compares his input also hashed wrt to the same prime number. If they are equal, he answers 1, else 0. Clearly, the answer is always correct when the hashes do not agree but the answer may be wrong even when they agree. This was used very cleverly for the string matching algorithm of Rabin-Karp (Section 8.1) where we showed that we can achieve this with high probability by using only $O(\log n)$ bits. In this case, Alice can also send the random prime number that is used to compute the hash within $O(\log n)$ bits. This implies an exponential improvement over the n bit scheme by using randomization. Clearly there is a provable gap between deterministic and randomized communication complexity. There are many interesting results known for communication complexity but our purpose here is to only highlight the relationship with streaming algorithms.

We hypothetically partition the stream into two halves, the first half and the second half. The first half can be thought of as Alice's part and the second half is with Bob. For any communication complexity problem, we can define an equivalent streaming problem as follows. We need to take care of a minor detail that the two input communication problem (x, y) should be thought of as a single input with x and y concatenated corresponding to the first and the second half. For example, the $Equality(x, y)$ is transformed into $Streamequal(x \cdot y)$ and is equal to 1 iff the first half is equal to the second half.

Alice simply simulates the streaming algorithm on her input and passes s bits to Bob where s is the space used by the corresponding streaming algorithm. If a streaming algorithm can compute this function correctly using s amount of space, then Bob should be able to successfully compute the function. Note that this applies both to deterministic and randomized setting. Therefore we can claim the following

Claim 16.3 *If the lower-bound for the one round communication complexity problem is s bits, then the corresponding one pass streaming algorithm has an $\Omega(s)$ space bound.*

Let us illustrate this technique on the problem of majority. For this, we define a problem known as the *Index* problem for Alice and Bob. Alice holds an n bit input string $X = \{0,1\}^n$ and Bob has an integer j, $1 \le j \le n$. Then $Index(X, j) = X[j]$, that is, Bob will output 0 or 1 depending on the value of the jth bit of X that is only known to Alice.

Claim 16.4 *The communication complexity of the Index problem is $\Omega(n)$.*

The proof of this claim is left as an Exercise Problem and can be argued along the lines of the *Equal* problem. Note that this problem is not symmetric like the *Equality* problem. If Bob was allowed to pass the bits, then it can be computed by Alice using $\log n$ communication. Another important property of the *Index* problem is that, even by using randomization, the communication complexity is $\Omega(n)$. The proof is based on Theorem 3.1 and requires a non-trivial construction of an appropriate distribution function – we leave it out from this discussion.

Let us now reduce the *Index* problem to a majority in streaming. We transform the bit sequence $X[i]$ into a sequence of integers σ_i as follows:

$$\sigma_i = 2i + X[i], \quad 1 \le i \le n$$

For example, 10011010 is transformed to 3,4,6, 9,11,12, 15, 16. Similarly, the input of Bob is transformed to the sequence $\sigma' = 2j, 2j, \ldots$ repeated n times. Then in the combined stream $\sigma \cdot \sigma'$ the integer $2j$ is a majority iff $X[i] = 0$. In the aforementioned example, suppose Bob has $j = 4$, then in the transformed stream

$$3, 4, 6, 9, 11, 12, 15, 16 || 8, 8, 8, 8, 8, 8, 8, 8$$

we do not have a majority since $X[4] = 1$. The reader can verify that if $X[4] = 0$, then there would be a majority element, namely, 8.

Further Reading

The study of streaming algorithms got formal recognition with the paper by Alon et al. [10], although there existed well-known works on space-bounded algorithms like those by Munro and Paterson [108] and the read-once paradigm implicit in the work of Misra and Gries [104]. The main technique of Misra and Gries has been re-discovered repeatedly in many later papers, implying the fundamental nature of this elegant technique. The Boyer–Moore voting algorithm was first discovered in 1980 and subsequently published much later [24].

Alon et al.'s [10] paper literally triggered a slew of fundamental results in the area of streaming algorithms which can be found in a survey book by Muthukrishnan [109]. The challenge in this model is typically more in the analysis and lower bounds rather than sophisticated algorithms because of the limitations in the model. The frequency-moments problem formalized by Alon et al. took a while to be satisfactorily settled [67] and interesting connections were discovered with metric embeddings. For lower bounds, there is a strong relationship between communication complexity [86] and streaming algorithms – see for example, Chakrabarti et al. [27] or Kalyansundaram and Schnitger [73]. Later papers extended the streaming model to multi-pass to understand the complexity of various challenging problems in this paradigm, in particular, many graph problems. See McGregor [100] for a nice survey.

Exercise Problems

16.1 Let f_i be the frequency of element i in the stream. Modify the Mishra–Gries algorithm (Fig. 16.3) to show that for a stream of length m, one can compute quantities $\hat{f}_i$ for each element i such that

$$f_i - \frac{m}{k} \leq \hat{f}_i \leq f_i$$

16.2 Recall the reservoir sampling algorithm described in Fig. 16.5. Prove by induction on i that after i steps, the random variable X is a uniformly chosen element from the stream $\{x_1, \ldots, x_i\}$.

16.3 Let $Y_1, \ldots, Y_t$ be t i.i.d. random variables. Show that the variance of $Z = \frac{1}{t} \cdot \sum_i Y_i$, denoted by $\sigma^2(Z)$, is equal to $\frac{1}{t} \cdot \sigma^2(Y_1)$.

16.4 Suppose $E_1, \ldots, E_k$ are k independent events such that each event occurs with probability at most $1/4$. Assuming $k \geq 4\log(1/\delta)$, prove that the probability that more than $k/2$ events occur is at most δ.

16.5 Let $a_1, a_2, \ldots, a_n$ be an array of n numbers in the range $[0,1]$. Design a randomized algorithm which reads only $O(1/\varepsilon^2)$ elements from the array and estimates the average of all the numbers in the array within the additive error of $\pm\varepsilon$. The algorithm should succeed with at least 0.99 probability.

16.6 Show that the family of hash functions H defined as

$$h(x) : a_{k-1}x^{k-1} + a_{k-2}x^{k-2} \ldots a_k \mod p$$

where $a_i \in_{\mathcal{U}} \{0, 1, 2, \ldots, p-1\}$ for some prime p is a k-independent universal hash family.

16.7 Consider a family of functions H where each member $h \in H$ is such that $h : \{0,1\}^k \to \{0,1\}$. The members of H are indexed with a vector $r \in \{0,1\}^{k+1}$. The value $h_r(x)$ for $x \in \{0,1\}^k$ is defined by considering the vector $x_0 \in \{0,1\}^{k+1}$ obtained by appending 1 to

x and then taking the dot product of $x0$ and $r$modulo 2 (i.e., you take the dot product of x_0 and r, and $h_r(x)$ is 1 if this dot product is odd, and 0 if it is even). Prove that the family H is three-wise independent.

16.8 For the algorithm given in Fig. 16.6 for estimating F_2, show that by maintaining t independent random variables and finally outputing the average of square of these values, Z,

$$\Pr[|Z - F_2| \geq \varepsilon F_k] \leq \frac{2}{\varepsilon^2 \cdot t}$$

16.9 Recall the setting for estimating the second frequency moment in a stream. There is a universe $U = \{e_1, \ldots, e_n\}$ of elements, and elements $x_1, x_2, \ldots$ arrive over time, where each x_t belongs to U. Now consider an algorithm which receives **two** streams: $S = x_1, x_2, x_3, \ldots$ and $T = y_1, y_2, y_3, \ldots$. Element x_t and y_t arrive at time t in two streams respectively. Let f_i be the frequency of e_i in the stream S and g_i be its frequency in T. Let G denote the quantity $\sum_{i=1}^n f_i g_i$.

- As in the case of second frequency moment, define a random variable whose expected value is G. You should be able to store X using $O(\log n + \log m)$ space only (where m denotes the length of the stream).
- Let $F_2(S)$ denote the quantity $\sum_{i=1}^n f_i^2$ and $F_2(T)$ denote $\sum_{i=1}^n g_i^2$. Show that the variance of X can be bounded by $O(G^2 + F_2(S) \cdot F_2(T))$.

16.10 You are given an array A containing n distinct numbers. Given a parameter ε between 0 and 1, an element x in the array A is said to be a near-median element if its position in the sorted (increasing order) order of elements of A lies in the range $[n/2 - \varepsilon n, n/2 + \varepsilon n]$. Consider the following randomized algorithm for finding a near-median: pick t elements from A, where each element is picked uniformly and independently at random from A. Now output the median of these t elements. Suppose we want this algorithm to output a near-median with probability at least $1 - \delta$, where δ is a parameter between 0 and 1. How big should we make t? Your estimate on t should be as small as possible. Give reasons.

16.11 **Sliding window model of streaming** Consider a variation of the conventional streaming model where we are interested in computing a function of only the last N entries (instead of from the beginning of the stream).

Given a 0–1 bit stream, design an algorithm to keep track of the number of 1s in the last N inputs using space s. Your answer can be approximate (in a multiplicative or additive manner) as a function of s. For instance, for $s = N$, we can get an exact answer.

16.12 Consider the problem of maintaining an approximate count of elements labeled $[0, 1, \ldots, n-1]$ as accurately as possible as an alternative to the Misra–Gries technique. The idea is to maintain a table T of hash values from where one can obtain an estimate of the required label.

The table T has dimensions $r \times k$, where the values of these parameters should be determined from the analysis. We will use r distinct universal hash functions (one for each row) where each hash function $h_i : \{0, 1, \ldots, n-1\} \to \{0, 1, k-1\}$, where $i \leq r$.

For each item $x \in [0, 1, \ldots, n-1]$, the location corresponding to $h_i(x)$ is incremented by 1 **for each** $i = 1, 2, \ldots, r$, that is, the locations $T(i, h_i(x))$, $i = 1, 2, \ldots, r$ will be incremented.

The query about the count of label j is answered as $F_j = \min_{1 \leq i \leq r} T(i, h_i(j))$.

(i) Show that $f_i \leq F_i$, where f_i is the actual count of items labeled i and F_i is an estimate.

(ii) Show how to choose r, k such that $\Pr[F_j \geq f_j + \varepsilon N_{-j}] \leq \delta$ for any given parameters $0 < \varepsilon, \delta < 1$ and $N_{-j} = \sum_i f_i - f_j$, the total count of all elements except j.

16.13 Densest interval problem Given a stream S of points x_i, $x_i \in \mathbb{R}$, $1 \leq i \leq m$ and a fixed length $r > 0$, we would like to find an interval $I = [s, s+r]$ such that $I \cap S$ is maximum. In other words, a placement of the length r interval that maximizes the number of points of S within the interval.

(i) Design a linear time sequential algorithm for this problem in a conventional model.

(ii) The general problem is quite difficult in the streaming setting, so we consider a special case where the points $x_1 < x_2 < \ldots$ are presented in a sorted order. Design an exact algorithm that outputs the densest interval using space $O(D)$ where D is the maximum density.

(iii) Given an approximation parameter ε, show how to output an interval I_z such that $D \geq |I_z \cap S| \geq (1-\varepsilon)D$ using space bounded by $O(\frac{\log n}{\varepsilon})$. Here I_z denotes the interval $[z, z+r]$.

(iv) Can you improve the space bound to $O(\frac{1}{\varepsilon})$?

Hint: Consider forming a sample of the sorted stream that is k apart and choose a suitable value of k that can be stored.

16.14 Prove Claim 16.4

Recurrences and Generating Functions

Consider a sequence $a_1, a_2, \ldots, a_n$ (i.e., a function with the domain as integers). A compact way of representing it is as an equation in terms of itself, a recurrence relation. One of the most common examples is the Fibonacci sequence specified as $a_n = a_{n-1} + a_{n-2}$ for $n \geq 2$ and $a_0 = 0$, $a_1 = 1$. The values a_0, a_1 are known as the *boundary conditions*. Given this and the recurrence, we can compute the sequence step by step, or better still, we can write a computer program. Sometimes, we would like to find the general term of the sequence. Very often, the running time of an algorithm is expressed as a recurrence and we would like to know the explicit function for the running time to make any predictions and comparisons. A typical recurrence arising from a *divide-and-conquer* algorithm is

$$a_{2n} = 2a_n + cn$$

which has a solution $a_n \leq 2cn\lceil \log_2 n \rceil$. In the context of algorithm analysis, we are often satisfied with an upper bound. However, to the extent possible, it is desirable to obtain an exact expression.

Unfortunately, there is no general method for solving all recurrence relations. In this chapter, we discuss solutions to some important classes of recurrence equations. In the second part, we discuss an important technique based on *generating functions* which are also important in their own right.

A.1 An Iterative Method – Summation

As starters, some recurrence relations can be solved by summation or *guessing* and verifying by induction.

Example A.1 *The number of moves required to solve the* Tower of Hanoi *problem with n disks can be written as*

$$a_n = 2a_{n-1} + 1$$

By substituting for a_{n-1}, this becomes

$$a_n = 2^2 a_{n-2} + 2 + 1$$

By expanding this till a_1, we obtain

$$a_n = 2^{n-1} a_1 + 2^{n-2} + \ldots + 1$$

This gives $a_n = 2^n - 1$ by using the formula for geometric series and $a_1 = 1$.

Example A.2 *For the recurrence*

$$a_{2n} = 2a_n + cn$$

we can use the same technique to show that $a_{2n} = \sum_{i=0} \log_2 n 2^i n / 2^i \cdot c + 2na_1$.

Remark *We made an assumption that n is a power of 2. In the general case, this may present some technical complications but the nature of the answer remains unchanged. Consider the recurrence*

$$T(n) = 2T(\lfloor n/2 \rfloor) + n$$

Suppose $T(x) = cx \log_2 x$ for some constant $c > 0$ for all $x < n$. Then, $T(n) = 2c \lfloor n/2 \rfloor \log_2 \lfloor n/2 \rfloor + n$. Therefore, $T(n) \leq cn \log_2(n/2) + n \leq cn \log_2 n - (cn) + n \leq cn \log_2 n$ for $c \geq 1$.

A very frequent recurrence equation that comes up in the context of divide-and-conquer algorithms (like Merge sort) has the form

$$T(n) = aT(n/b) + f(n) \quad a, b \text{ are constants and } f(n) \text{ is a positive monotonic function}$$

Theorem A.1 *For the following different cases, the aforementioned recurrence has the following solutions*

- *If $f(n) = O(n^{\log_b a - \varepsilon})$ for some constant ε, then $T(n)$ is $\Theta(n^{\log_b a})$.*
- *If $f(n) = O(n^{\log_b a})$, then $T(n)$ is $\Theta(n^{\log_b a} \log n)$.*
- *If $f(n) = O(n^{\log_b a + \varepsilon})$ for some constant ε, and if $af(n/b)$ is $O(f(n))$, then $T(n)$ is $\Theta(f(n))$.*

Example A.3 *What is the maximum number of regions induced by n lines in the plane? If we let L_n represent the number of regions, then we can write the following recurrence*

$$L_n \leq L_{n-1} + n \quad L_0 = 1$$

Again by the method of summation, we can arrive at the answer $L_n = \frac{n(n+1)}{2} + 1$.

Example A.4 *Let us try to solve the recurrence for Fibonacci, namely*

$$F_n = F_{n-1} + F_{n-2} \quad F_0 = 0, \ F_1 = 1$$

If we try to expand this in the way that we have done previously, it becomes unwieldy very quickly. Instead we "guess" the following solution

$$F_n = \frac{1}{\sqrt{5}} \left(\phi^n - \bar{\phi}^n \right)$$

where $\phi = \frac{(1+\sqrt{5})}{2}$ and $\bar{\phi} = \frac{(1-\sqrt{5})}{2}$. This solution can be verified by induction. Of course, it is far from clear how one can magically guess the right solution. We shall address this later in the chapter.

A.2 Linear Recurrence Equations

A recurrence of the form

$$c_0 a_r + c_1 a_{r-1} + c_2 a_{r-2} + \ldots + c_k a_{r-k} = f(r)$$

where c_i are constants is called a *linear recurrence equation* of order k. Most of the examples in this chapter fall under this class. If $f(r) = 0$, then it is *homogeneous linear recurrence*.

A.2.1 Homogeneous equations

We will first outline the solution for the homogeneous class and then extend it to the general linear recurrence. Let us first determine the number of solutions. It appears that we must know the values of $a_1, a_2, \ldots, a_k$ to compute the values of the sequence according to the recurrence. In the absence of this, there can be different solutions based on different boundary conditions. Given the k boundary conditions, we can *uniquely* determine the values of the sequence. Note that this is not true for a *non-linear* recurrence like

$$a_r^2 + a_{r-1} = 5 \text{ with } a_0 = 1$$

This observation (of unique solution) makes it somewhat easier for us to guess some solution and verify.

Let us guess a solution of the form $a_r = A\alpha^r$, where A is some constant. This may be justified from the solution of Example A.1. By substituting this in the homogeneous linear recurrence and simplification, we obtain the following equation

$$c_0\alpha^k + c_1\alpha^{k-1} + \ldots + c_k = 0$$

This is called the *characteristic equation* of the recurrence relation and this degree k equation has k roots, say $\alpha_1, \alpha_2, \ldots, \alpha_k$. If these are all distinct, then the following is a solution to the recurrence

$$a_r^{(h)} = A_1\alpha_1^r + A_2\alpha_2^r + \ldots + A_k\alpha_k^r$$

which is also called the *homogeneous solution to linear recurrence*. The values of $A_1, A_2, \ldots, A_k$ can be determined from the k boundary conditions (by solving k simultaneous equations).

When the roots are not unique, i.e., some roots have multiplicity, then for multiplicity m, $\alpha^n, n\alpha^n, n^2\alpha^n, \ldots, n^{m-1}\alpha^n$ are the associated solutions. This follows from the fact that if α is a multiple root of the characteristic equation, then it is also the root of the derivative of the equation.

A.2.2 Inhomogeneous equations

If $f(n) \neq 0$, then there is no general methodology. Solutions are known for some particular cases; they are known as *particular* solutions. Let $a_n^{(h)}$ be the solution by ignoring $f(n)$ and let $a_n^{(p)}$ be a particular solution; then, it can be verified that $a_n = a_n^{(h)} + a_n^{(p)}$ is a solution to the non-homogeneous recurrence.

The following is a table of some particular solutions

d a constant	B
dn	$B_1 n + B_0$
dn^2	$B_2 n^2 + B_1 n + B_0$
ed^n, e,d are constants	Bd^n

Here B, B_0, B_1, B_2 are constants to be determined from initial conditions. When $f(n) = f_1(n) + f_2(n)$ is a sum of the aforementioned functions, then we solve the equation for $f_1(n)$ and $f_2(n)$ separately and add them in the end to obtain a particular solution for $f(n)$.

A.3 Generating Functions

An alternative representation for a sequence $a_1, a_2, \ldots, a_i$ is a polynomial function $a_1 x + a_2 x^2 + \ldots + a_i x^i$. Polynomials are very useful objects in mathematics, in particular as 'placeholders.' For example, if we know that two polynomials are equal (i.e., they evaluate to the same value for all x), then all the corresponding coefficients must be equal. This follows from the well-known property that a degree d polynomial has no more than d distinct roots (unless it is the zero polynomial). The issue of convergence is not important at this stage but will be relevant when we use the method of differentiation.

Example A.5 *Consider the problem of changing a Rs 100 note using notes of the following denomination -50, 20, 10, 5, and 1. Suppose we have an infinite supply of each denomination; then, we can represent each of these using the following polynomials where the coefficient corresponding to x^i is non-zero if we can obtain a certain sum using the given denomination.*

$$P_1(x) = x^0 + x^1 + x^2 + \ldots$$

$$P_5(x) = x^0 + x^5 + x^{10} + x^{15} + \ldots$$

$$P_{10}(x) = x^0 + x^{10} + x^{20} + x^{30} + \ldots$$

$$P_{20}(x) = x^0 + x^{20} + x^{40} + x^{60} + \ldots$$

$$P_{50}(x) = x^0 + x^{50} + x^{100} + x^{150} + \ldots$$

For example, we cannot have 51 to 99 using Rs 50, so all those coefficients are zero.

By multiplying these polynomials, we obtain

$$P(x) = E_0 + E_1 x + E_2 x^2 + \ldots + E_{100} x^{100} + \ldots + E_i x^i$$

where E_i is the number of ways the terms of the polynomials can combine such that the sum of the exponents is 100. Convince yourself that this is precisely what we are looking for. However, we must still obtain a formula for E_{100} or more generally E_i, which is the number of ways of changing a sum of i.

Note that for the polynomials $P_1, P_5, \ldots, P_{50}$, the following holds

$$P_k(1 - x^k) = 1 \quad for \ k = 1, 5, .., 50 \ giving$$

$$P(x) = \frac{1}{(1-x)(1-x^5)(1-x^{10})(1-x^{20})(1-x^{50})}$$

We can now use the observations that $\frac{1}{1-x} = 1 + x^2 + x^3 \ldots$ and $\frac{1-x^5}{(1-x)(1-x^5)} = 1 + x^2 + x^3 \ldots$. So the corresponding coefficients are related by $B_n = A_n + B_{n-5}$, where A and B are the coefficients of the polynomials $\frac{1}{1-x}$ and $\frac{1}{(1-x)(1-x^5)}$. Since $A_n = 1$, this is a linear recurrence. Find the final answer by extending these observations.

Let us try the method of generating functions on the Fibonacci sequence.

Example A.6 *Let the generating function be $G(z) = F_0 + F_1 z + F_2 z^2 + \ldots + F_n z^n$, where F_i is the ith Fibonacci number. Then, $G(z) - zG(z) - z^2 G(z)$ can be written as the infinite sequence*

$$F_0 + (F_1 - F_2)z + (F_2 - F_1 - F_0)z^2 + \ldots + (F_{i+2} - F_{i+1} - F_i)z^{i+2} + \ldots = z$$

for $F_0 = 0, F_1 = 1$. Therefore, $G(z) = \frac{z}{1-z-z^2}$. This can be worked out to be

$$G(z) = \frac{1}{\sqrt{5}} \left(\frac{1}{1-\phi z} - \frac{1}{1-\bar{\phi} z} \right)$$

where $\bar{\phi} = 1 - \phi = \frac{1}{2} \left(1 - \sqrt{5} \right)$.

A.3.1 Binomial theorem

The use of generating functions necessitates computation of the coefficients of power series of the form $(1+x)^\alpha$ for $|x| < 1$ and *any* α. For that, the following result is very useful – the coefficient of x^k is given by

$$C(\alpha, k) = \frac{\alpha \cdot (\alpha-1) \ldots (\alpha-k+1)}{k \cdot (k-1) \ldots 1}$$

This can be seen from an application of Taylor's series. Let $f(x) = (1+x)^\alpha$. Then, from Taylor's theorem, expanding around 0 for some z,

$$f(z) = f(0) + zf'(0) + \alpha \cdot z + z^2 \frac{f''(0)}{2!} + \ldots + z^k \frac{f^{(k)}(0)}{k!} \ldots$$

$$= f(0) + 1 + z^2 \frac{\alpha(\alpha-1)}{2!} + \ldots + C(\alpha, k) + \ldots$$

Therefore, $(1+z)^\alpha = \sum_{i=0}^{\infty} C(\alpha, i) z^i$, which is known as the binomial theorem.

A.4 Exponential Generating Functions

If the terms of a sequence is growing too rapidly, that is, the nth term exceeds x^n for any $0 < x < 1$, then it may not converge. It is known that a sequence converges iff the sequence $|a_n|^{1/n}$ is bounded. Then, it makes sense to divide the coefficients by a rapidly growing function like $n!$. For example, if we consider the generating function for the number of permutations of n *identical* objects

$$G(z) = 1 + \frac{p_1}{1!} z + \frac{p_2}{2!} z^2 + \ldots + \frac{p_i}{i!} z^i$$

where $p_i = P(i, i)$, then $G(z) = e^z$. The number of permutations of r objects when selected out of (an infinite supply of) n kinds of objects is given by the exponential generating function (EGF)

$$\left(1 + \frac{p_1}{1!} z + \frac{p_2}{2!} z^2 + \ldots \right)^n = e^{nx} = \sum_{r=0}^{\infty} \frac{n^r}{r!} z^r$$

Example A.7 *Let D_n denote the number of derangements of n objects. Then, it can be shown that $D_n = (n-1)(D_{n-1} + D_{n-2})$. This can be re-written as $D_n - nD_{n-1} = -(D_{n-1} - (n-2)D_{n-2})$. Iterating this, we obtain $D_n - nD_{n-1} = (-1)^{n-2}(D_2 - 2D_1)$. Using $D_2 = 1, D_1 = 0$, we obtain*

$$D_n - nD_{n-1} = (-1)^{n-2} = (-1)^n.$$

Multiplying both sides by $\frac{x^n}{n!}$, and summing from $n = 2$ to ∞, we obtain

$$\sum_{n=2}^{\infty} \frac{D_n}{n!} x^n - \sum_{n=2}^{\infty} \frac{nD_{n-1}}{n!} x^n = \sum_{n=2}^{\infty} \frac{(-1)^n}{n!} x^n$$

If we let $D(x)$ represent the exponential generating function for derangements, after simplification, we get

$$D(x) - D_1 x - D_0 - x(D(x) - D_0) = e^{-x} - (1-x)$$

or $D(x) = \frac{e^{-x}}{1-x}$.

A.5   Recurrences with Two Variables

For selecting r out of n distinct objects, we can write the familiar recurrence

$$C(n,r) = C(n-1,r-1) + C(n-1,r)$$

with boundary conditions $C(n,0) = 1$ and $C(n,1) = n$.

The general form of a linear recurrence with constant coefficients that has two indices is

$$C_{n,r}a_{n,r} + C_{n,r-1}a_{n,r-1} + \ldots + C_{n-k,r}a_{n-k,r} + \ldots + C_{0,r}a_{0,r} + \ldots = f(n,r)$$

where $C_{i,j}$ are constants. We will use the technique of generating functions to extend the one variable method. Let

$$A_0(x) = a_{0,0} + a_{0,1}x + \ldots + a_{0,r}x^r$$

$$A_1(x) = a_{1,0} + a_{1,1}x + \ldots + a_{1,r}x^r$$

$$A_n(x) = a_{n,0} + a_{n,1}x + \ldots + a_{n,r}x^r$$

Then we can define a generating function with $A_0(x), A_1(x)A_3(x)\ldots$ as the sequence – the new indeterminate can be chosen as y.

$$A_y(x) = A_0(x) + A_1(x)y + A_2(x)y^2 + \ldots + A_n(x)y^n$$

For this example, we have

$$F_n(x) = C(n,0) + C(n,1)x + C(n,2)x^2 + \ldots C(n,r)x^r + \ldots$$

$$\sum_{r=0}^{\infty} C(n,r)x^r = \sum_{r=1}^{\infty} C(n-1,r-1)x^r + \sum_{r=0}^{\infty} C(n-1,r)x^r$$

$$F_n(x) - C(n,0) = xF_{n-1}(x) + F_{n-1}(x) - C(n-1,0)$$

$$F_n(x) = (1+x)F_{n-1}(x)$$

or $F_n(x) = (1+x)^n C(0,0) = (1+x)^n$ as expected.

Bibliography

[1] Dimitris Achlioptas. Database-friendly random projections: Johnson-lindenstrauss with binary coins. *Journal of Computer and System Sciences*, 66(4):671–687, June 2003.

[2] Leonard M. Adleman. On constructing a molecular computer. In *DNA Based Computers, Proceedings of a DIMACS Workshop, Princeton, New Jersey, USA, April 4, 1995*, pages 1–22, 1995.

[3] A. Aggarwal and J. S. Vitter. The input/output complexity of sorting and related problems. *Communications of the ACM*, pages 1116–1127, 1988.

[4] Alok Aggarwal, Bowen Alpern, Ashok K. Chandra, and Marc Snir. A model for hierarchical memory. In *Proceedings of the 19th Annual ACM Symposium on Theory of Computing (STOC)*, pages 305–314. ACM, 1987.

[5] Manindra Agrawal, Neeraj Kayal, and Nitin Saxena. Primes is in P. *Annals of Mathematics*, 2:781–793, 2002.

[6] Alfred V. Aho and Margaret J. Corasick. Efficient string matching: An aid to bibliographic search. *Communications of the ACM*, 18(6):333–340, June 1975.

[7] Alfred V. Aho, John E. Hopcroft, and Jeffrey D Ullman. *The design and analysis of computer algorithms*. Addison-Wesley, Reading, 1974.

[8] Ravindra K. Ahuja, Thomas L. Magnanti, and James B. Orlin. *Network Flows: Theory, Algorithms, and Applications*. Prentice-Hall, Inc., Upper Saddle River, NJ, USA, 1993.

[9] M. Ajtai, J. Komlós, and E. Szemerédi. An 0(n log n) sorting network. In *Proceedings of the 15th Annual ACM Symposium on Theory of Computing (STOC)*, pages 1–9, 1983.

[10] Noga Alon, Yossi Matias, and Mario Szegedy. The space complexity of approximating the frequency moments. *Journal of Computer and System Sciences*, 58(1):137–147, 1999.

[11] Ingo Althöfer, Gautam Das, David P. Dobkin, Deborah Joseph, and José Soares. On sparse spanners of weighted graphs. *Discrete and Computational Geometry*, 9:81–100, 1993.

[12] L. Arge, M. T. Goodrich, M. Nelson, and N. Sitchinava. Fundamental parallel algorithms for private-cache chip multiprocessors. In *Proceedings of the 20th ACM Symposium on Parallelism in. Algorithms and Architectures (SPAA)*, pages 197–206, 2008.

[13] Sanjeev Arora and Boaz Barak. *Computational Complexity: A Modern Approach.* Cambridge University Press, New York, NY, USA, 2009.

[14] Sanjeev Arora, Carsten Lund, Rajeev Motwani, Madhu Sudan, and Mario Szegedy. Proof verification and the hardness of approximation problems. *J. ACM*, 45(3):501–555, May 1998.

[15] Surender Baswana and Sandeep Sen. A simple and linear time randomized algorithm for computing sparse spanners in weighted graphs. *Random Structures and Algorithms*, 30:532–563, 2007.

[16] K. E. Batcher. Sorting networks and their application. *Proc. AFIPS 1968 SJCC*, 32:307–314, 1968.

[17] Paul Beame, Stephen A. Cook, and H. James Hoover. Log depth circuits for division and related problems. *SIAM Journal on Computing*, 15(4):994–1003, 1986.

[18] Paul Beame and Johan Hastad. Optimal bounds for decision problems on the crcw pram. *Journal of the ACM*, 36(3):643–670, July 1989.

[19] Jon Louis Bentley. Multidimensional binary search trees used for associative searching. *Communications of the ACM*, 18(9):509–517, September 1975.

[20] Mark de Berg, Otfried Cheong, Marc van Kreveld, and Mark Overmars. *Computational Geometry: Algorithms and Applications.* Springer-Verlag TELOS, Santa Clara, CA, USA, 3rd ed. edition, 2008.

[21] Binay K. Bhattacharya and Sandeep Sen. On a simple, practical, optimal, output-sensitive randomized planar convex hull algorithm. *Journal of Algorithms*, 25(1):177–193, 1997.

[22] Manuel Blum, Robert W. Floyd, Vaughan Pratt, Ronald L. Rivest, and Robert E. Tarjan. Time bounds for selection. *Journal of Computer and System Sciences*, 7(4):448–461, August 1973.

[23] Stephen Boyd and Lieven Vandenberghe. *Convex Optimization.* Cambridge University Press, New York, NY, USA, 2004.

[24] Robert S. Boyer and J. Strother Moore. MJRTY: A fast majority vote algorithm. In *Automated Reasoning: Essays in Honor of Woody Bledsoe*, Automated Reasoning Series, pages 105–118. Kluwer Academic Publishers, 1991.

[25] A. Bykat. Convex hull of a finite set of points in two dimensions. *Information Processing Letters*, 7:296 – 298, 1978.

[26] J. Lawrence Carter and Mark N. Wegman. Universal classes of hash functions (extended abstract). In *Proceedings of the 9th Annual ACM Symposium on Theory of Computing (STOC)*, pages 106–112, 1977.

[27] Amit Chakrabarti, Subhash Khot, and Xiaodong Sun. Near-optimal lower bounds on the multi-party communication complexity of set disjointness. In *Proceedings of the 18th IEEE Annual Conference on Computational Complexity (CCC)*, pages 107–117, 2003.

[28] Timothy M. Chan, Jack Snoeyink, and Chee-Keng Yap. Primal dividing and dual pruning: Output-sensitive construction of four-dimensional polytopes and three-dimensional voronoi diagrams. *Discrete & Computational Geometry*, 18(4):433–454, 1997.

[29] Bernard Chazelle. A minimum spanning tree algorithm with inverse-ackermann type complexity. *Journal of the ACM*, 47(6):1028–1047, 2000.

[30] G. Blelloch R. Chowdhury P. Gibbons V. Ramachandran S. Chen and M. Kozuch. Provably good multicore cache performance for divide-and-conquer algorithms. In *Proceedings of the 19th ACM-SIAM Symposium on Discrete Algorithms (SODA)*, pages 501–510, 2008.

[31] V. Chvátal. *Linear Programming*. Series of books in the mathematical sciences. W.H. Freeman, 1983.

[32] Kenneth L. Clarkson and Peter W. Shor. Application of random sampling in computational geometry, II. *Discrete & Computational Geometry*, 4:387–421, 1989.

[33] Richard Cole. Parallel merge sort. *SIAM Journal on Computing*, 17(4):770–785, August 1988.

[34] Richard Cole and Vijaya Ramachandran. Resource oblivious sorting on multicores. In *Proceedings of the 37th International Colloquium on Automata, Languages and Programming (ICALP)*, pages 226–237, 2010.

[35] Stephen A. Cook. The complexity of theorem-proving procedures. In *Proceedings of the 3rd Annual ACM Symposium on Theory of Computing (STOC)*, STOC '71, pages 151–158, 1971.

[36] J. W. Cooley and J. W. Tukey. An algorithm for the machine computation of the complex fourier series. *Mathematics of Computation*, 19:297–301, April 1965.

[37] Thomas H. Cormen, Charles E. Leiserson, Ronald L. Rivest, and Clifford Stein. *Introduction to Algorithms*. The MIT Press, 2nd edition, 2001.

[38] David E. Culler, Richard M. Karp, David Patterson, Abhijit Sahay, Eunice E. Santos, Klaus Erik Schauser, Ramesh Subramonian, and Thorsten von Eicken. Logp: A practical model of parallel computation. *Communications of the ACM*, 39(11):78–85, November 1996.

[39] Sanjoy Dasgupta and Anupam Gupta. An elementary proof of a theorem of johnson and lindenstrauss. *Random Structures and Algorithms*, 22(1):60–65, January 2003.

[40] Sanjoy Dasgupta, Christos H. Papadimitriou, and Umesh Vazirani. *Algorithms*. McGraw-Hill, Inc., New York, NY, USA, 1 edition, 2008.

[41] Rene De La Briandais. File searching using variable length keys. In *Papers Presented at the the March 3-5, 1959, Western Joint Computer Conference*, IRE-AIEE-ACM '59 (Western), pages 295–298. ACM, 1959.

[42] Reinhard Diestel. *Graph Theory (Graduate Texts in Mathematics)*. Springer, August 2005.

[43] Martin Dietzfelbinger, Anna Karlin, Kurt Mehlhorn, Friedhelm Meyer auf der Heide, Hans Rohnert, and Robert E. Tarjan. Dynamic perfect hashing: Upper and lower bounds. *SIAM Journal on Computing*, 23(4):738–761, 1994.

[44] James R. Driscoll, Neil Sarnak, Daniel D. Sleator, and Robert E. Tarjan. Making data structures persistent. *Journal of Computer and System Sciences*, 38(1):86–124, February 1989.

[45] Herbert Edelsbrunner. *Algorithms in Combinatorial Geometry*. Springer-Verlag, Berlin, Heidelberg, 1987.

[46] Taher El Gamal. A public key cryptosystem and a signature scheme based on discrete logarithms. In *Proceedings of CRYPTO 84 on Advances in Cryptology*, pages 10–18, 1985.

[47] M. J. Fischer and M. S. Paterson. String-matching and other products. Technical report, Massachusetts Institute of Technology, Cambridge, MA, USA, 1974.

[48] R. Floyd. Permuting information in idealized two-level storage. *Complexity of Computer Computations*, pages 105–109, 1972.

[49] Robert W. Floyd and Ronald L. Rivest. Expected time bounds for selection. *Communications of the ACM*, 18(3):165–172, March 1975.

[50] Steven Fortune and James Wyllie. Parallelism in random access machines. In *Proceedings of the 10th Annual ACM Symposium on Theory of Computing (STOC)*, pages 114–118. ACM, 1978.

[51] Edward Fredkin. Trie memory. *Communications of the ACM*, 3(9):490–499, September 1960.

[52] Michael L. Fredman, János Komlós, and Endre Szemerédi. Storing a sparse table with 0(1) worst case access time. *Journal of the ACM*, 31(3):538–544, June 1984.

[53] Michael L. Fredman and Robert Endre Tarjan. Fibonacci heaps and their uses in improved network optimization algorithms. *Journal of the ACM*, 34(3):596–615, 1987.

[54] Michael L. Fredman and Dan E. Willard. Surpassing the information theoretic bound with fusion trees. *Journal of Computer and System Sciences*, 47(3):424–436, 1993.

[55] Alan M. Frieze, Ravi Kannan, and Santosh Vempala. Fast monte-carlo algorithms for finding low-rank approximations. *Journal of the ACM*, 51(6):1025–1041, 2004.

[56] Matteo Frigo, Charles E. Leiserson, Harald Prokop, and Sridhar Ramachandran. Cache-oblivious algorithms. In *Proceedings of the 40th Annual Symposium on Foundations of Computer Science (FOCS)*, pages 285–298, 1999.

[57] Michael R. Garey and David S. Johnson. *Computers and Intractability; A Guide to the Theory of NP-Completeness*. W. H. Freeman & Co., New York, NY, USA, 1990.

[58] Andrew V. Goldberg and Satish Rao. Beyond the flow decomposition barrier. *Journal of the ACM*, 45(5):783–797, 1998.

[59] R.L. Graham. An efficient algorith for determining the convex hull of a finite planar set. *Information Processing Letters*, 1(4):132 – 133, 1972.

[60] Dan Gusfield. *Algorithms on Strings, Trees, and Sequences: Computer Science and Computational Biology*. Cambridge University Press, New York, NY, USA, 1997.

[61] Harary. *Graph Theory*. Perseus Books, Reading, MA, 1999.

[62] Daniel S. Hirschberg. Algorithms for the longest common subsequence problem. *Journal of the ACM*, 24(4):664–675, October 1977.

[63] John E. Hopcroft and Jeff D. Ullman. *Introduction to Automata Theory, Languages, and Computation*. Addison-Wesley Publishing Company, 1979.

[64] John E. Hopcroft and Jeffrey D. Ullman. Set merging algorithms. *SIAM Journal on Computing*, 2(4):294–303, 1973.

[65] Ellis Horowitz, Sartaj Sahni, and Sanguthevar Rajasekaran. *Computer Algorithms*. Silicon Press, Summit, NJ, USA, 2nd edition, 2007.

[66] Piotr Indyk and Rajeev Motwani. Approximate nearest neighbors: Towards removing the curse of dimensionality. In *Proceedings of the 30th Annual ACM Symposium on Theory of Computing (STOC)*, pages 604–613, 1998.

[67] Piotr Indyk and David P. Woodruff. Optimal approximations of the frequency moments of data streams. In *Proceedings of the 37th Annual Symposium on Theory of Computing (STOC)*, pages 202–208, 2005.

[68] Joseph JáJá. *An Introduction to Parallel Algorithms*. Addison Wesley Longman Publishing Co., Inc., Redwood City, CA, USA, 1992.

[69] R.A. Jarvis. On the identification of the convex hull of a finite set of points in the plane. *Information Processing Letters*, 2(1):18 – 21, 1973.

[70] Hong Jia-Wei and H. T. Kung. I/o complexity: The red-blue pebble game. In *Proceedings of the 13th Annual ACM Symposium on Theory of Computing (STOC)*, pages 326–333, New York, NY, USA, 1981. ACM.

[71] W. B. Johnson and J. Lindenstrauss. Extensions of lipschitz mappings into hilbert space. *Contemporary Mathematics*, 26:189– 206, 1984.

[72] Adam Kalai. Efficient pattern-matching with don't cares. In *Proceedings of the 13th Annual ACM-SIAM Symposium on Discrete Algorithms (SODA)*, SODA '02, pages 655–656, 2002.

[73] Bala Kalyanasundaram and Georg Schnitger. The probabilistic communication complexity of set intersection. *SIAM Journal on Discrete Mathematics*, 5(4):545–557, 1992.

[74] David R. Karger. Global min-cuts in rnc, and other ramifications of a simple min-cut algorithm. In *Proceedings of the 4th Annual ACM/SIGACT-SIAM Symposium on Discrete Algorithms (SODA)*, pages 21–30, 1993.

[75] David R. Karger, Philip N. Klein, and Robert Endre Tarjan. A randomized linear-time algorithm to find minimum spanning trees. *Journal of the ACM*, 42(2):321–328, 1995.

[76] David R. Karger and Clifford Stein. A new approach to the minimum cut problem. *Journal of the ACM*, 43(4):601–640, 1996.

[77] R. Karp. Reducibility among combinatorial problems. In R. Miller and J. Thatcher, editors, *Complexity of Computer Computations*, pages 85–103. Plenum Press, 1972.

[78] Richard M. Karp and Michael O. Rabin. Efficient randomized pattern-matching algorithms. *IBM J. Res. Dev.*, 31(2):249–260, March 1987.

[79] S. Khuller and Y. Matias. A simple randomized sieve algorithm for the closest-pair problem. *Information and Computation*, 118(1):34 – 37, 1995.

[80] David G. Kirkpatrick and Raimund Seidel. The ultimate planar convex hull algorithm? *SIAM Journal on Computing*, 15(1):287–299, 1986.

[81] Jon Kleinberg and Eva Tardos. *Algorithm Design*. Addison-Wesley Longman Publishing Co., Inc., Boston, MA, USA, 2005.

[82] Donald E. Knuth. *Seminumerical Algorithms*, volume 2 of *The Art of Computer Programming*. Addison-Wesley, Reading, Massachusetts, second edition, 1981.

[83] Donald E. Knuth. *Sorting and Searching*, volume 3 of *The Art of Computer Programming*. Addison-Wesley, Reading, Massachusetts, second edition, 1981.

[84] Donald E. Knuth, James H. Morris Jr., and Vaughan R. Pratt. Fast pattern matching in strings. *SIAM Journal on Computing*, 6(2):323–350, 1977.

[85] Jnos Komls, Yuan Ma, and Endre Szemerdi. Matching nuts and bolts in o(n log n) time. *SIAM Journal on Discrete Mathematics*, 11(3):347–372, 1998.

[86] Eyal Kushilevitz and Noam Nisan. *Communication complexity*. Cambridge University Press, 1997.

[87] Richard E. Ladner and Michael J. Fischer. Parallel prefix computation. *Journal of the ACM*, 27(4):831–838, October 1980.

[88] E. Lawler. *Combinatorial optimization - networks and matroids*. Holt, Rinehart and Winston, New York, 1976.

[89] F. Thomson Leighton. *Introduction to Parallel Algorithms and Architectures: Array, Trees, Hypercubes*. Morgan Kaufmann Publishers Inc., San Francisco, CA, USA, 1992.

[90] Frank Thomson Leighton, Bruce M. Maggs, Abhiram G. Ranade, and Satish Rao. Randomized routing and sorting on fixed-connection networks. *Journal of Algorithms*, 17(1):157–205, 1994.

[91] L.A. Levin. Universal sequential search problems. *Probl. Peredachi Inf.*, 9:115–116, 1973.

[92] Leonid A Levin. Average case complete problems. *SIAM Journal on Computing*, 15(1):285–286, February 1986.

[93] Harry R. Lewis and Christos H. Papadimitriou. *Elements of the Theory of Computation*. Prentice Hall PTR, 2nd edition, 1997.

[94] L. Lovasz and M. D Plummer. *Matching Theory*. Elsevier, 1986.

[95] Rabin M. Probabilistic algorithm for testing primality. *Journal of Number Theory*, 12(1):128–138, 1980.

[96] Aleksander Madry. Computing maximum flow with augmenting electrical flows. In *IEEE 57th Annual Symposium on Foundations of Computer Science (FOCS)*, pages 593–602, 2016.

[97] Udi Manber and Gene Myers. Suffix arrays: A new method for on-line string searches. In *Proceedings of the 1st Annual ACM-SIAM Symposium on Discrete Algorithms (SODA)*, pages 319–327. Society for Industrial and Applied Mathematics, 1990.

[98] Edward M. McCreight. A space-economical suffix tree construction algorithm. *Journal of the ACM*, 23(2):262–272, April 1976.

[99] Edward M. McCreight. Priority search trees. *SIAM Journal on Computing*, 14(2):257–276, 1985.

[100] Andrew McGregor. Graph stream algorithms: a survey. *SIGMOD Record*, 43(1):9–20, 2014.

[101] K. Mehlhorn. *Data Structures and Algorithms III: Multi-dimensional Searching and Computational Geometry*, volume 3 of *EATCS Monographs on Theoretical Computer Science*. Springer, 1984.

[102] Kurt Mehlhorn. *Data structures and algorithms. Volume 1 : Sorting and searching*, volume 1 of *EATCS Monographs on Theoretical Computer Science*. Springer, 1984.

[103] G Miller. Riemann's hypothesis and tests for primality. *Journal of Computer and System Sciences*, 13(3):300–317, 1976.

[104] Jayadev Misra and David Gries. Finding repeated elements. *Science of Computer Programming*, 2(2):143–152, 1982.

[105] Michael Mitzenmacher and Eli Upfal. *Probability and Computing: Randomized Algorithms and Probabilistic Analysis*. Cambridge University Press, New York, NY, USA, 2005.

[106] Rajeev Motwani and Prabhakar Raghavan. *Randomized Algorithms*. Cambridge University Press, 1995.

[107] Ketan Mulmuley. A fast planar partition algorithm, I (extended abstract). In *Proceedings of the 29th Annual IEEE Symposium on Foundations of Computer Science (FOCS)*, pages 580–589, 1988.

[108] J. I. Munro and M. S. Paterson. Selection and sorting with limited storage. In *Proceedings of the 19th Annual Symposium on Foundations of Computer Science (SFCS)*, SFCS '78, pages 253–258, 1978.

[109] S. Muthukrishnan. Data streams: Algorithms and applications. *Foundations andTrends in Theoretical Computer Science*, 1(2):117–236, August 2005.

[110] Jaroslav Nesetril, Eva Milková, and Helena Nesetrilová. Otakar boruvka on minimum spanning tree problem translation of both the 1926 papers, comments, history. *Discrete Mathematics*, 233(1-3):3–36, 2001.

[111] Michael A. Nielsen and Isaac L. Chuang. *Quantum Computation and Quantum Information*. Cambridge University Press, 2000.

[112] James B. Orlin. Max flows in o(nm) time, or better. In *ACM 45th Annual Symposium on Theory of Computing Conference (STOC)*, pages 765–774, 2013.

[113] Rasmus Pagh and Flemming Friche Rodler. Cuckoo hashing. *Journal of Algorithms*, 51(2):122–144, May 2004.

[114] Christos M. Papadimitriou. *Computational complexity*. Addison-Wesley, Reading, Massachusetts, 1994.

[115] David Peleg and A. A. Schaffer. Graph spanners. *Journal of Graph Theory*, 13:99–116, 1989.

[116] Yehoshua Perl, Alon Itai, and Haim Avni. Interpolation search: a log logn search. *Communications of the ACM*, 21(7):550–553, July 1978.

[117] F. P. Preparata and S. J. Hong. Convex hulls of finite sets of points in two and three dimensions. *Communications of the ACM*, 20(2):87–93, February 1977.

[118] Franco P. Preparata and Michael I. Shamos. *Computational Geometry: An Introduction*. Springer-Verlag, Berlin, Heidelberg, 1985.

[119] William Pugh. Skip lists: A probabilistic alternative to balanced trees. *Communications of the ACM*, 33(6):668–676, June 1990.

[120] Sanguthevar Rajasekaran and Sandeep Sen. A generalization of the 0-1 principle for sorting. *Information Processing Letters*, 94(1):43–47, 2005.

[121] John H. Reif. Depth-first search is inherently sequential. *Information Processing Letters*, 20:229–234, 06 1985.

[122] John H. Reif. *Synthesis of Parallel Algorithms*. Morgan Kaufmann Publishers Inc., San Francisco, CA, USA, 1st edition, 1993.

[123] John H. Reif and Leslie G. Valiant. A logarithmic time sort for linear size networks. *Journal of the ACM*, 34(1):60–76, January 1987.

[124] Rüdiger Reischuk. A fast probabilistic parallel sorting algorithm. In *Proceedings of the 22nd Annual IEEE Symposium on Foundations of Computer Science (FOCS)*, pages 212–219. IEEE Computer Society, 1981.

[125] R. L. Rivest, A. Shamir, and L. Adleman. A method for obtaining digital signatures and public-key cryptosystems. *Communications of the ACM*, 21(2):120–126, February 1978.

[126] Sheldon M. Ross. *Introduction to Probability Models*. Academic Press, San Diego, CA, USA, sixth edition, 1997.

[127] Neil Sarnak and Robert E. Tarjan. Planar point location using persistent search trees. *Communications of the ACM*, 29(7):669–679, July 1986.

[128] Isaac D. Scherson and Sandeep Sen. Parallel sorting in two-dimensional VLSI models of computation. *IEEE Transactions on Computers*, 38(2):238–249, 1989.

[129] Raimund Seidel and Cecilia R. Aragon. Randomized search trees. *Algorithmica*, 16(4/5):464–497, 1996.

[130] Sandeep Sen. Some observations on skip-lists. *Information Processing Letters*, 39(4):173–176, 1991.

[131] Sandeep Sen, Siddhartha Chatterjee, and Neeraj Dumir. Towards a theory of cache-efficient algorithms. *Journal of the ACM*, 49(6):828–858, November 2002.

[132] Adi Shamir. Factoring numbers in o(log n) arithmetic steps. *Information Processing Letters*, 8(1):28–31, 1979.

[133] Y. Shiloach and Uzi Vishkin. An o (logn) parallel connectivity algorithm. *Journal of Algorithms*, 3:57 – 67, 1982.

[134] Peter W. Shor. Polynomial-time algorithms for prime factorization and discrete logarithms on a quantum computer. *SIAM Journal on Computing*, 26(5):1484–1509, 1997.

[135] Marc Snir. Lower bounds on probabilistic linear decision trees. *Theoretical Computer Science*, 38:69 – 82, 1985.

[136] Robert Solovay and Volker Strassen. A fast monte-carlo test for primality. *SIAM Journal on Computing*, 6(1):84–85, 1977.

[137] H. S. Stone. Parallel processing with the perfect shuffle. *IEEE Transactions on Computers*, 20(2):153–161, February 1971.

[138] Robert E. Tarjan and Jan van Leeuwen. Worst-case analysis of set union algorithms. *Journal of the ACM*, 31(2), March 1984.

[139] Robert Endre Tarjan. Efficiency of a good but not linear set union algorithm. *Journal of the ACM*, 22(2):215–225, 1975.

[140] Robert Endre Tarjan. Sensitivity analysis of minimum spanning trees and shortest path trees. *Information Processing Letters*, 14(1):30–33, 1982.

[141] Mikkel Thorup. Integer priority queues with decrease key in constant time and the single source shortest paths problem. *Journal of Computer and System Sciences*, 69(3):330–353, 2004.

[142] Mikkel Thorup and Uri Zwick. Approximate distance oracles. *Journal of Association of Computing Machinery*, 52:1–24, 2005.

[143] E. Ukkonen. On-line construction of suffix trees. *Algorithmica*, 14(3):249–260, Sep 1995.

[144] L. G. Valiant. A bridging model for multi-core computing. In *Proceedings of the 16th Annual European Symposium on Algorithms (ESA)*, pages 13–28, 2008.

[145] Leslie G. Valiant. Parallelism in comparison problems. *SIAM Journal on Computing*, 4(3):348–355, 1975.

[146] Leslie G. Valiant. A bridging model for parallel computation. *Communications of the ACM*, 33(8):103–111, August 1990.

[147] Peter van Emde Boas. Preserving order in a forest in less than logarithmic time. In *Proceedings of the 16th Annual Symposium on Foundations of Computer Science (FOCS)*, pages 75–84. IEEE Computer Society, 1975.

[148] Vijay V. Vazirani. *Approximation Algorithms*. Springer Publishing Company, Incorporated, 2010.

[149] Andrew J. Viterbi. Error bounds for convolutional codes and an asymptotically optimum decoding algorithm. *IEEE Transactions on Information Theory*, 13(2):260–269, 1967.

[150] Jeffrey Scott Vitter. *Algorithms and Data Structures for External Memory*. Now Publishers Inc., Hanover, MA, USA, 2008.

[151] Jean Vuillemin. A data structure for manipulating priority queues. *Communications of the ACM*, 21(4):309–315, April 1978.

[152] P. Weiner. Linear pattern matching algorithms. In *14th Annual Symposium on Switching and Automata Theory (SWAT)*, pages 1–11, Oct 1973.

[153] Hassler Whitney. On the abstract properties of linear dependence. *American Journal of Mathematics*, 57(3):509–533, 1935.

[154] David P. Williamson and David B. Shmoys. *The Design of Approximation Algorithms*. Cambridge University Press, New York, NY, USA, 2011.

[155] A. C. C. Yao. Probabilistic computations: Toward a unified measure of complexity. In *18th Annual Symposium on Foundations of Computer Science (sfcs 1977)*, pages 222–227, Oct 1977.

[156] Andrew C-C. Yao. Separating the polynomial-time hierarchy by oracles. In *Proceedings of the 26th Annual IEEE Symposium on Foundations of Computer Science (FOCS)*, pages 1–10, 1985.

[157] Andrew Chi-Chih Yao. Should tables be sorted? *Journal of the ACM*, 28(3):615–628, July 1981.

[158] F. Frances Yao. Speed-up in dynamic programming. *SIAM Journal on Algebraic Discrete Methods*, 3(4):532–540, 1982.

Index

Amortized
  path compression, 73
  potential, 46, 165
  union find, 68
Approximation
  3 coloring, 253
  Greedy, 76
  knapsack, 251
  max-cut, 254
  PTAS, 250
  set cover, 252
  TSP, 253
  vertex cover, 250

Closest Pair
  backward analysis, 115
  expected runtime, 145
  ric, 150
Computational models
  external memory, 10
  parallel, 11–12

PRAM, 11
  RAM, 10
Convex hull
  definition, 137
  Graham scan, 140
  Jarvis march, 139
  Merging, 141
  parallel, 299
  quickhull, 142
  relation to sorting, 141
  union and intersection, 141
Cooley–Tukey
  butterfly graph, 176
  FFT, 175
  roots of unity, 174–175

DFA
  prefix computation, 291
  string matching, 179
Dynamic programming
  context free parsing, 95

Fibonacci series, 92
function approximation, 99
independent set in trees, 102
Knapsack, 94–95
longest monotonic subsequence, 97
Viterbi maximum likelihood, 100

Fibonacci
  matrix recurrence, 2
  number, 1
  recursive definition, 1
Flow
  augmenting path, 212
  circulation, 222–223
  disjoint path, 216–217
  Edmond-Karp, 214
  Ford Fulkerson, 213–214, 219
  Halls theorem, 220
  Max-flow, 212
  Max-flow-min-cut theorem, 212
  project planning, 224,
  residual graph, 210, 212–213
  Whitneys theorem, 191

Gradient descent
  algorithm, 79, 84–85
  convergence, 81
  convex function, 78
  perceptron, 84
  point location, 146
Graph
  All pairs shortest path, 193
  augmenting path, 212
  Bellman Ford, 194–195, 198
  bi connectivity, 191
  bipartite matching, 217
  DFS, 184, 188
  Dijkstras algorithm, 195

global mincut, 201
Max-flow, 212
parallel connectivity, 304
path decomposition, 209
spanners, 184, 198
s-t cut, 201
strong components, 188
topological sort, 187
Turan's theorem, 281
Greatest Common Divisor (gcd)
  Euclid's algorithm, 34–35
  Extended Euclid, 35–36
Greedy
  Boruvka's algorithm, 295
  generic greedy, 61
  gradient descent, 77
  half-matching, 63
  Kruskal's algorithm, 64, 66, 69, 74
  Matroid, 61, 74
  maximum spanning tree, 64
  minimum set cover, 252
  Prim's algorithm, 74
  Routing, 301
  Scheduling, 65

Heaps
  binary, 43
  Binomial, 44–45

Johnson Lindenstrauss
  dimension reduction, 258
  lemma, 259
  random projection, 259–262
  using normal distribution, 260–261

Knapsack
  approximation, 251
  dynamic programming, 244

formulation, 251
tree search and pruning strategy, 55

Linear algebra
   Clustering, 271
   document classification, 157
   Frobenius norm, 268
   Gaussian elimination, 262
   low rank approximation, 267
   spectral norm, 267
   SVD theorem, 265–267
   unitary matrix, 266, 274
Lower bound
   communication complexity, 337
   convex hull, 141
   element distinctness, 49
   external memory permutation, 314
   external sorting, 314
   parallel comparison tree, 281
   randomized, 48
   sorting: average, 48
   sorting: worst case, 48

Matroid
   exchange property, 61
   greedy, 74
   rank property, 61
Memory hierarchy
   Cache model, 316
   lower bound, 337
   matrix multiplication, 311–312
   merge sort, 313
   models, 308
   oblivious, 309, 317
   transpose, 310–311
MST
   Boruvka, 76
   Kruskal, 74

Matroid, 64
   Prim, 74
   red-blue rule, 75
Multidimensional Search
   Interval trees, 129
   k-d trees, 132–135
   persistent data structure, 146
   priority search, 135
   Range trees, 129
Multiplication
   cache efficient, 311
   integers, 176
   matrix, 311–312
   recursive, 3
   Schonage-Strassen, 176

NP completeness
   3-SAT, 240–241, 247
   co-NP, 247
   Cook-Levin theorem, 235
   Definition, 230, 236
   hardness, 249
   independent set, 240
   knapsack, 244
   polynomial reducibility
   PSPACE, 247
   RP, PP, BPP, 247
   Satisfiability, 235
   set cover, 252
   three coloring, 242, 253
   vertex cover, 241

Parallel
   Connectivity, 298
   DFA simulation, 290
   extremal selection, 304
   interconnection networks, 278, 300
   List ranking, 292

load balancing, 279–280, 294
lower bound: PCT
models, 300
partition sort, 285
PRAM, 300
prefix computation, 299
quickhull, 299
routing on mesh, 301
shear sort, 284–285
sorting, 282
Point location
binary search, 148
persistent data structure, 146
planar subdivision, 148
Polynomial
Convolution, 172
evaluation, 171
FFT, 171
Interpolation, 171–172
Lagrange's formula, 172
Probability
Conditional, 30
density function, 19
independent, 18–19
inequality:Chebychev, 23
inequality:Chernoff
inequality:Markov, 21–24
measure, 17
random number generation, 26
random permutation, 29
random variable, 27
space, 17

Quicksort
expected runtime, 145
lower bound, 47–49
parallel, 285
partition, 8

partition sort, 285
quickhull, 142
recurrence, 53
running time, 125

Randomized Algorithms
backward analysis, 115
closest pair, 149
expectation, 9
game tree evaluation, 57–58
high probability, 30, 202
incremental construction, 29, 149, 151
Las Vegas, 10, 122, 248
Monte Carlo, 10, 122, 203, 248
Permutations, 29
random mate, 294
tail estimates, 113
RSA
complexity, 36
encryption and decryption, 36

Scheduling
Greedy, 65–66
using topological sort, 187
Searching
AVL, 109
Loglogn, 122
perfect hashing, 121
Priority search, 135
Semi-dynamic, 45
Skip Lists, 116
Treaps, 114, 135
universal hashing, 117, 121
Selection
expected runtime, 39
k-th smallest, 37, 39
median of medians, 39
random splitter, 38

Sorting
  external memory, 313
  integer, 313
  lexicographic, 41
  merge sort, 313–314
  parallel, 282
  quicksort, 8
  radix, 41–43
  stable, 41, 43
  0-1 principle, 283, 303
Streaming
  Boyer-Moore majority, 325
  distinct elements, 327
  frequency moment, 331
  frequency threshold, 325
  lower bound, 337
  median-of-median, 39
  Mishra-Gries, 327

  Model, 337
  reservoir sampling, 332
  second frequency, 335
String matching
  hashing, 159
  KMP, 161, 165
  Rabin Karp, 157
  Tries and suffix trees, 167–168
  using convolution, 181
  wildcard in pattern, 180
  wildcard in pattern and string, 179

Union Find
  Array, 68
  MST, 64
  path compression, 70, 73
  tree based, 46
  union by rank, 68, 70